13th Topical Conference on Gas Utilization 2013

Topical Conference at the 2013 AIChE Spring Meeting and 9th Global Congress on Process Safety

San Antonio, Texas, USA
28 April – 2 May 2013

ISBN: 978-1-62748-454-1

Printed from e-media with permission by:

Curran Associates, Inc.
57 Morehouse Lane
Red Hook, NY 12571

CURRAN ASSOCIATES INC. proceedings.com

Some format issues inherent in the e-media version may also appear in this print version.

Printed by Curran Associates, Inc. (2013)

AIChE
3 Park Avenue
New York, NY 10016-5991

Phone: (203) 702-7660
Fax: (203) 775-5177

www.aiche.org

Additional copies of this publication are available from:

Curran Associates, Inc.
57 Morehouse Lane
Red Hook, NY 12571 USA
Phone: 845-758-0400
Fax: 845-758-2634
Email: curran@proceedings.com
Web: www.proceedings.com

TABLE OF CONTENTS

() Rebirth of Manufacturing in the U.S.: What are the Implications for Chemical Engineers?

Dan Romasko, Tesoro Corporation, San Antonio, TX

Technology advances have increased production of oil and natural gas from reserves that were previously unprofitable. This rapid increase in production is restructuring the energy industry and providing more cost advantaged feedstocks and fuel to the nation's manufacturing base, particularly the petrochemical industry. This improved cost structure leads to increased competitiveness in the global markets and supports growth in the domestic industry. Along with this growth comes the need for skilled engineers at a time when the demographic changes point toward a declining supply of engineering talent.

(25a) Current Myths about LNG

Christopher Caswell[1], David Coyle[2], Charles Durr[3], Heinz Kotzot[3] and **Jeffery Gaspard**[2], (1)Gas Monetization, KBR, Houston, TX, (2)KBR, Houston, TX, (3)KBR, houston, TX

Since the startup of the first baseload LNG plant (Camel) in 1964, there have been several generations of LNG plant designs. Each generation can be defined by a noticeable step change in elements of the overall plant configuration. These steps may include developments in equipment, changes in cooling media, breakthroughs in train throughput, or combinations of changes leading to new industry standards or benchmarks. Since the first LNG projects with train capacities of less than 1 Mt/a, generations of technical innovations have led to operating trains of nearly 8 Mt/a, and front end engineering designs (FEED) exceeding this current benchmark.

Regardless of the state of the LNG industry, there are often misconceptions regarding the application of new technologies, determination of process configurations, or implementation of project execution strategies to reduce project costs and shorten schedules. These misconceptions or innocent mistakes, if perpetuated at conferences or reflected in FEED, can be perceived as reality. These "design and execution myths" will result in less accurate cost estimates or increased technical and execution risk for new projects.

This paper will review LNG technology and project execution myths that could derail the innovation necessary to develop the next generation of LNG projects. By openly discussing these myths, we can stimulate constructive debate on the challenges facing today's projects.

(25b) Boil-off Gas System Design for Integrated Liquefaction and Regasification Facilities

Bindupriya Bandari, Bechtel Oil, Gas & Chemicals, Inc., Houston, TX and Dipanjan Bhattacharya, Bechtel Oil and Gas, Houston, TX

BOG system design for integrated liquefaction and re-gasification LNG facilities

Dipanjan Bhattacharya

Senior Process Engineer

Bechtel Oil, Gas, and Chemicals, Inc.

Bindu Bandari

Process Engineering Specialist

Bechtel Oil, Gas, and Chemicals, Inc.

Designing boil-off gas (BOG) system for a LNG facility that offers both liquefaction and re-gasification capabilities is inherently complex and challenging. Steady state simulation tools are commonly employed to design the BOG system for independent liquefaction or regasification facilities. But their application is limited when working with a complex network of multiple jetties, multiple LNG storage tanks, multiple boil-off gas compressors and multiple modes of operation.

This topic discusses a case study based on a dynamic simulation model of an integrated facility that is converted from an existing LNG import terminal to a bi-directional facility. The design challenges are compounded by the fact that the model has to evaluate the hydraulic adequacy of the existing systems for an entirely new operating envelope.

A well-developed integrated dynamic simulation model can be used for hydraulic analysis, compressor selection, what-if-analysis and preliminary tuning of the controllers and sizing of control valves. The what-if-analysis can be performed to evaluate transient operating conditions, system performance and response; to various operating modes - ship loading mode, normal operation and upset scenarios such as compressor trip

This case study illustrates that a rigorous dynamic model is an effective means to analyze a complex BOG system for an integrated LNG facility.

(25d) Removal of Heavy Hydrocarbons from Lean Natural Gas

Yu-Nan Liu, Air Products & Chemicals, Allentown, PA and **Gowri Krishnamurthy**, Air Products and Chemicals, Inc., Allentown, PA

Removal of heavy hydrocarbons, such as C_6+ from natural gas is essential prior to cryogenic liquefaction in order to prevent freeze-out of these components in the liquefier. Traditionally, cryogenic methods such as passing feed gas through a scrub column or a front-end NGL extraction unit have been employed for this purpose. However, these conventional approaches may not be feasible for certain feed gas streams such as lean natural gas containing low amounts of C_2-C_4, yet significant amounts of C_6+. Lately, several sources of natural gas, such as domestic pipelines, shale gas, coal-bed methane, etc., report lean feed gas of this nature. Adsorption-based separation processes are an alternative method to strip trace heavy hydrocarbons from natural gas. In this paper, we discuss important considerations for evaluating such adsorption systems as well as possible integration with conventional cryogenic methods for the effective removal of heavy hydrocarbons from natural gas.

"Qatar: the largest LNG exporter in less than ten years"

By J C Kuo, Hao Dong, Marco Vaccaro and Dr. K H Wang.

University of Houston

Abstract

Natural gas is one of the most important energy sources. Due to its large reserves, there is enough supply to last over two hundred years for human consumption. Due to supply constraints and location, natural gas is often transported as liquefied natural gas (LNG). Qatar first started exporting LNG to Spain in 1997 and became the world's largest LNG exporter within a decade. This amazing growth is due to the development of the North Field which is the world's largest non-associated gas field. Qatar is now exporting LNG all over the world and represents over 30% of total LNG exporting. As the world energy demand expands, Qatar will continue to play a major role in LNG exports.

This paper will cover the following:

1) An introduction and background on Qatar
2) Specifics on Qatar's natural gas reserves and production
3) General information on Liquefied Natural Gas
4) Details on Liquefied Natural Gas in Qatar
5) Future and conclusions of Liquefied Natural Gas in Qatar

(1) Introduction to Qatar

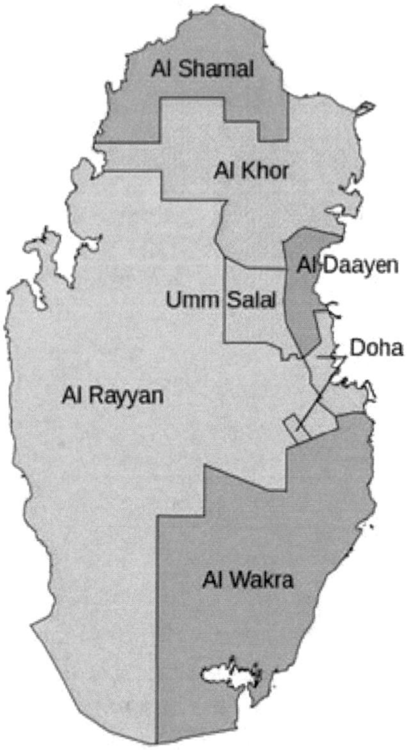

Figure 1 Qatar Municipality Map after 2004 [4]

Qatar is located in the Middle East. Qatar's sole land border is with Saudi Arabia, and the rest of the territory is surrounded by the Persian Gulf. The country spans 4,416 square miles with a population of 1,759,000. Although this population is small it is very diverse. The Qatari ethnic groups include: Arabic 40%, Indian 20%, Nepali 13%, Filipino 10%, Pakistani 7%, Sri Lankan 5%, and other 5%. Key languages spike in Qatar are: Arabic, English, Malayalam, Hindi, Tamil, Kannada, Tagalog, Urdu, and Punjabi. Three quarters of the country is Muslim while other religions make up the final 25%. The geography in Qatar is mostly desert, where water is scarce and agriculture is minimal. [2]

Qatar was once controlled by the Sheikhs of Bahrain until 1867. The country has been ruled by the Al Thani family since the mid-nineteenth century. Qatar was a British protectorate noted mainly for pearling and fishing in 1916. Once oil was discovered in the 1940's, it became one of the region's wealthiest states due to its enormous oil and natural gas production. Qatar became independent in 1971. Since then Qatar has built intimate military ties with the United States during the Persian Gulf War in 1992, and signed a defense pact with United States in 1994. [3]

Qatar now has commercial ties with the United States. Historical trade volumes have grown from $738 million in 2003 to $3.2 billion in 2009. Over the same period, the U.S has become one of the largest import partners for Qatar. Qatar attracted an estimated $100 billion in energy investment, approximately $60 to $70 billion coming from the United States.

Qatar's government is an elected, unicameral, federal republic. It is divided into seven municipalities, as shown by Figure 1. Qatar is a member of the Organization of the Petroleum Exporting Countries (OPEC), Gulf Cooperation Council (GCC), and the Oil & League. 60% of Qatar's gross domestic product is from oil and natural gas revenues. Qatar's GDP is $172.98 billion and has the world's largest GDP per capita at $109,000. In 2011, Qatar consumed approximately 147,000 bbl/d of oil equivalent, 79% of this total was attributable to natural gas. [4]

Qatar Petroleum (QP) controls all aspects of Qatar's oil sector, including exploration, production, refining, transport, and storage. It also holds the rights to all petroleum resources in the Qatari territories. QP often enlists foreign company involvement such from ExxonMobil, Chevron, Shell, and Total.

Qatar has the world's largest and second largest natural gas companies, Qatargas and Rasgas. Both are located in Doha, the capital of Qatar. Doha is a very large and developed city. Doha contains more than 60% of the Qatar's population. Due to the overcrowding in Doha, Qatar is building a new city nearby named Lusail.

(2) Qatar's natural gas reserves and production

The Persian Gulf's coastal areas are the world largest single source of crude oil. The Safaniyi Oil Field and the North field represent the world's largest offshore oilfield and non-associated natural gas field respectively..

Qatar has proved natural gas reserves of 896 Tcf, which represents 14% of the world's natural gas reserves. Qatar's natural gas reserves rank third behind Russia and Iran, Figure 2. Qatar produces 4.12 Tcf/y, of which it exports 3.35 Tcf/y.

Qatar is now exporting nature gas all over the world. This includes Asia, Europe, South and North America. Because of the development of the North Field, Qatar became the world's largest LNG exporter in 2006. Qatar exported over 80 MTPA of LNG in 2011, around 70% of which was LNG. Qatar currently exports 2 Bcf/d of natural gas to UAE and Oman through the Dolphin pipeline, which designed capacity, is 3.2 Bcf/d. [5]

Qatar also utilizes gas-to-liquid facilities for some of its excess gas production. Two world class GTL facilities include the Oryx GTL plant and the Pearl GTL project which is the largest GTL plant in the world . Qatar is one of the only three countries along with South Africa, Malaysia to have operational GTL facilities. Pearl is the largest GTL plant in the world and the first integrated GTL operation.

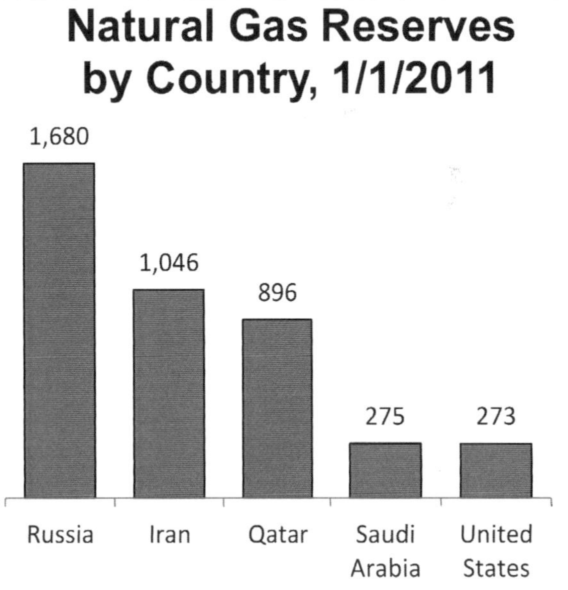

Figure 2 World Natural Gas Reserves TCF, by Country, 2011 [6]

(3) Introduction to Liquefied Natural Gas

The demand for natural gas has risen by 28% from 2001 to 2011.[7] However, this demand is not being met by supply in some remote parts of the world. This is mainly due to the difficulties of transporting natural gas through terrains such as mountains, and oceans. Liquefied Natural Gas provides a solution to this problem by significantly reducing the volume of gas and using tankers for transport. Along with volume reductions LNG provides a much purer gas, consistent of more methane than typical natural gas. LNG capital costs are significant, however shipping costs are very competitive compared to other forms of gas transport.

The LNG process includes liquefaction, transport and regasification. Liquefaction consists of cleaning upstream gas through field facilities, followed by cooling the gas to -160℃. Gas liquefaction is assisted with refrigerants that match the cooling curves of the gas. There are many different methods used for gas liquefaction. Today commercial liquefaction is dominated by only a few licensed technologies. The Propane-MR (PMR) and AP-X of Air Products (APCI) process, and its derivatives, accounts for about 75% of the market share. The Optimized Cascade (OC) process of

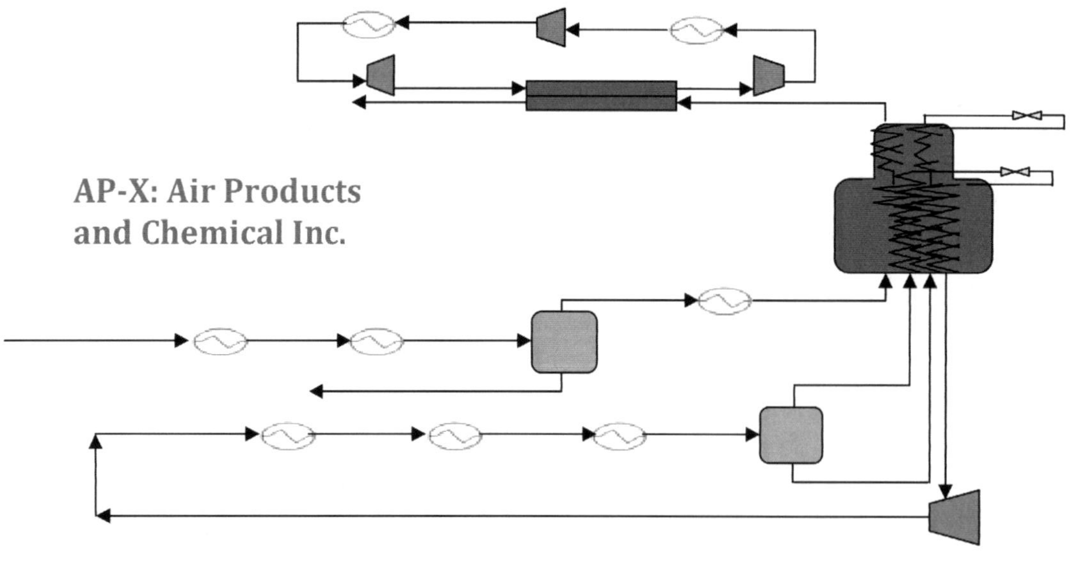

Figure 3 The AP-X Process of Air Products [17]

ConocoPhillips

(COP) represents another 20%. The other processes, such as the Multi-Fluid Cascade (MFC) of Linde and Single Mixed Refrigerant (SMR), make up the balance.[11] All of these processes reduce 54.6 Mscf of natural gas to one metric ton of LNG. This is a 600:1 volume reduction. The Optimized Cascade (OC) process of ConocoPhillips (COP) is shown in Figure 4. Mathematical examples of the cooling process are

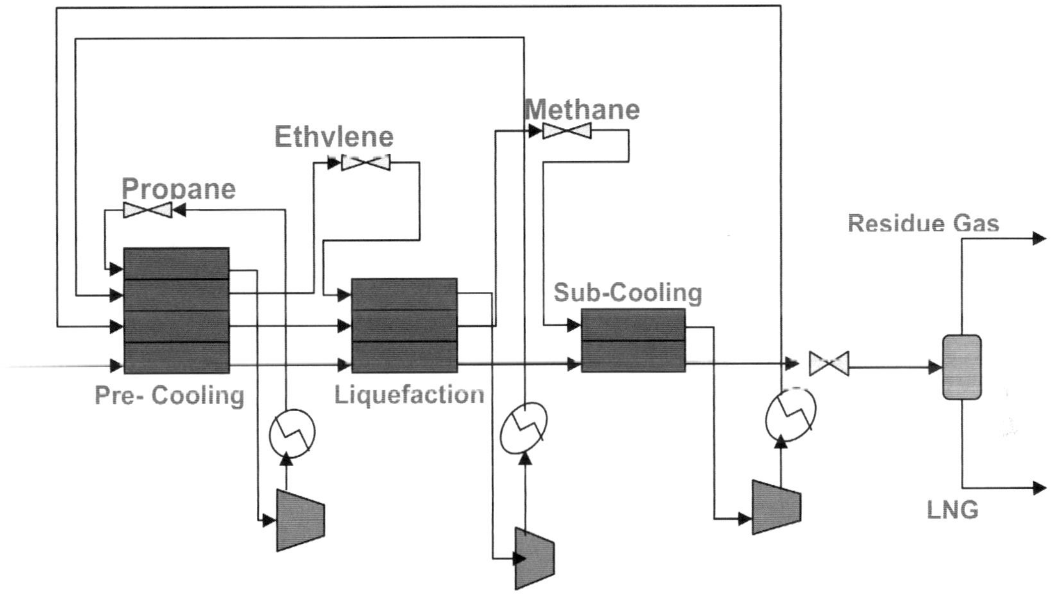

Figure 4 The Optimized Cascade (OC) Process of ConocoPhillips [17]

demonstrated in Wang, X., and Economides, M.J.: "Advanced Natural Gas Engineering".[1]

LNG train capacity has been increasing over the past thirty years in order to capture economies of scale. The capacity of an LNG plant is determined by many factors. The first consideration is the size of the gas reservoir which can support several large LNG trains for a reasonable time period. The second consideration is the technology limitations such as the size of gas turbine, compressors and piping. The third consideration is the marketability of the product.

LNG transport typically involves large carriers containing containment systems. As with the liquefaction process, there are many different LNG carrier designs. These carriers contain spherical or membrane type holding systems. Carriers allow for remote sea travel which eliminates geographical issues to natural gas transport.

Gas regasification begins once the LNG is offloaded from the carrier. Heat is added to the LNG to get an initial fraction of natural gas. Once all of the gas is in its natural form, it is odorized and sent to gas pipelines.

Natural gas liquefaction was developed by Michael Faraday in 1873 and by 1912 the first liquefied natural gas plant was built in West Virginia. The first LNG tanker was built from a converted World War II freighter in 1959. The United Kingdom became the first LNG importer when they started trading with Algeria in 1964.

As of 2011, there are 96 liquefaction trains and 89 regasification terminals around the world. This represents 279 MTPA in liquefaction capacity and 608 MTPA in regasification capacity. The global LNG carrier fleet currently consists of over 360 LNGC.[16] In total, 2011 LNG import/exports exceeded 240 MT. This was a 10% increase from 2010.[14] Recent increases in LNG demand were fueled by the severe earthquake and tsunami in Japan. A breakdown of the growth of LNG and exports by country is given below in figures 5, and 6.

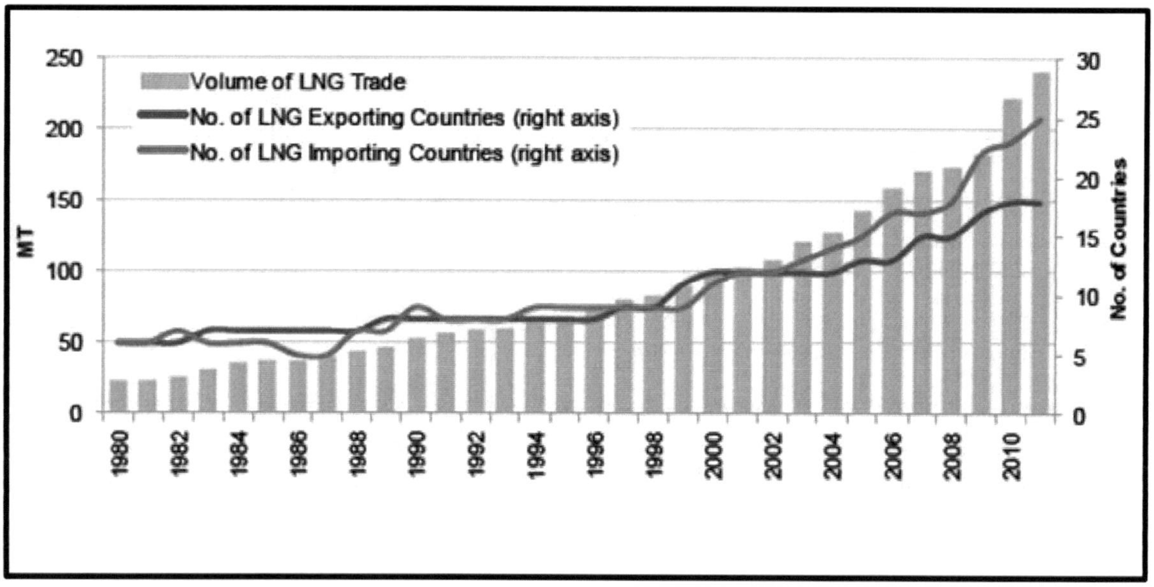

Australia will have signaficant growth in the near future due to the new projects such as Gorgon, Wheatstone and other LNG base load plants.

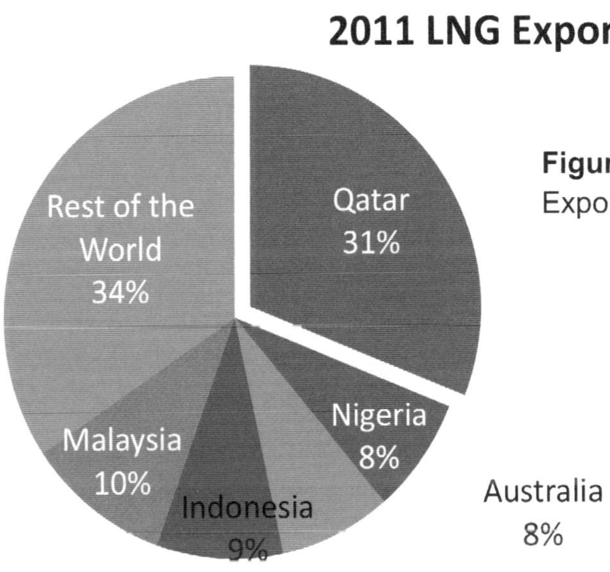

Figure 6 2011 LNG Exports by Country [7]

(4) Details on Liquefied Natural Gas in Qatar

LNG began in Qatar with the establishment of the Qatargas company in 1984. Qatargas was founded simply as an LNG company. By 1996, Qatargas produced and shipped its first supply of LNG to Japan. Qatargas has grown and into four separate

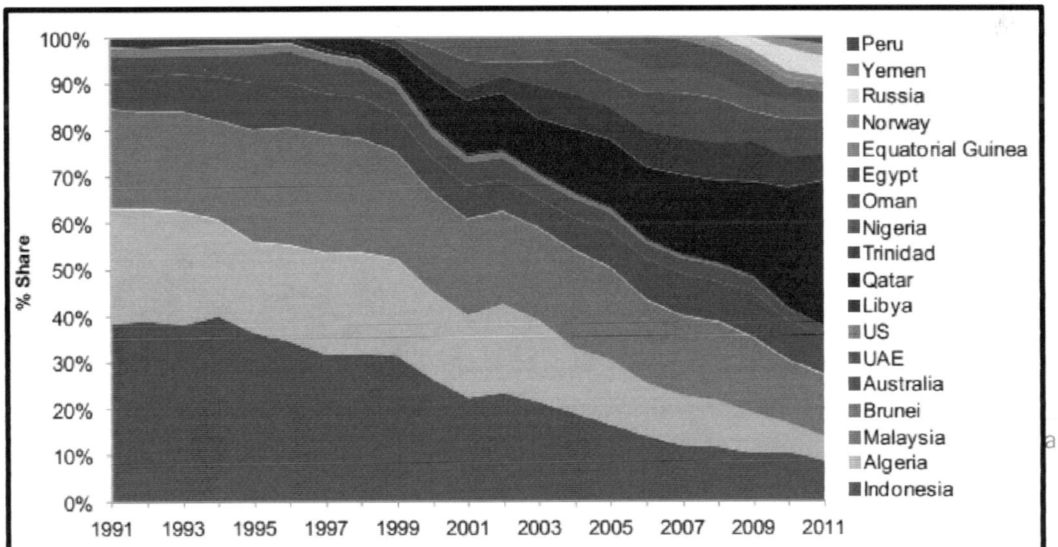

ventures since 1996 and now supplies 42 MTPA of LNG. Figure 7 shows a chart of the growth of LNG in Qatar and other countries since 1991. As you can see Qatar did not get into LNG until after other countries had shown success in the industry.
In Qatar, the typical LNG base load plant from feed gas to transportation is shown in Figure 8. Gas come from the offshore or onshore fields and then will go through Acid gas removal, Mercury removal, water removal/dehydration, heavy hydrocarbon removal, and liquefaction. Finally the LNG is put on a LNGC (LNG carrier) for ship transportation.

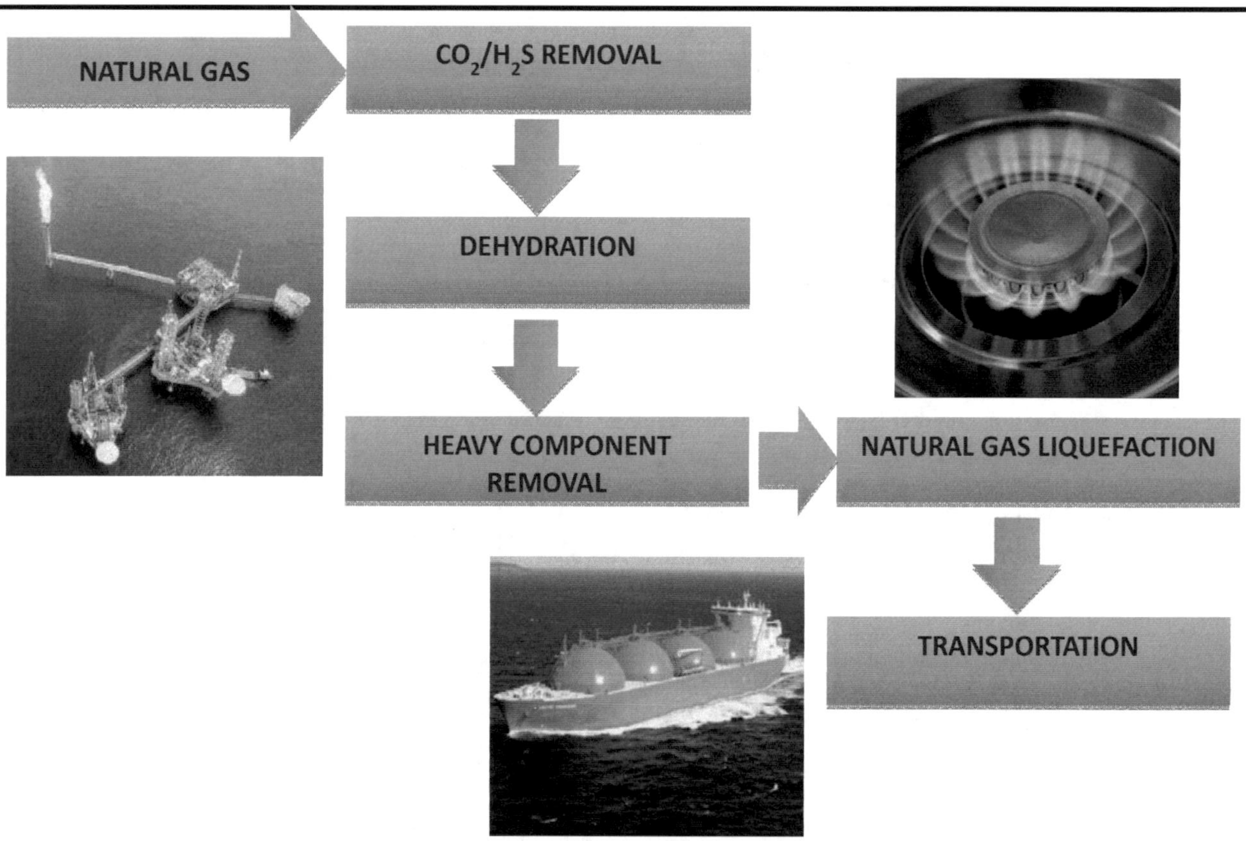

Figure 8 LNG from feed gas to transportation [17]

Qatar only uses APCI process trains. Qatar has 14 LNG trains, including six "mega-trains" that produce 7.8 MTPA. Trains 6 and 7 of Ras Laffan Liquefied Natural Gas Company (RasGas), and trains 4, 5, 6 and 7 of Qatar gas.

Qatar's LNG capacity has grown from zero to about 80 MTPA in less than 10 years. The offshore gas supply is the only bottle neck for increased production. Therefore, most of the LNG trains have pre-investment for more capacity. Debottlenecking has already been successful on Qatar gas trains 1, 2 and 3, where it has increased their combined capacity from 6.6 to 10.2 MTPA (54.5% increase).

Each mega-train will be able to give about 1.6 MTPA of additional capacity. Qatar could add 7.2 to 10.0+ MTPA at short notice due to their pre-investment in the early stage. Typically without major pre-investment, the maximum additional capacity is around 5-7+%.

Before 2006, there were approximately 155 LNG carriers in operation with a total capacity of about 18 million cubic meters. In 2010, Qatar added another 54 ships. 9 of which are conventional LNGC (120 km3), 31 of these ships are Q-Flex (205 km3) and 14 are Q-Max (250 km3). This is a total added capacity of 11 million cubic meters.

Qatar's LNG Infrastructure, January 2011			
Unit	**Liquefaction Capacity**	**Start-up**	**Primary Market(s)**
RasGas Facilities			
Trains 1& 2	2 x 3.2 MMt (320 Bcf)	Aug. 1999	South Korea
Train 3	4.7 MMt (230 Bcf)	Feb. 2004	India
Train 4	4.7 MMt (230 Bcf)	Aug. 2005	Europe
Train 5	4.7 MMt (230 Bcf)	Nov. 2006	Europe & Asia
Train 6	7.8 MMt (380 Bcf)	Jul. 2009	Asia, Europe & N. America
Train 7	7.8 MMt (380 Bcf)	Feb. 2010	Asia, Europe & N. America
QatarGas Facilities			
Trains 1-3	3 x 3.2 MMt (480 Bcf)	Dec. 1996	Japan & Spain
Train 4	7.8 MMt (380 Bcf)	Apr. 2009	UK, Europe & Asia
Train 5	7.8 MMt (380 Bcf)	Sep. 2009	UK, Europe & Asia
Train 6	7.8 MMt (380 Bcf)	Nov. 2010	Asia & N. America
Train 7	7.8 MMt (380 Bcf)	Mar. 2011	Asia & N. America
Source: RasGas, Qatargas, media reports			

Table 1 LNG Trains in Qatar [4]

The Q vessels are a new breed of LNGC that are being considered as an industry breakthrough in size and design. The tankers will enable more efficient transportation of natural gas throughout the world. The Q-Flex and the even larger Q-Max are a new generation of LNG mega ships with about 40% lower energy requirements and carbon emissions than conventional vessels due to the economies of scale created by their size and the efficiency of the engines.

Alongside excellent operational servicing, these vessels are also known for their excellent constructional features. Specifically, the usage of silicon ensures the protection and safety of the marine ecology. The price of a Q-Max increased from $300 million in 2005 to about $500 million currently because steel and raw material prices have risen.

(5) Future of LNG in Qatar and Conclusion

Qatar only has a population of 1.8 million including the expats that work there. This is only about 0.025% of the world's population. However, they have built enough LNG base load plants and LNG carriers to support about 7.2% of the world's total energy consumption. Furthermore, Qatar became the world's no. 1 LNG exporter in 2006. This was achieved in less than a 10 year time span.

Qatar's North Field, located in the Persian Gulf, is one of the largest non-associated natural gas fields in the world. This is a very good geographic petroleum location in terms of development and transportation. Qatar's natural gas reserves are no. 3 in the world and could supply LNG for about 225 years with current reserve and production. Russia's natural gas reserves are currently the largest in the world. However, many reservoirs are located in very severe climate conditions. Therefore, natural gas development cost is higher and scheduling takes longer. Iran has the second largest natural gas reserves. However, their current political standing and relationships with the United States and European countries is unsteady. The most dominate LNG technology providers, APCI (Air Products) and COP (ConocoPhillips), are both US companies.

Qatar has signed a defense pact with United States. They also have excellent relationship with the US and most industrialized countries. Furthermore they work with most of the major international oil companies. This includes ExxonMobil, Chevron, Shell, Total and ConocoPhillips in many projects. Typical engineering and construction of any major LNG based plant takes four to five years to complete. Additional trains will take one and a half to three years. Nevertheless, with excellent planning and the large oil companies' expertise, Qatar was able to build an LNG empire in a short period of time.

There are no plans to build new LNG facilities in the near-term future in Qatar. So any additional capacity increases will be the result of improvements to their existing facilities. Most LNG trains have pre-investment for additional capacity. Each mega-train will be able to give about 1.20 to 1.66+ MTPA of additional capacity. Qatar could add 7.2 to 10.0+ MTPA at short notice due to the pre-investment in the early stage.

Currently, Qatar exports LNG to numerous countries all over the world. This includes Spain, Turkey, Italy, U.S., France, South Korea, India, China, and the U.K. All of these countries share a strong relationship with Qatar. The European countries are dependent on LNG due to their location and their production of natural gas. The Asian countries are booming economically. This means that they will need lots of energy in the future.

As we can see from Table 2, Japan is the largest LNG importer in the world, but the country is much diversified in selection. Japan imports average 10% from each country such as Australia, Indonesia, Malaysia, Russia, and Qatar.

Korea is the second largest LNG importer in the world with more than 20% coming from Qatar. U.K., Spain, India, and Taiwan have a similar situation to Korea. They import a lot of LNG, but Qatar is already their major exporter.

China has grown by an average of ten percent per year over the past two decades, and became the second-largest economy and energy user in 2010 after the U.S. [18] China's bill for LNG has surged in the past four years as it feeds its booming economy and cuts reliance on coal for power generation. [19] China consumed 2,432 MT of energy in 2011 However, only 4% was from Natural gas compare to 26% in the U.S. Also, the nature gas price is high in China compared to the rest of the world. Average natural gas price in China is $282/T, which is about three times the U.S. price. Even though China is the fifth largest LNG importer, it has the most growth potential in the market. Qatar has all of the key essentials to meet this demand. This includes: reserves, management skill, relationships, hardware and track record.

Importer	LNG Import From Qatar, MT	Total LNG Import, MT	Percentage
Japan	7.91	78.8	10.04%
Korea	7.58	35.8	21.17%
UK	10.57	18.6	56.83%
Spain	4.20	17.1	24.56%
China	1.27	12.8	9.92%
India	8.05	12.7	63.39%
Taiwan	2.88	12.2	23.61%
France	1.77	10.7	16.54%
Italy	4.56	6.4	71.25%
US	0.94	5.9	15.93%
Turkey	1.46	4.6	31.74%
Belgium	4.51	4.5	100.22%
Argentina	0.18	3.2	5.63%
Mexico	0.72	2.9	24.83%
Chile	0.12	2.8	4.29%
Canada	0.18	2.4	7.50%
Kuwait	0.00	2.4	0.00%
Portugal	0.06	2.2	2.73%
UAE	0.12	1.2	10.00%
Greece	0.03	1	3.00%
Dom. Rep.	0.00	0.7	0.00%
Thailand	0.00	0.7	0.00%
Brazil	0.38	0.6	63.33%
Netherlands	0.00	0.6	0.00%
Puerto Rico	0.00	0.5	0.00%
Total	57.49	241.3	

Table 2 LNG Imports by Country, 2011 [9]

References

1. Wang, X., and Economides, M.J., “Advanced Natural Gas Engineering”. Gulf Publishing Company (2009).
2. Central Intelligence Agency, World Wildlife Fund, US State Department, Qatar, January 8, 2012. <http://www.eoearth.org/article/Qatar?topic=49460>
3. Qatar, <http://www.infoplease.com/ipa/A0107901.html>
4. Wikipedia, Qatar, <http://en.wikipedia.org/wiki/Qatar>
5. EIA, Qatar Energy Data, Statistics and Analysis - Oil, Gas, Electricity, Coal, 2011
6. Oil & Gas Journal, Jan. 1, 2011 and EIA Natural Gas Navigator (U.S. only), Dec. 31, 2009
7. 2011 BP Statistical Study, BP, <http://www.bp.com/sectionbodycopy.do ?categoryId=7500&contentId=7068481>
8. Dominion, 2012, <https://www.dom.com/business/gas-transmission/cove-point/history-of-lng.jsp>
9. International Gas Union, PETRONAS, “World LNG Report 2011”, (2012)
10. Saggas, <http://www.saggas.com/en/caracteristicas-tecnicas/proceso-de-regasificacion/>
11. Pek, J. J. B., van Driel, A., de Jong, E. C. J. N., Nagelvoort, R. K., “Large Capacity LNG Plant Development”, paper presented at LNG 14, Doha, Qatar, March 21-24 (2004).
12. Avidan, A., Martinez, B., Varnell, W., “Study Evaluates Design Considerations of Larger, More Efficient Liquefaction Plants”, Oil & Gas Journal, issue of August 23 (2003).
13. Bauer, H., “A Novel Concept”, Hydrocarbon Engineering, issue of May (2002).
14. Dam, W., Ho, S.-M., “Unusual Design Considerations Driver Selection of Sakhalin LNG Plant Facilities”, Oil & Gas Journal, issue of October 1 (2001).
15. Grootjans, H. F, Nagelvoort, K. N., Vink, K., J., “Liquefying a Stream enriched in Methane”, U. S. Patent 6,370,910 B1 (2002).
16. Grootjans, H. F, “Compression Apparatus for Gaseous Refrigerant”, U. S. Patent 6,637,238 B2 (2003).
17. Aduku, A., Harris, O., Rivera, V. Bagajewicz, M., “Evaluation of LNG Production Technologies”, University of Oklahoma
18. Saltanat Berdikeeva, China Turns to Natural Gas to Fuel their Economic Growth, (2012) <http://oilprice.com/Energy/Natural-Gas/China-Turn-to-Natural-Gas-to-Fuel-their-Economic-Growth.html>

(45a) Pilot and Full-Scale Application of Technologies for Mercury Control

Noah Meeks, Research & Environmental Affairs, Southern Company Services, Inc., Birmingham, AL

With the final MATS rule going into effect in 2015, electric utilities must now consider mercury control as a needed technology in their emissions train. This talk wiill cover recent full-scale injections of activated carbon and alkali injection, as well as pilot-scale studies of non-halogenated carbons or non-carbon sorbents. We will also discuss recent mixed matrix membrane pilot studies. **(Most recent data will be included in the abstract closer to the meeting)

(45b) Acid Tunneling Vs Acidizing : A Case Study of Rudatain Oilfield-Kuwait

Adel Elsharkawy, Petroleum Engineering, Kuwait University, Kuwait, Kuwait and Mohamed Al-Hadher, Pet Eng, Kuwait Oil Company, Safat, Kuwait

Acid Tunneling Vs Acidizing : A Case Study of Rudatain Oilfield-Kuwait

By Mohammed Al-Hadher and Adel Elsharkawy

Kuwait University, College of Engineering & Petroleum

Po Box 5969 Safat 13060 Kuwait

Abstract

Acidizing is one of the most important techniques used in oil industry in order to enhance the reservoirs and increase oil production. There are three types of acidizing techniques which are matrix acid stimulation, acid fracturing, and acid tunneling. In this paper a comparison these types of acidizing is discussed. The case study of RA-0007 acid tunneling completed in Kuwait Oil Company is presented. It represents the first dendritic with extensive worm holes created by acid tunneling for maximum reservoir contact. The whole effect is aptly demonstrated by the change in PI of the well before and after the acid tunneling as mentioned before. Analysis of the well designed and implemented job showed that two tunnels were created successfully in MA-C formation through a two days rig-less operation. The results indicated an improvement in the well productivity index (PI) which had been increased from 1.3 to 3.87 bpd/Psi.

(45e) Sustained Casing Pressure Modeling and Well Integrity

Tony Rocha-Valadez, Mary Kay O'Connor Process Safety Center, Artie McFerrin Department of Chemical Engineering, Texas A&M University, College Station, TX, A. Rashid Hasan, Harold Vance Department of Petroleum Engineering, Texas A&M University, College Station, TX, C. Shah Kabir, EPT, Hess Corporation, Houston, TX and Sam M. Mannan, Artie McFerrin Department of Chemical Engineering, Mary Kay O'Connor Process Safety Center, Texas A&M University, College Station, TX

SUSTAINED CASING PRESSURE MODELING AND WELL INTEGRITY

T. Rocha-Valadez[1*], A.R. Hasan[2], C.S. Kabir [3] and S.M. Mannan[1]

[1]Mary Kay O'Connor Process Safety Center, Artie McFerrin Department of Chemical Engineering, Texas A&M University, College Station, Texas 77843-3122, US., email: *tonyrochav@tamu.edu, mannan@tamu.edu.

[2] Harold Vance Department of Petroleum Engineering, Texas A&M University, College Station, Texas 77843-3122, US, email: rashid.hasan@pe.tamu.edu.

[3]Hess Corporation, 1501 McKinney Street, Houston, TX 77010, US, email: skabir@hess.com.

With an increasing demand for hydrocarbons and other sources of energy, the methods for delivering such energy needs have become complex. In the case of the oil and gas industry, exploration and production have gone to deeper waters and deeper horizons than ever before, resulting in drilling several miles into the earth's core. Deeper well penetration means encountering increasingly high formation pressures and temperatures. For example, in deepwater drilling, well construction and production management deal with significant heat transfer between the produced fluids and its surroundings. The high-rate production and the attendant heat transfer demand close attention in terms of operational safety and potential environmental impact. One phenomenon that occurs in producing wells is sustained casing pressure (SCP), which presents a risk to both process safety and the environment.

In oilfield terminology, the casing is a steel pipe cemented to the surrounding formation. Multiple concentric casings are involved in a typical well construction. The casing immediately surrounding the production tubing is known as the production casing, which is most vulnerable to thermal and mechanical stresses and fluid leakage. The cement sheaths provide zonal isolation, support the casing, and protect it from corrosive fluids. Ideally, pressures monitored at the wellhead for multiple casings should change minimally after being bled off through a needle valve at the surface. When the pressure rebuilds after being bled off, that casing is said to exhibit SCP [1].

SCP occurs because of stresses that the system experiences owing to fluids production up the tubing. Besides the normal source of SCP, a poor cement job may allow hydrocarbon migration from the producing or other horizons and accumulation into one or more of the casings. The risk of not attaining well integrity can range from the activation of rupture discs to the blowout of hydrocarbons. Because of the safety and environmental dangers of SCP, regulations require remedial treatments for a casing string that exhibits significant SCP problems. Unfortunately, available methodologies to test for SCP are often qualitative and empirical in nature [2, 3], rendering them inapplicable in many situations. No practical systematic testing procedure has emerged to quantify SCP.

In this paper, we present a model for SCP that is rooted in the transport processes of the system. The model assumes gas leaking through the cement barrier, and then migrating up the annular-liquid column. It accounts for the fluid-temperature profile along the wellbore and liquid compressibility. Some robust assumptions about the transport processes allow us to write the governing equation as a first-order, linear-differential equation that is solved with appropriate boundary conditions. The resulting algebraic expression for the casing pressure rise as a function of time paves the way for all applications. A comparative study of the model's performance with field data suggests excellent agreement for wells where the casing pressure increase is owes gas influx into the annulus. The proposed model could form the basis for standardization of SCP testing. This study also offers some test procedures to assess SCP-related safety concerns.

References

1. Bourgoyne Jr., A.T.J., S.L. Scott, and W. Manowski, ***A Review of Sustained Casing Pressure(SCP) Ocurring on the OCS***, 2000, Final Report Submiteted to MMS.

2. Xu, R. and A.K. Wojtanowicz, ***Diagnosis of Sustained Casing Pressure from Bleed-off/Buildup Testing Patterns.*** SPE, 2001: p. Productions and Operations Symposium.

3. Wojtanowicz, A.K., S. Nishikawa, and R. Xu, ***Diagnosis and Remediation of Sustained Casing Pressure in Wells***, July 2001, MMS.

(45f) Evaluation of Particle Growth in an Electrochemical Fluidised Bed Reactor

Emmanuel Ehirim, Rivers State University of Science and Technology, Port Harcourt, Nigeria

The process of electrodeposition within electrochemical reactor results in the increase of particle sizes, which increases the volume of the bed. This work entails a mathematical model to describe the effect of this phenomenon on the behaviour of a three-dimensional fluidised bed electrode operating under the condition of electron transfer. The model incorporates the pseudo-stationary behaviour in the reactor since the bed increases in volume due to the particle size growth. The Orthogonal Collocation method was used to solve the developed differential equations. The overpotential distribution and the reaction rate within the reactor have been calculated. The principal results are the effects of the electron transfer operation on the superficial displacement with time. The effects of total applied current, electrode geometry and the bed porosity on the behaviour of the system have been analysed. The result shows a very gradual increase in the bed height.

(45g) Debottlenecking Gas Dryers and Desiccant Adsorbers

Benjamin A. Schmitt, Hydrocarbon Processing Industry (HPI) - Business Unit, Johnson Screens, Inc., New Brighton, MN and Michael Ekholm, Engineering, Johnson Screens, Inc., New Brighton, MN

Debottlenecking Gas Dryers and Desiccant Adsorbers

Though Maximization of the Useable Bed Volume with Johnson Screens' SSG™

Benjamin Schmitt, PE

Senior Mechanical Engineer, Technology

Michael Ekholm, PE

Applications and Product Design Manager

Johnson Screens Inc. St. Paul, MN

AIChE Spring Meeting, April 2013

13th Topical Conference

San Antonio, TX

Introduction:

The following presentation of the Shaped Support Grid (SSG™) releases additional supporting data and updates to the initial presentation given by Robert G. Norell at the 2010 AIChE Spring Conference.

Limitations on ethylene dryers and desiccant adsorbers can create bottlenecks in the production process which can lead to a need for increased dryer and adsorber capacity. Capital intensive investments may be necessary in order to eliminate the bottleneck which may include the addition of an additional vessel. With limits on the plot plan, this may not be a workable solution. Adding to the problem is the amount of time consumed in engineering, planning and construction, which can have negative effects on the operation of the ethylene unit. Increased desiccant volume and operational efficiency can increase the unit's production with minimal capital expenditures. Johnson Screens', patent pending, SSG™ can add the increased process efficiency and cycle life necessary to extend the operating capability of the unit.

Extending the operation, though debottlenecking, can be achieved by maximizing the time before breakthrough of the media bed. There are three primary functions that maximize the breakthrough time:

Proper gas distribution in both normal operation and regeneration conditions which maximizes the bed utilization.

Increased molecular sieve or desiccant bed volume are achieved through the elimination of traditional bed supports comprised of support grids and beams. The support interface for the screen configuration also allows for thermal expansion typically present in vessels with a thermal cycle as part of the process. The design allows for thermal cycling without the risks of seal or structural failure inherent in other solutions.

Minimizing the bed volume consumed by the distribution and support system increases the net desiccant volume lost when utilizing alternative bed supports.

Johnson Screens ***SSG*** is designed to provide the industry with an alternative, cost effective, approach to increasing capacity. Supporting data and preliminary case studies will be depicted to quantify the benefits that are achievable through the implementation of the ***SSG***.

Conclusions:

The use of a Johnson Screens ***SSG*** provides a reliable immediate solution to reduce capacity bottlenecks. The ***SSG*** increases the bed volume, when compared to traditional bed support arrangements comprised of grids and beams. The increase in desiccant volume will allow for increases in feed flow rate and/or adsorbtion cycle time. The ***SSG*** maximizes the flow distribution during desiccant regeneration, maximizing the utilization of the complete cross section of the vessel. This solution can be implemented in both new and existing vessels, providing consistency in design between planned increases in performance of current vessels and future expansions.

The modular design of the ***SSG*** allows for easy installation and maintenance. Connection of the ***SSG*** to the vessel is performed within the outlet nozzle, eliminating all welding on the pressure vessel and eliminating any need for recertification of the pressure vessel after installation. The design allows for ease of dismantling for inspection of the vessel head during routine maintenance and inspections. The low profile will also reduce interference with any required media exchanges. The design package of the ***SSG*** provides the end user with a holistic solution that fulfills debottlenecking needs with limited capital expenditures.

Biography of Speaker:

Benjamin Schmitt is a Senior Mechanical Engineer in Johnson Screens' Hydrocarbon Processing Industry Technology Group. Ben graduated from North Dakota State University in 2004 with a Bachelor of Science degree in Mechanical Engineering. He currently working toward his Master's in Business Administration from Hamline University. Ben is member of AIChE and is a registered professional engineer in the state of Minnesota.

AIChE 2013 Spring Meeting Poster board

Design Considerations for effective liquid removal In Suction Drums

Presented By:

Kanti Patel, PE

Compressor down again?

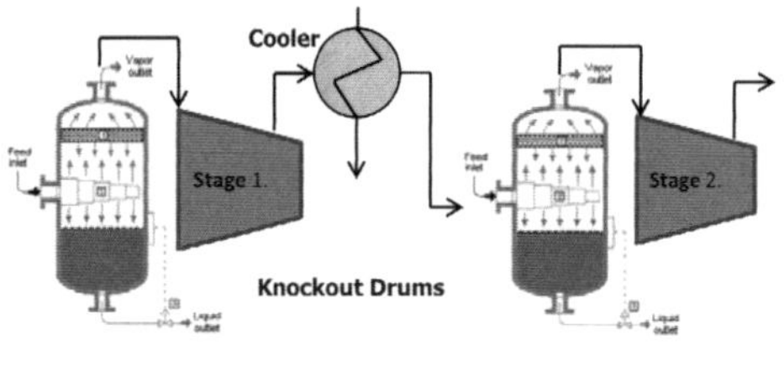

Is liquid carryover the culprit?

<u>We have heard it all before</u>:

-Premature compressor shutdown.

-Production time lost.

-High yearly compressor maintenance costs.

-Impeller damage due to pitting corrosion.

-Corrosion due to salt deposits.

Common Design Problems with Suction Drums which promotes liquid carryover

1. Choosing the correct mist eliminator (M.E. equipment)
2. Knowing where to install the M.E. in a suction drum.
3. Consider a high mist loading in the gas when selecting your M.E.
4. Fouling considerations.
5. M.E. disengaging space, bad gas distribution & liquid slug in the feed.
6. One micron and great droplets, versus submicron mist removal.
7. High viscosity liquids.
8. Dislocated M.E. due to pressure surges.
9. Through put exceeds original design capacity.
10. Capacity range too large for suction drum.

AMACS
PROCESS TOWER INTERNALS

Choosing the correct Mist eliminator equipment

- Proper sizing

+ Use Souders-Brown equation to find optimum velocity. $\mathbf{Vd = K} \cdot \sqrt{\frac{Pl}{Pg} - 1}$
+ Select the correct vapor velocity factor K. based on the operating pressure, and type of equipment.
  - Lower K-factor 0.01 for every 100 psi
  - K factor to select when using:

Mesh Pad	K= 0.1 - 0.55
Chevron Vane	K= 0.45 - 0.65
MultiPocket® Vane	K= 0.75 – 1.0

AMACS
PROCESS TOWER INTERNALS

Choosing the correct Mist eliminator equipment cont.

- Selection based on droplet size.

Table 2. Droplet sizes (water in air) typically captured with 99.9% efficiency by mist eliminator elements of various types

Element type	Size range microns
Fiber candles or panels	0.1 and larger
Mesh with co-knit yarn	2.0 μm and larger
0.006-inch knitted mesh	5.0 μm and larger
0.011-inch knitted mesh	10 μm and larger
AMACS MultiPocket ® Vane	10 μm and larger
Conventional vane arrays	15 μm and larger

AMACS
PROCESS TOWER INTERNALS

Where to install the M.E. in the suction drum

- Improperly installed M.E. can cause severe liquid carryover problems.
- General accepted spacing guidelines.

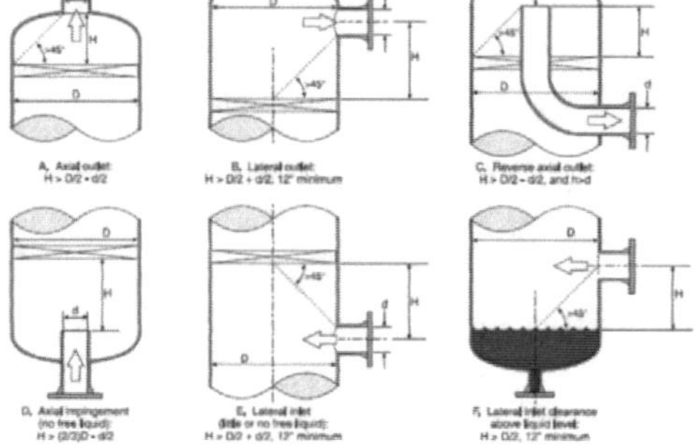

AMACS
PROCESS TOWER INTERNALS

Consider high mist loading when selecting your M.E. equipment

+ High mist loading in the gas stream can flood your M.E. equipment and cause liquid carryover.

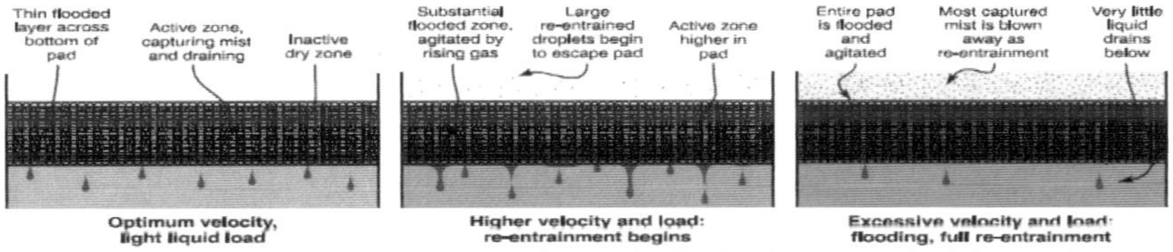

Figure 4. Envisioning mesh pad performance degradation as liquid and gas loads increase (vertical cross-sections)

AMACS
PROCESS TOWER INTERNALS

Consider high mist loading when selecting your M.E. equipment cont.

+ Limited drainage capabilities
  - mesh M.E. <0.5 gal/FT2/min
  - Vane type M.E. <2 gal/FT2/min

+ Select a combination of vane & mesh mist eliminators

AMACS
PROCESS TOWER INTERNALS

Fouling Applications

+ Fouling of mesh mist eliminator will cause liquid carryover.

+ For mild fouling install a water spray system below the mist eliminator for occasional inline cleaning (based on Delta-P)

+ For consistent fouling problems, install a Vane mist eliminator upstream of the mesh mist eliminator.
  - Vanes remove solids and bulk liquids therefore, protecting the downstream mesh mist eliminator

AMACS
PROCESS TOWER INTERNALS

Fouling Applications cont.

+ For consistent fouling problems where the solids are "sticky" install a water spray system below the vane mist eliminator.

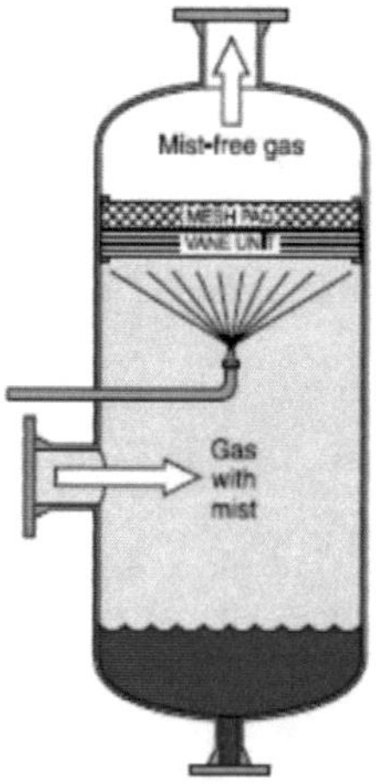

AMACS
PROCESS TOWER INTERNALS

Bad Gas distribution, limited disengaging space and liquid slugs.

+ Bad gas distribution, limited disengaging space in a suction drum can cause severe liquid carryover.

+ CFD analysis can assist in locating high velocity areas.

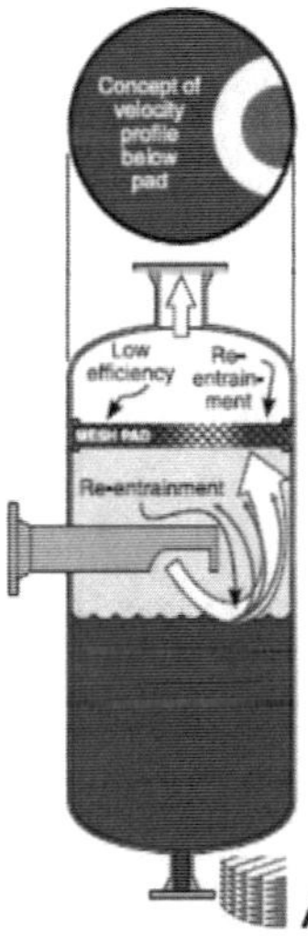

AMACS
PROCESS TOWER INTERNALS

Bad Gas distribution, limited disengaging space and liquid slugs cont.

+ Use inlet diffuser to force uniform gas distribution i suction drum.

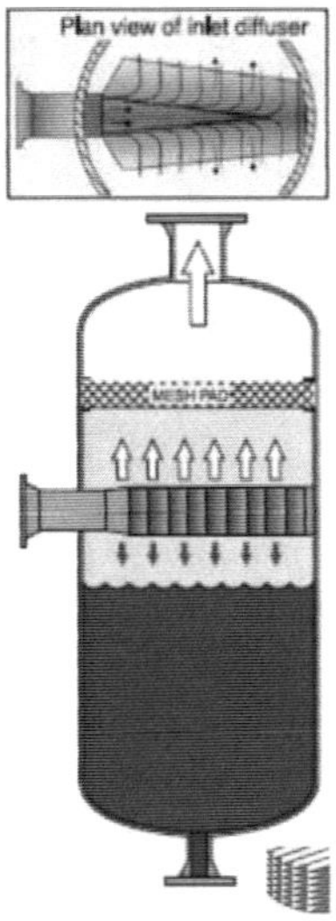

AMACS
PROCESS TOWER INTERNALS

Inlet Feed Device

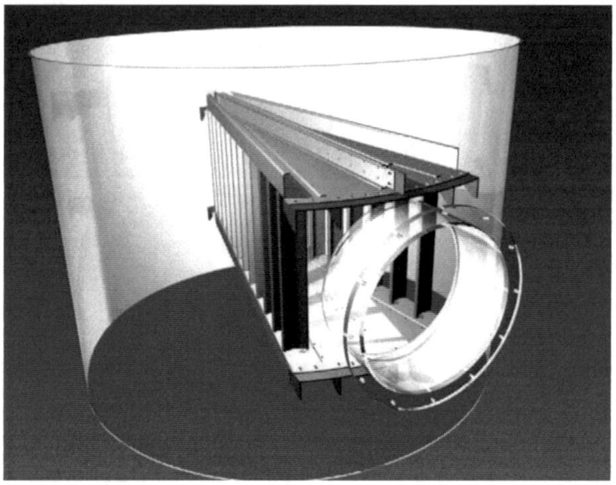

AMACS
PROCESS TOWER INTERNALS

Inlet Feed Device

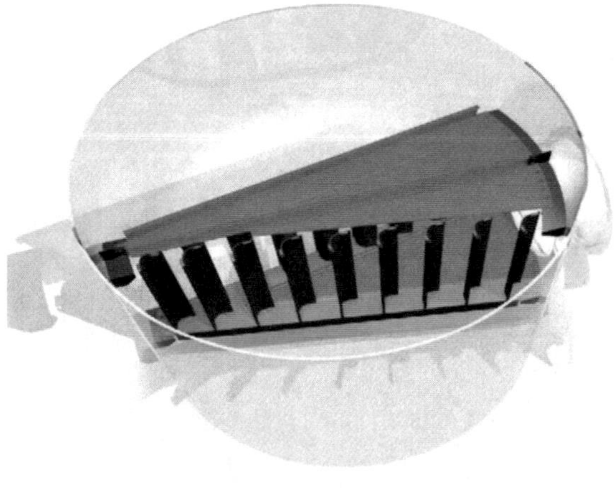

AMACS
PROCESS TOWER INTERNALS

Bad Gas distribution, limited disengaging space and liquid slugs cont.

+ Use a perforated distribution plate downstream of the M.E. for additional distribution assistance.

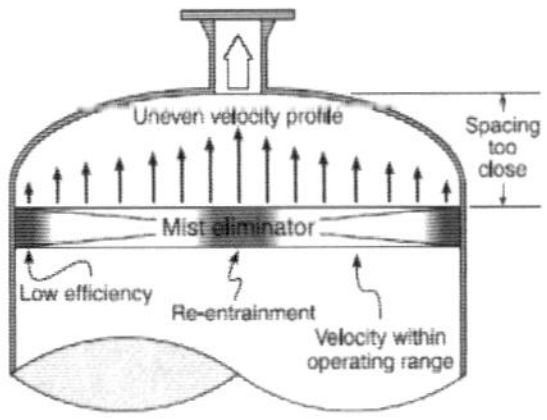

Figure 7. Example of mist eliminator performance degradation due to uneven velocity profile

+ Inlet diffusers also handle incoming liquid slug which otherwise can cause flooding of the Mist eliminator.

AMACS
PROCESS TOWER INTERNALS

Know what mist size droplets You are dealing with

+ Know if your mist is > 1 micron or submicron.
+ When gas travels from A to B, condensation in the pipeline will cause a fine mist (fog or submicron droplets)

  Examples:

  - Incoming gas from an offsite location
  - Fuel gas to furnaces/heaters with low NOx burners

AMACS
PROCESS TOWER INTERNALS

Know what mist size droplets You are dealing with cont.

+ Submicron mist is not captured by conventional M.E.

+ Submicron mist requires a coalescing filter, typically
installed in a two stage filter separator.

AMACS PROCESS TOWER INTERNALS

Know what mist size droplets You are dealing with cont.

<u>1^{c} stage;</u> submicron droplets are coalesced to > 10 micron droplets.

<u>2^{c} stage;</u> greater than > 10 micron droplets are ejected from the coalescing filter and captured by a vane, which captures droplets > 8 micron.

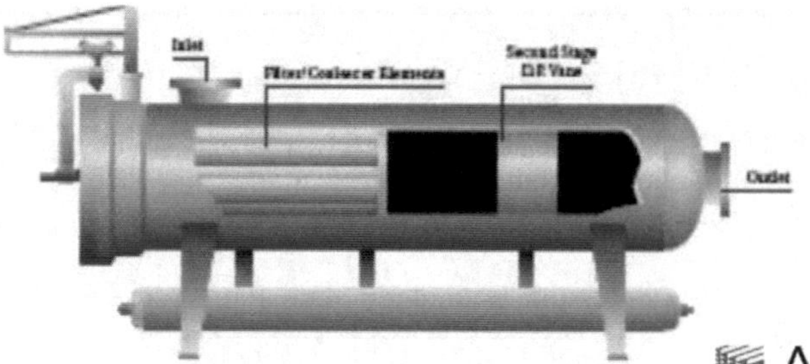

AMACS PROCESS TOWER INTERNALS

High viscosity liquid as mist

+ Viscous liquids drain slower out of the M.E. and could cause localized flooding.
+ Use a mesh mist eliminator with an open drainage layer on the bottom to improve liquid drainage.
+ Use a lower K- value (lower face viscosity) for the M.E. To allow additional drainage time.
+ Use a vane type M.E. upstream of the mesh mist eliminator, to remove the bulk of the viscous liquid, so the M.E. can effectively handle the residual liquids.

Dislocated M.E. due to System pressure surges

+ M.E. sections can be dislodged into the downstream equipment, which then creates massive carryover.
+ Standard M.E. equipment often gets installed and secured to the support ring with tie wire.

Dislocated M.E. due to System pressure surges cont.

+ Where system pressure surges are common, there are more secure ways to install the M.E.

  - dual support rings
  - through bolts
  - install a pressure relief door

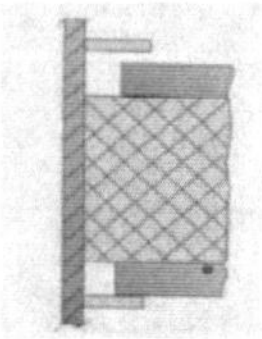

Dual support rings

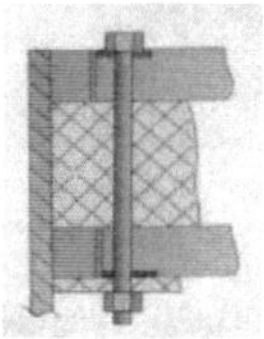

Through bolts

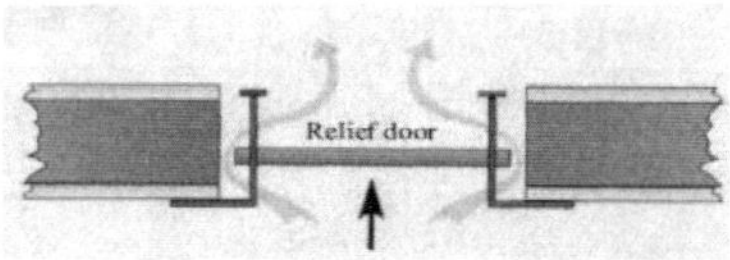

AMACS
PROCESS TOWER INTERNALS

When throughput exceeds original capacity

+ There are various solutions to increase capacity in an existing vessel.

  - Patented High capacit
    mist eliminator 0% - 1

AMACS
PROCESS TOWER INTERNALS

When throughput exceeds original capacity cont.

-Patented MISTFIX™
cylindrical mist eliminator 0% - 25%

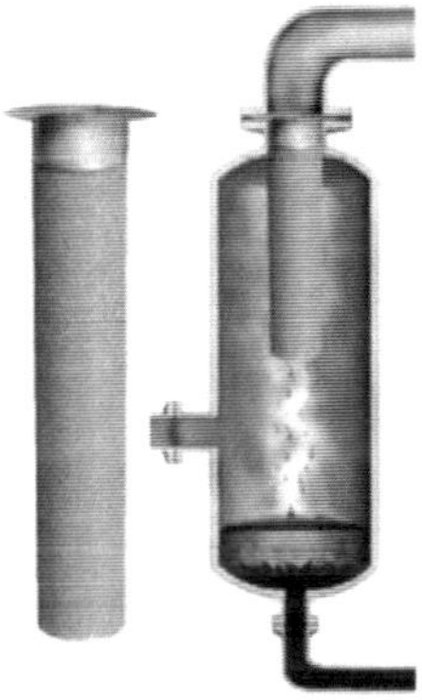

When throughput exceeds original capacity cont.

- High capacity Patented MultiPocket® vane. 0%-50%

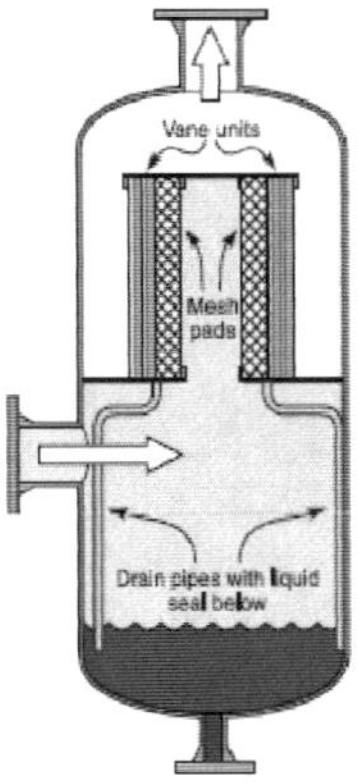

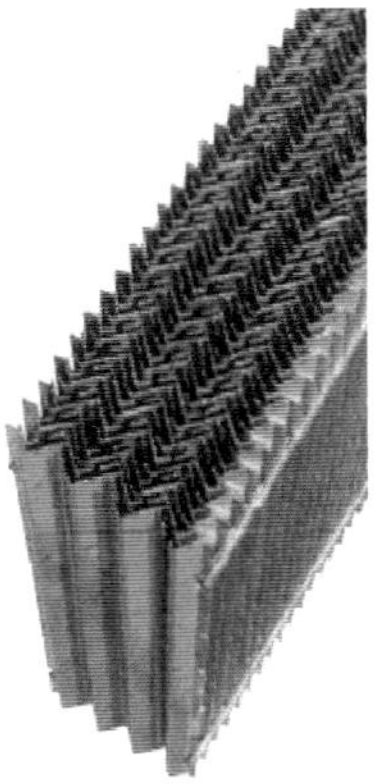

AMACS
PROCESS TOWER INTERNALS

When throughput exceeds original capacity cont.

+ MistFix™ is installed through the gas outlet, all other M.E. equipment comes in sections, installed through the manway.
+ With increased capacity comes increased bad gas distribution.
    - Use an inlet diffuser.
+ To avoid welding new support rings and housing for the high capacity new equipment, use double expansion rings, a compression fit.

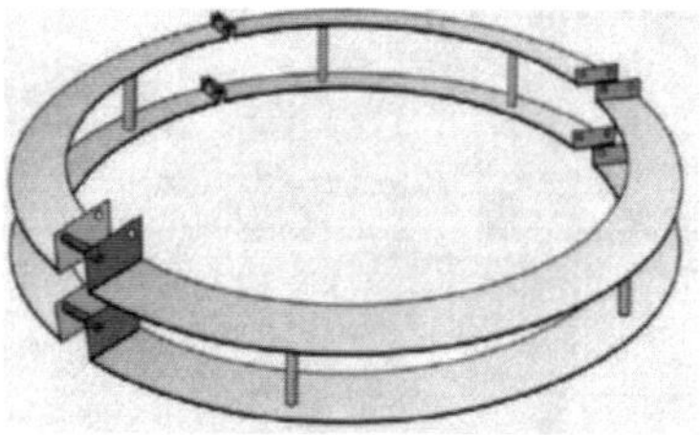

AMACS
PROCESS TOWER INTERNALS

Support Rings

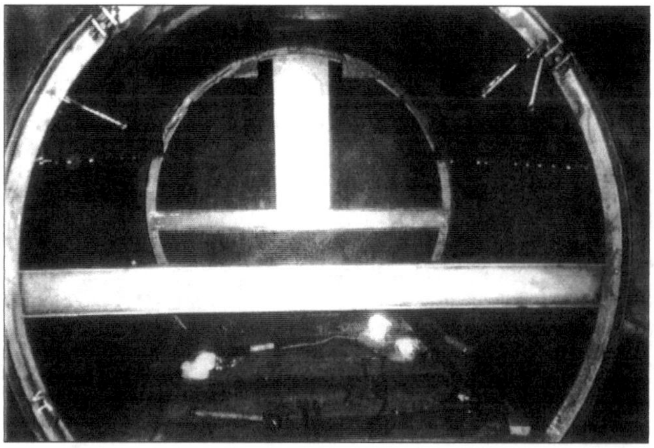

AMACS
PROCESS TOWER INTERNALS

Support Rings

AMACS
PROCESS TOWER INTERNALS

When the capacity range is too large for a suction drum

+ Typical M.E. equipment will operate between

  30% – 110% capacity.

+ When a suction drum often operates below 30% liquid carryover may occur.

AMACS
PROCESS TOWER INTERNALS

When the capacity range is too large for a suction drum Cont.

To facilitate a capacity range of 10% - 110% consider a 2 phase M.E. in the same drum.

- one that operates 30% - 1
- second that operates between 10% - 30%

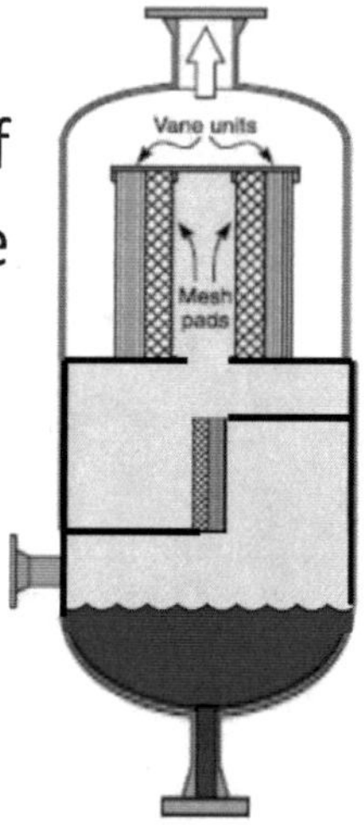

(45i) Renewable Feedstock for Steam Crackers: Catalytic Upgrading of Crude Tall Oil (CTO) Into Bio-Naphtha

Jinto Anthonykutty, Juha Linnekoski, Antero Laitinen and Ali Harlin, VTT Technical Research Centre of Finland, Espoo, Finland

Renewable feedstocks from bio resources are better alternatives to petroleum derived naphtha for the production of olefins and aromatics by steam cracking or reforming.[1, 2] Mostly, these renewable feedstocks (bio-naphtha) rich in n-alkanes are produced by catalytic upgrading methods via hydrotreating/hydroprocessing; in which oxygenates are removed from bio-oil by various deoxygenation processes. At present, hydrotreated vegetable (HVO) oil is produced in many countries on industrial scale using existing petroleum refinery methods[3]. However, employing higher priced vegetable oils could be a major concern in these processes and the main challenge is therefore finding an inexpensive and non-food chain affecting feedstock. Tall oil, obtained as a by-product from Kraft pulping process, is abundant and cheap;and hence, meets the criteria of a low cost starting material for upgrading[4]. Moreover, the utilization of tall oil in conventional oil refineries could serve as an eco-friendly method to convert bio-wastes into bio-naphtha[1].

In the present research approach, catalytic upgrading (hydrotreating) of crude tall oil (CTO) into bio-naphtha suitable for a steam cracker is discussed. CTO, which contains mainly fatty acids (42.9 wt. %) and resin acids (25.7 wt. %) was obtained from Stora-Enso Kaukopää (Finland) mill and used as received. A commercial NiMo catalyst was employed and the experiments were conducted in a continuous, down flow, fixed bed reactor at different process conditions (WHSV 1-3h^{-1}, Pressure 5 MPa and Temperature 350 - 450 ^{o}C). After the stabilization period (6 h.), the liquid sample was collected and separated into organic and aqueous phases.The product distribution in organic phase was analyzed with a Shimadzu GC/MS equipped with a Zebron MS 30m × 0.25 mm × 0.25µm column.

This current approach reports that CTO produces mainly paraffins, cycloalkanes and aromatics during upgrading process. Paraffinic hydrocarbons mainly comprise n-octadecane (20-25 wt. %) and n-heptadecane (15-20 wt. %), are obtained in maximum yield (55- 60 wt. %) in low temperature reactions. The yield of paraffinic hydrocarbons increased as a function of increasing space time in low temperature reactions. 18-Norabietane and other mono-, di- and tri- cycloalkanes are obtained from resin acids in CTO, and these cyclic structures are converted to aromatics such as naphthalene and phenanthrene in high temperature reactions (>400 ^{o}C). Stigmastane, apparently produced from sterols in CTO are also present as a major three ring component in the product stream. Additionally, gaseous products obtained and the yield increased as a function of temperature. The trend observed with the distribution of gaseous products reveal that decarboxylation and cracking reactions are favored at higher temperatures, while hydrogenation and hydrodeoxygenation (HDO) are suppressed.

The elemental composition of CTO shows that it contains relatively high sulfur content (1800 mg/kg), could be useful to keep the catalyst's activity with time on stream (TOS). However, the considerable amount (87 mg/kg) of alkali metals (Na, K, Ca and P) in CTO might be critical for the catalyst activity with TOS, will be discussed in this contribution.

The product composition obtained with CTO is assessed in comparison with compositions and characteristics of refinery naphtha feeds, and will be discussed in detail. Furthermore, this contribution will discuss the achieved deoxygenation rate and success criteria for the product composition of upgraded CTO.

References

1. Gornay, J.; Coniglio, L.; Billaud, F.; Wild, G. Steam Cracking and Steam Reforming of Waste Cooking Oil in a Tubular Stainless Steel Reactor with Wall Effects. ***energy&fuels***, **2009**, 23(11), 5663-5676.
2. Pyl, S.P.; Schietekat, C.M.; Reyniers, M.F.; Abhari, R.; Marin, G.B.; Van Geem, K.M. Biomass to olefins: Cracking of renewable naphtha. ***Chemical Engineering Journal***, **2011**, 176-177, 178-187.
3. Sunde, K.; Brekke, A.; Solberg, B. Environmental Impacts and Costs of Hydrotreated Vegetable Oils, Transesterified Lipids and Woody BTL- A Review. ***Energies***, **2011**, 4(6), 845-877.
4. Coll, R.; Udas, S.; Jacoby, W.A. Conversion of the Rosin Acid Fraction of Crude Tall Oil into Fuels and Chemicals. ***energy&fuels***, **2001**, 15, 1166-1172.

AIChE Paper Number 45j
High Efficiency Ethylene Cracking Furnaces

Rajaram Ramesh

Product Development Manager - Ethylene
Technip
Netherlands

Marco W. M. van Goethem, PhD

Product Development Manager – Modeling and Testing
Technip
Netherlands

25th Ethylene Producers Conference
AIChE 2013 Spring National Meeting
San Antonio, Texas, April 29 – May 3, 2013

Abstract

Environmental and economic pressures generate a driving force to produce products more efficiently. A direct effect of more efficient processes is that the emission of green house gases, particularly carbon dioxide decreases. In households nowadays "high efficiency furnaces" are available with efficiencies in the range of 94-97%. This paper describes the outcome of a feasibility study on applying the high efficiency furnace concept to steam cracking furnaces. The high efficiency is achieved by enhanced heat recovery from the flue gases before they are released to the atmosphere at ca. 65°C.

Technip studied the application of this concept for cracking furnaces based on gas as well as liquid feedstocks. Natural gas was used as basis for the energy price. Since energy prices have a strong impact on project economics, the evaluations were done for an EU average natural gas price and a US natural gas price.

Two alternative heat sinks have been considered: preheating of combustion air and preheating of cold hydrocarbon feed. Both are viable for liquid cracking furnaces. In the case of gas cracking furnaces feed preheating results in a pinch and this makes it not viable.

Capital costs are based on Technip's cost database and quotations from air preheater vendors. Special attention has been paid to the material selection of the affected convection section banks.

This study shows that high efficiency ethylene cracking furnaces are economically viable. For Europe the study indicates an attractive payback period. In the US the payback period is longer due to the lower energy prices. If energy prices increase in the future, the concept's economic viability will become better. Inclusion of CO2 credits will make the payback period more attractive.

Furnace Efficiency

In this paper furnace efficiency has been calculated as net thermal efficiency as interpreted by API560[1]. Furnace efficiency is determined by drawing an envelope around the furnace and including in it the radiant section, convection section and any heat exchanger equipment that recover heat from the flue gas (air preheater, economizer, steam superheater, etc.). The efficiency is the ratio of the useful heat absorbed to the fired heat of the fuel and any other external heat inputs. Ambient temperature is a reference temperature for efficiency calculations and any heat inflow or outflow above this reference is taken into consideration as a heat input or heat loss.
Furnace efficiency can be determined as follows:

$$\frac{heat\ absorption\ in\ radiant\ section\ +\ convection\ section}{(firing\ rate\ \times\ fuel\ heating\ value) + sensible\ heat\ of\ fuel\ and\ air\ above\ ambient\ temp. + other\ external\ heat\ inputs}$$

Alternately, $\quad$ Furnace efficiency $= 100\% -$ Total heat loss, where:

$$\text{Total heat loss} = \frac{stack\ loss\ +\ radiation\ heat\ loss}{(firing\ rate\ \times\ fuel\ heating\ value) + sensible\ heat\ of\ fuel\ and\ air\ above\ ambient\ temp. + any\ external\ heat\ inputs}$$

In a furnace where fuel and combustion air enter the furnace at ambient temperature and there are no other external heat inputs, the first expression can be simplified to the following:

$$Furnace\ efficiency = \frac{heat\ absorption\ in\ radiant\ section\ +\ convection\ section}{fuel\ firing\ rate\ \times\ fuel\ heating\ value}$$

Please note that net thermal efficiency, as defined by API 560, is based on the lower heating value of the fuel.

Study objectives

Technip undertook this study with the goal to reduce fuel firing in cracking furnaces by at least two percent. Although this figure does not look very ambitious, it must be remembered that since efficiencies in current designs are already in the mid nineties (upper part of the technology S-curve), further improvements will not be easy and can only be incremental.

Since stack loss is the main source of inefficiency (radiation heat loss is comparatively smaller and more difficult / expensive to reduce) it was decided to make stack loss reduction as the main focus of the study.

By a relatively straight forward calculation it can be shown that the two percent efficiency rise target would require the flue gas to be cooled to approximately sixty or seventy degrees Celsius.

The main challenge in reducing stack temperature to this level is finding an appropriate heat sink for the recovered heat. In a cracking furnace there are two suitable heat sinks, combustion air and hydrocarbon feed, and both were studied. The following sections discuss the results for liquid furnaces and gas furnaces.

Basis of Study

We have done the study on a naphtha furnace and an ethane furnace. We have chosen these two feedstocks since a large number of furnaces operate on these two feeds and we can draw conclusions that can be applied approximately to other liquid and gas feeds. The study has been performed on the following bases:

- The reference liquid and gas furnace designs used in this study are based on actual furnaces from Technip's past projects. The furnaces were simulated using Technip's proprietary tool SPYRO which is used in the design and simulation of cracking furnaces.
- The budgetary capital cost need to achieve the target furnace efficiency was calculated as an incremental cost. It was estimated from vendor quotes and Technip's internal cost database.
- The cost estimates are for a new project, not a revamp, and are for a West European location.
- Operating costs are based on cost of base utilities and derived utilities. Base utilities are natural gas and demineralized water. These are taken from market data.
- Since natural gas prices are very different in different parts of the world, the operating costs and economic analysis have been done for European and United States gas prices.
- EU natural gas price was fixed at €25.1 /MWh of higher heating value (HHV) which is equivalent to €28.0 /MWh of lower heating value (LHV). The prices are based on EUROSTAT data for 1st half of 2010 and were obtained from the publications of the Department of Energy & Climate Change in the UK. They correspond to the EU-15 average for Large Industrial Consumers (i.e. 2.13 pence/kWh HHV).
- US Natural gas prices were obtained from the US Department of Energy. The average value was 6 \$/MM BTU (HHV) or 17 €/MWh (LHV) for 1st half of 2010. The average HHV used is 1029 BTU per cubic foot or 1.029 MM BTU per thousand cubic feet.
- Cost of demineralized water was assumed at € 2.00 /ton.
- The prices of other utilities (boiler feed water, steam and electricity) have been calculated from those of the two base utilities. Electricity was priced assuming 35% overall efficiency of power generation.
- The economic calculations do not include any CO2 credit for the fuel saved.
- Assumed exchange rate: 1 US dollar = 0.75 euros
- The plant is assumed to operate 8400 hours per year

The resultant utility prices used in the study are summarized below.

UTILITY COSTS

		EU basis	US basis	
FUEL as NATURAL GAS	Base	**28.0**	**17.0**	€ / MWh
DEMINERALISED WATER	Base	2.0	1.5	€ / ton
VHP STEAM (net value)	Derived	23.7	14.4	€ / ton
MP STEAM (net value)	Derived	13.9	8.5	€ / ton
LP STEAM	Derived	9.7	5.9	€ / ton
ELECTRICAL POWER	Derived	80.1	48.6	€ / MWh

Study Results

Naphtha Furnace using Air Preheat

Combustion air can be preheated in a plate type air preheater (APH). These are cross flow type exchangers and have low pressure drops. Combustion air is at a lower temperature than any of the process or utilities streams entering the convection section; therefore, from the standpoint of efficiency, APHs are ideally suited to be the last heat recovery section before the stack.

Fig 1 shows the Base Case Naphtha furnace while Fig 3 shows the same furnace with an APH. Please also refer to Table 1 for a performance comparison of the APH design with the Base Case naphtha furnace. The following points are of interest:

1) The overall efficiency of the furnace will improve from 94.4 to 97.0% for the APH option. This is mainly due to the lower stack temperature but also partly due to the smaller excess air, which is explained below.

2) This option will require a forced draft combustion air fan to overcome the pressure drop in the APH. As a result the furnace operation will change from induced draft mode to balanced draft. The availability of air in forced draft will enable more accurate control of the excess air and it will be possible to operate the furnace with lower excess air (5% instead of 8%). This leads to a lower flue gas flow, a lower stack loss and therefore better efficiency.

3) There will be a drop of 4.8% in fuel firing which is greater than the efficiency increase. This is because of the lower absorbed heat. The furnace is designed and controlled in such a way as to supply all the required process heat and to absorb the surplus heat into utilities to the extent possible. When the furnace operates more efficiently and the heat demanded by the process media remains unchanged, as in the APH case, there will be less surplus heat available for VHP steam. As a result the VHP steam generation will go down from 79.7 to 76.0 t/h.

4) The heat transfer surface in some parts of the APH will be below the water dew point of the flue gas (around 60°C). This can lead to water condensation and create two possible issues: 1) accumulation of water in the APH and 2) corrosion due to the presence of liquid water, oxygen and carbon dioxide.

   Two measures have been included in the design to counter these problems:
   - Placement of unit such that flue gas flows downward to aid easy draining
   - Use of corten steel to resist corrosion

5) We have evaluated the impact of the air preheat option on NOx emissions using API 535[2]. The rise in air temperature and the drop in excess air will have opposing effects on the NOx production. The net effect is that the NOx concentration will be close to the Base Case.

   If a DeNOx unit is present, as is often the case in the US, the APH option may help to slightly extend the life of the DeNOx catalyst. This is because the total NOx load will be smaller than the Base Case due to the smaller flue gas flow.

Table 2 summarizes the results of the economic analysis of this option. We have used payback period to compare different options. Payback period is the ratio of the incremental capital cost above the Base Case CAPEX divided by the savings in yearly operating cost.

1) The capital cost (CAPEX) of the APH option includes the cost of the APH, the combustion air fan, extra ducting and structural steel. The operating cost (OPEX) savings is the savings in furnace fuel but this must be offset against the value of the lost VHP steam production (which notionally must be made up in a utility boiler elsewhere) and the additional electricity for the two fans.

2) It can be seen that the payback period is three years based on European gas prices and five years based on American gas prices.

Naphtha Furnace using Feed preheat

Naphtha is normally stored at ambient temperature and is preheated to 60°C using quench water before it is routed to the furnaces. Most cracking furnaces are equipped with a Feed Preheater (FPH) bank in the convection section. If flue gas is utilized to preheat naphtha from ambient conditions the stack temperature can be lowered to the desired range of sixty to seventy degrees Celsius. Till now designers and operators have not taken full advantage of this mainly because of the higher CAPEX and fear of dew point corrosion.

One of the consequences of this feature will be that the quench water duty previously absorbed by the naphtha will now shift to the air coolers in the quench water circuit. The additional electrical power that the air cooler fans will require will be about 10 KW per furnace.

Please refer to Table 3 for an overview of the performance of a naphtha furnace with and without an extended feed preheater. Please note that the Base Case naphtha furnace is the same as the one considered in the APH option.

1) The overall efficiency of the furnace can be improved from 94.4 to 96.0% for this option.

2) Unlike the APH option a combustion air fan will **not** be required. The furnace will continue to operate in induced draft mode.

3) The fuel firing and the VHP steam generation will drop marginally.

4) The upper tube rows of the FPH will be below the water dew point of the flue gas and some condensation will occur. As a precaution we have considered that the top four rows will be of stainless steel.

5) The flue gas NOx concentration will be unaffected by this option. However the total hourly NOx emissions and load on the DeNOx unit, if any, will decrease to the same extent as the fuel firing, which is around 0.6%.

Table 4 summarizes the results of the economic analysis of this option.

1) The capital cost (CAPEX) of this option covers the incremental cost of the extended FPH, extra ducting and structural steel. The OPEX savings is the reduction in furnace fuel after adjusting for the reduction in VHP steam production and the additional electricity for the ID fan and the Quench Water Air Cooler fans.

2) In the Base Case the naphtha feed piping between the external feed preheater and the convection section is insulated for heat conservation. In the FPH option naphtha will be fed at ambient temperature to the furnace and so these lines do not have to be insulated. This will yield some capital cost saving but this has not been taken into account in our analysis.

3) The payback period excluding CO2 credit will be three and a half years with European natural gas and five and a half years with US natural gas.

Ethane Furnace using Air Preheat

Fig 2 shows the Base Case Ethane furnace that was used in the study. The APH arrangement on an ethane furnace will be quite similar to the one on the naphtha furnace. Table 5 shows the impact that the APH would have on the furnace performance. The main points are as follows:

1) The overall efficiency of the furnace will improve from 91.7 to 96.5% for the APH option. This is mainly due to the lower stack temperature but also partly due to the smaller excess air.

2) In this furnace also the APH will require a forced draft combustion air fan. This will allow the excess air to be reduced from 8% to 5% which will result in lower stack losses and a better efficiency.

3) The fuel firing will go down by 11.0%, from 8.2 to 7.3 t/h. The absorbed heat will drop from 116.8 to 110.6 MW. As a result of these the VHP steam generation will go down from 97.0 to 88.8 t/h.

4) A portion of the APH will operate below the flue gas water dew point. Consequently the unit must be designed to prevent water accumulation and corrosion.

5) The impact on NOx was evaluated here also. The rise in air temperature will have a greater impact on NOx than the fall in excess air. We estimate that as a result the flue gas NOx concentration will increase slightly (around two percent). This can be a cause for concern if the furnace does not have a DeNOx unit and is operating close to the limits imposed by authorities. However since the total firing will go down, the hourly NOx production will drop. Plants can consider discussing with authorities to obtain a higher allowable NOx limit,

   If the furnace has a DeNOx unit, the chemical load on the unit will reduce, which should help to obtain a longer catalyst life.

The operating cost savings and payback for this option have been summarized in Table 6.

1) The capital cost includes the cost of the APH, the combustion air fan, extra ducting and structural steel. The operating cost savings is the savings in fuel firing with an adjustment for the value of the reduced VHP steam generation and the extra electricity for the ID fan and FD fan.

2) The payback period with natural gas priced at European rates will be one and a half years. The corresponding figure for US gas prices will be three years.

Ethane Furnace using Feed Preheat

Using feed at ambient temperature does not significantly reduce the stack temperature in a gas furnace. This is because the feed preheat curve has a steeper slope than the flue gas cooling curve, causing a pinch on the hot side of the Feed Preheater. Fig 4 shows the heating and cooling curves of ethane, boiler feed water and flue gas in the convection section. The hatched shape on the plot shows the maximum additional heat duty that can be recovered into feed, over and above the Base Case. The plot clearly shows that when the ethane feed temperature is lowered from the base value of 55°C to 30°C, the stack temperature will drop from 170°C to about 150°C. The flue gas cannot be cooled any more than this modest amount.

For this reason the feed preheat option was not investigated any further.

Conclusions and Summary

The key results from the different study options are listed together below.

Cracking furnace feed	Naphtha		Naphtha		Ethane	
Cold sink	Combustion air		Hydrocarbon feed		Combustion air	
	EU nat. gas	US nat. gas	EU nat. gas	US nat. gas	EU nat. gas	US nat. gas
OPEX savings	€ 0.60 mill	€ 0.36 mill	€ 0.17 mill	€ 0.10 mill	€ 1.27 mill	€ 0.77 mill
Payback time	3 years	5 years	3.5 years	5.5 years	1.5 years	3 years
Salient features of each option	Small number of additional equipment		No additional equipment		Small number of additional equipment	
	Higher CAPEX		Lower CAPEX		Higher CAPEX	
	More OPEX savings		Less OPEX savings		Best OPEX savings	
	Slight effect on NOx concentration		No effect on NOx		Around 2% higher NOx concentration	

All three scenarios have attractive payback periods and will be worth considering in high energy cost countries such as Europe and Asia. In US and other energy rich nations, the economic incentive will be much lower. In the US, where ethane cracking is common and increasing even more, improvement of furnace efficiency with air preheat can be worth considering as the payback time is less than 3 years.

If the options are compared against each other, in our opinion the Air preheating option has more potential, irrespective of feedstock.

The Feed Preheating option in liquid furnaces is cheaper but it also gives proportionally modest savings. Moreover, as earlier mentioned, the feed preheat duty is being shifted from one waste heat source (quench water) to another (flue gas). This option will not lead to a net improvement or cash savings in the plant as a whole unless there is a shortage of quench water and there are other, better uses for the saved quench water duty

Acknowledgements

The authors wish to express their thanks to their colleagues Mr Ashok Laghate, Ms Susan Cooper, Mr Sachin Singare and others in Technip Benelux who contributed to this study.

References

1) API 560: Fired Heaters for General Refinery Service, 4th edition

2) API 535: Burners for Fired Heaters in General Refinery Service, 2nd edition

Table 1
Naphtha Furnace with Combustion Air Preheating – Performance Summary

		Base Case	Air Preheat Case
Naphtha feed	t/h	60.8	60.8
Ambient conditions	-	15°C, 85% R.H.	15°C, 85% R.H.
Air temp. to burners	°C	15	77
Stack temperature	°C	108	61
Excess air	%	8	5
Absorbed heat	MW	113.2	110.4
Overall efficiency	%	94.4	97.0
Fuel flow	t/h	8.4	8.0
Fired heat	MW	119.9	113.9
VHP steam production	t/h	79.66	76.01
NOx concentration		Base	Close to Base

Table 2
Economics of Naphtha Furnace with Combustion Air Preheating

		Air Preheat Case (EU location)	Air Preheat Case (US location)
Savings in fuel firing	million €/yr	1.42	0.86
Cost of lost VHP steam capacity	million €/yr	0.73	0.44
Cost of additional electricity	million €/yr	0.10	0.06
Net OPEX saving	million €/yr	0.60	0.36
Payback	Years	3	5

Table 3

Naphtha Furnace with Feed Preheating – Performance Summary

		Base Case	Feed Preheat Case
Naphtha feed	t/h	60.8	60.8
Feed temperature	-	60°C	30°C
Combustion air temp.	°C	15	15
Stack temperature	°C	108	71
Excess air	%	8	8
Absorbed heat	MW	113.2	114.3
Overall efficiency	%	94.4	96.0
Fuel flow	t/h	8.4	8.35
Fired heat	MW	119.9	119.1
VHP steam production	t/h	79.66	79.63
NOx concentration		Base	Base

Table 4

Economics of Naphtha Furnace with Feed Preheating

		Feed Preheat Case (EU location)	Feed Preheat Case (US location)
Savings in fuel firing	million €/yr	0.19	0.12
Cost of lost VHP steam capacity	million €/yr	< 0.01	< 0.01
Cost of additional electricity	million €/yr	0.01	< 0.01
Net OPEX saving	million €/yr	0.17	0.10
Payback	Years	3.5	5.5

Table 5

Ethane Furnace with Air Preheating – Performance Summary

		Base Case	Air Preheat Case
Ethane feed	t/h	51.9	51.9
Ambient conditions	-	15°C, 85% R.H.	15°C, 85% R.H.
Air temp. to burners	°C	15	129
Stack temperature	°C	169	64
Excess air	%	10	5
Absorbed heat	MW	116.8	110.6
Overall efficiency	%	91.7	96.5
Fuel flow	t/h	8.2	7.3
Fired heat	MW	127.3	114.5
VHP steam production	t/h	97.0	88.75
NOx concentration		Base	Base + 2%

Table 6

Economics of Ethane Furnace with Air Preheating

		Air Preheat Case (EU location)	Air Preheat Case (US location)
Savings in fuel firing	million €/yr	3.01	1.83
Cost of lost VHP steam capacity	million €/yr	1,64	1.00
Cost of additional electricity	million €/yr	0.10	0.06
Net OPEX saving	million €/yr	1.27	0.77
Payback	Years	1.5	3

Fig 1
Naphtha cracking furnace

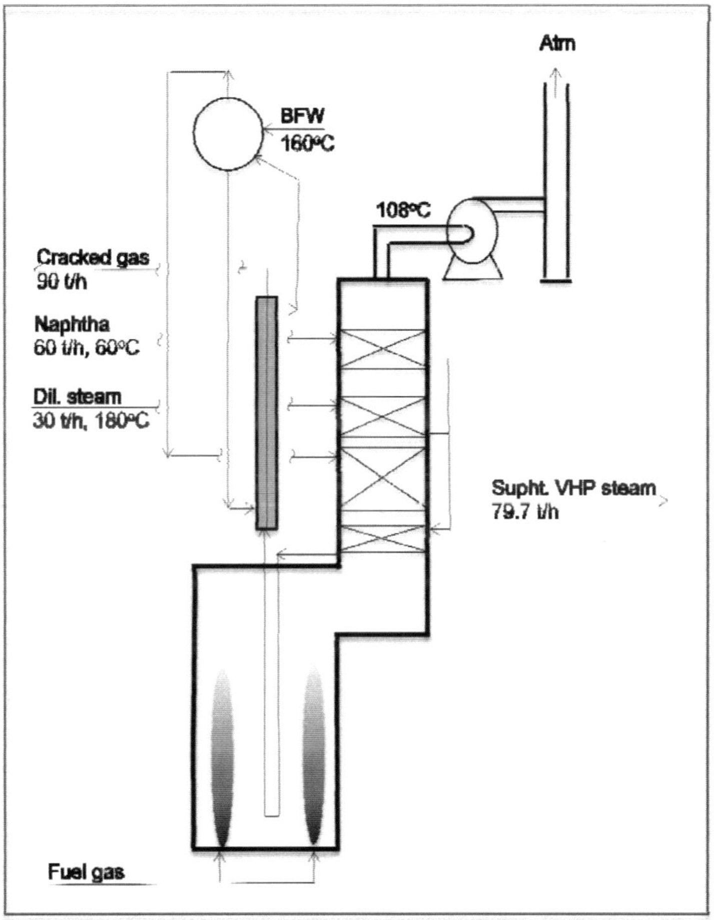

Fig 2
Ethane cracking furnace

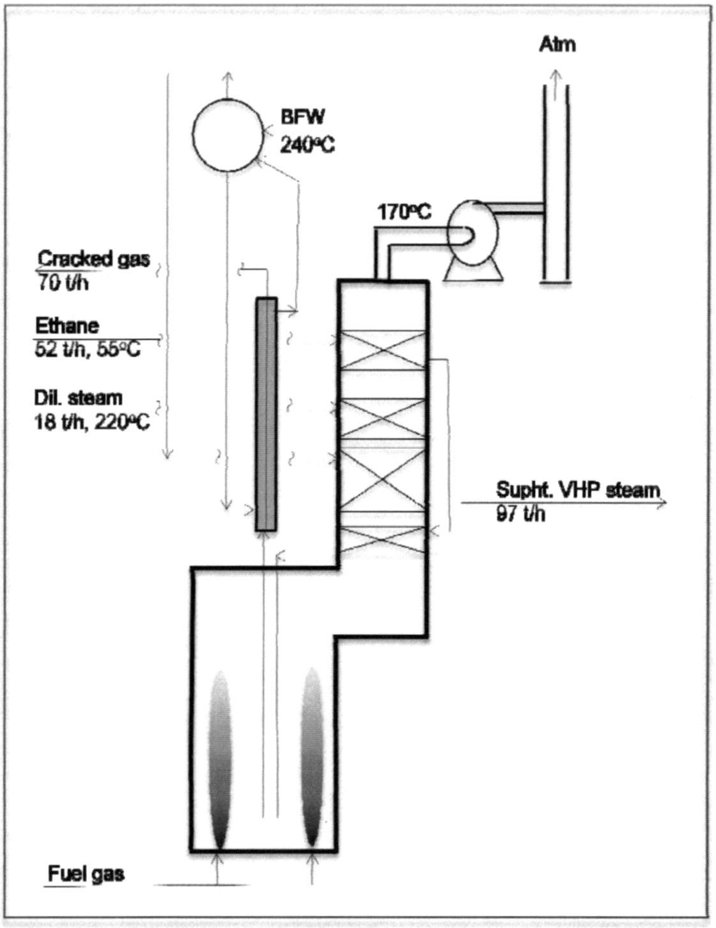

Fig 3

Naphtha cracking furnace – Air Preheater option

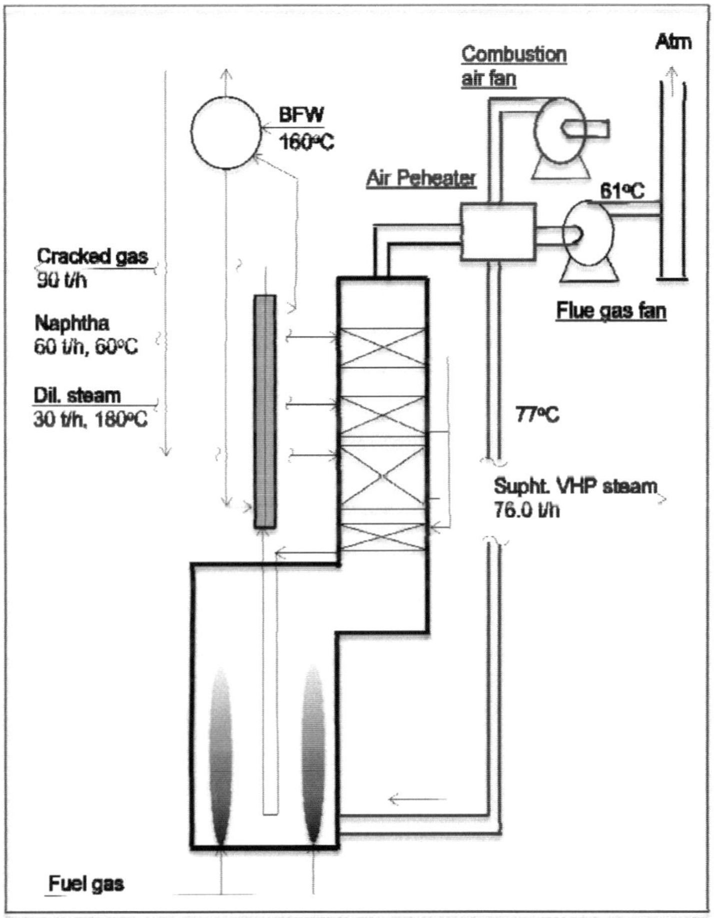

Fig 4

Ethane cracking furnace – Feed Preheater option

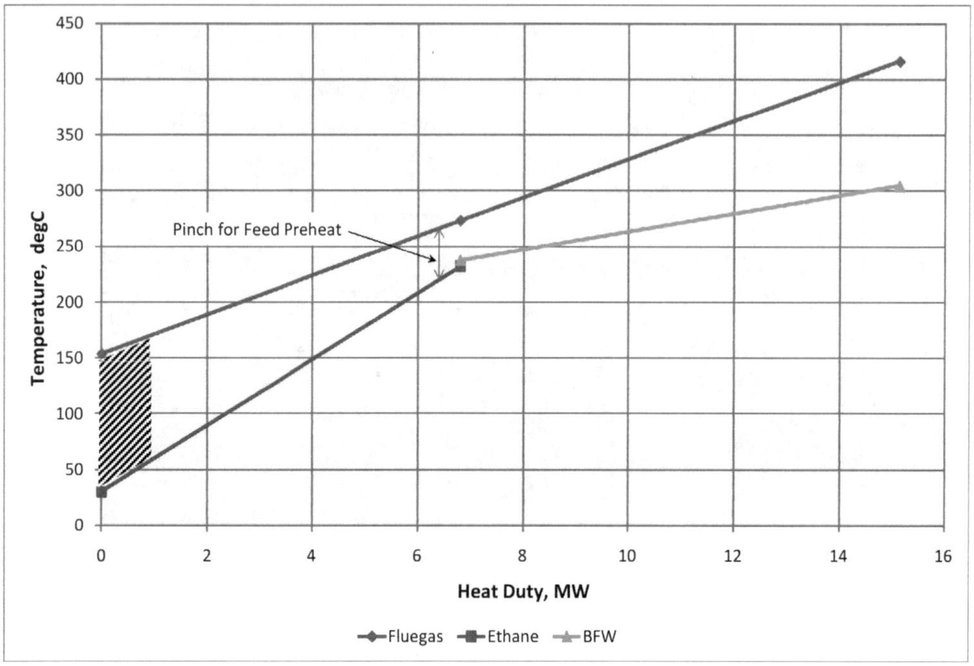

Modernize Your Control Functionality
Easy Steps to Start Process Optimization

13AIChE
Spring Meeting, San Antonio, TX

Presenters

- Laurie Ben
- James Beall

EMERSON
Process Management

Agenda

- Introduction
- Examples – Best Practices & Lessons Learned
  - Split Range Duty
  - Smith Predictor
  - Gain Scheduling/Adaptive Control
  - Signal Characterization
- Summary

Introduction

- Modernizing Control System – Decision Done Check!
- 'Do I' or 'Don't I' take advantage of new technology
  - Implemented older cumbersome techniques?
  - Complex custom strategies that handle automation challenges?
  - Hard to implement & maintain?
- Standard features that are field proven, practical, and easy to implement
  - Easier to implement & maintain
  - More efficient and effective control strategies
  - Achieve improvements in process performance
  - Lays the foundation for optimization

Agenda

- Introduction
- Examples – Best Practices & Lessons Learned
  - Split Range Duty
    - Control Challenge
    - Legacy Solution
    - Newer Technology Solution
    - Summary
  - Smith Predictor
  - Gain Scheduling/Adaptive Control
  - Signal Characterization
- Summary

Split Range Duty – Control Challenge

- Use steam and hot oil as energy source for distillation column for temperature controller.
- Hot oil is "free energy" and should be maximized
- Control Scenarios
  - Maximize hot oil and supplement with steam
  - Base load steam (if meter bad) and supplement with hot oil
  - Base load hot oil flow (if bad meter) and supplement with steam
- Constraints
  - Minimum and maximum hot oil valve position and flow rate
  - Minimum steam flow rate

Split Range Duty – Legacy Solution

– Old legacy system implemented this feature using a velocity PID with incremental output and a "micro expert rules system"

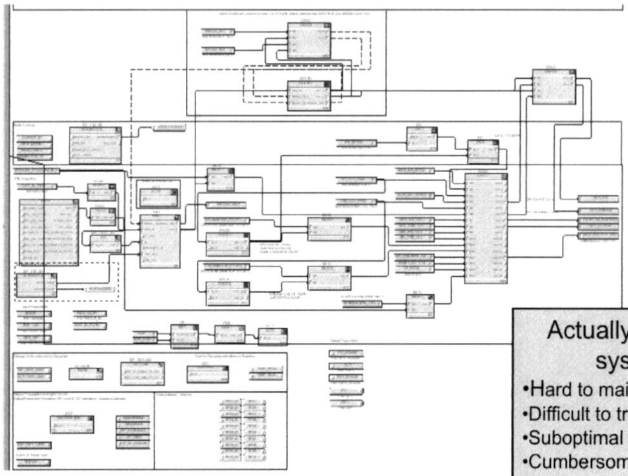

Actually Duplicated in New system, but still…

- Hard to maintain and modify
- Difficult to troubleshoot
- Suboptimal performance
- Cumbersome/Kludge/Ugly

Split Range Duty – New Technology Solution

Embedded MPC Function Block provides all control functionality! EASY!!!

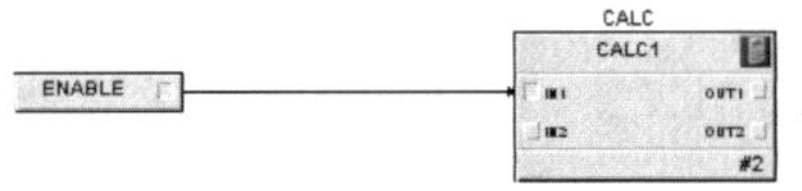

Simple calculations to calculate limits

Split Range Duty – New Technology Solution

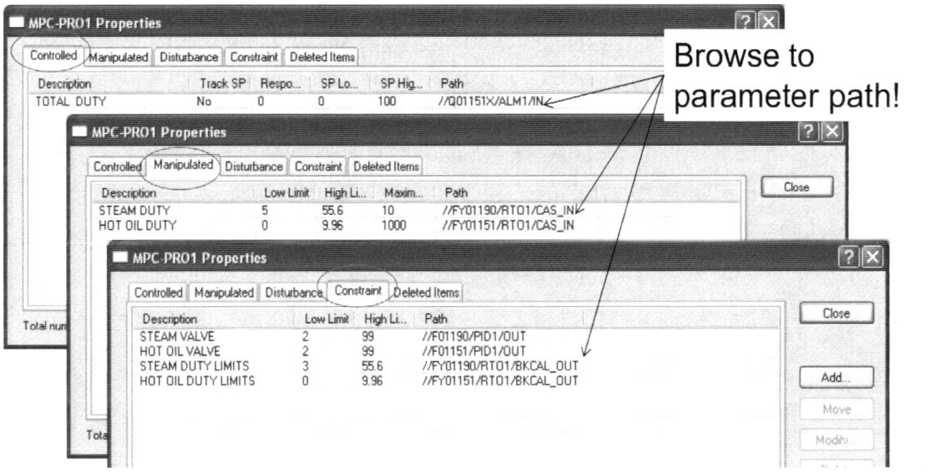

Split Range Duty – New Technology Solution

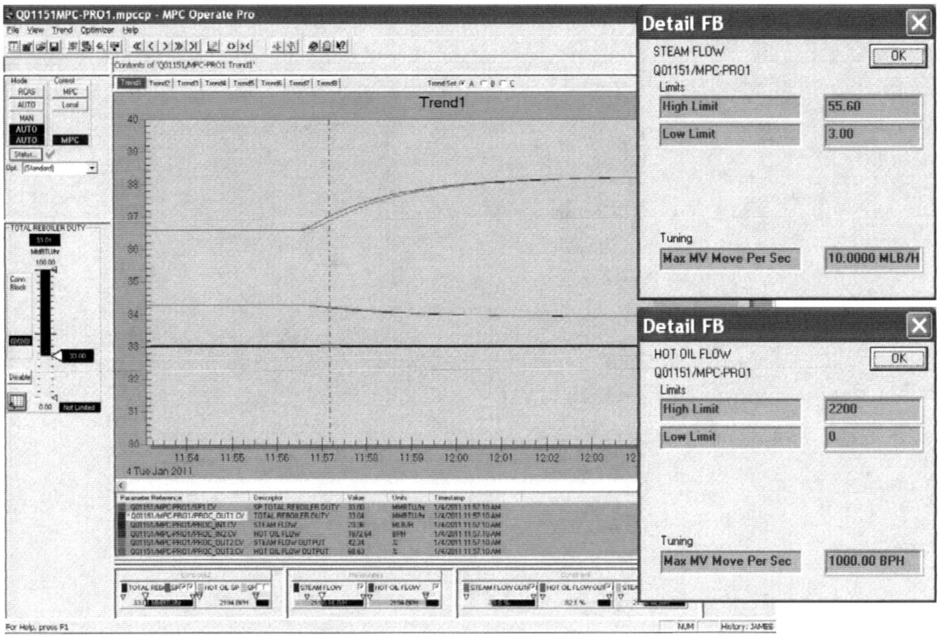

Summary Split Range Duty Control

Embedded MPC (New Technology) vs Legacy Solution

- Better performing solution
  - Bumpless transfers and constraint control
- Easier to Maintain
  - Change limits and settings
  - Simpler implementation
  - Easy to understand, train others, document
- Easily expands to handle future constraints
- Reduce energy costs!

Agenda

- Introduction
- Examples – Best Practices & Lessons Learned
  - Split Range Duty
  - Smith Predictor
    - Control Challenge
    - Legacy Solution
    - Newer Technology Solution
    - Summary
  - Gain Scheduling/Adaptive Control
  - Signal Characterization
- Summary

Smith-Predictor PID – Control Challenge

- Control of processes that are "dead time dominant" (dead time > time constant)
- Typically custom configured
  - some legacy control systems have preconfigured options, e.g. PROVOX
- Algorithm uses a process model
  - (gain, dead time and time constant)
- Challenge: as process dynamics change from model used, performance degrades rapidly

Smith-Predictor PID – Legacy Solution

- Traditional Smith-Predictor

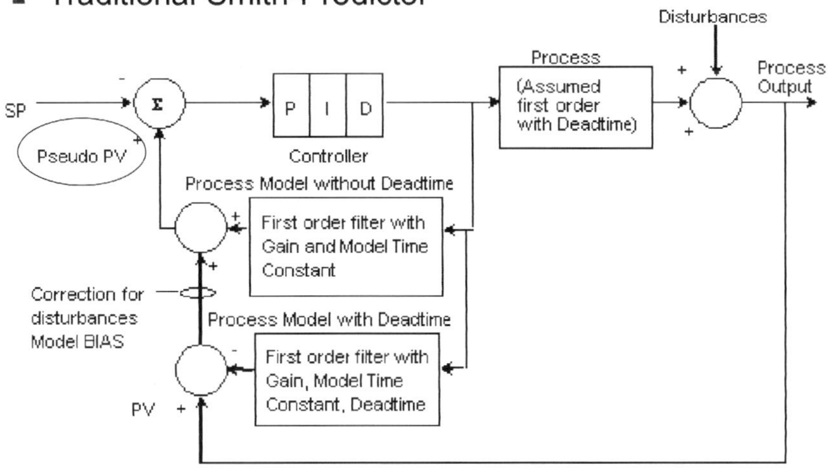

•High maintenance when process dynamics change
•Performance degrades rapidly as process dynamics change from configured model

Smith-Predictor PID – New Technology Solution #1

Easy to implement the Smith-Predictor but…

- High Maintenance when process dynamics change
- Performance degrades rapidly as process dynamics change from configured model

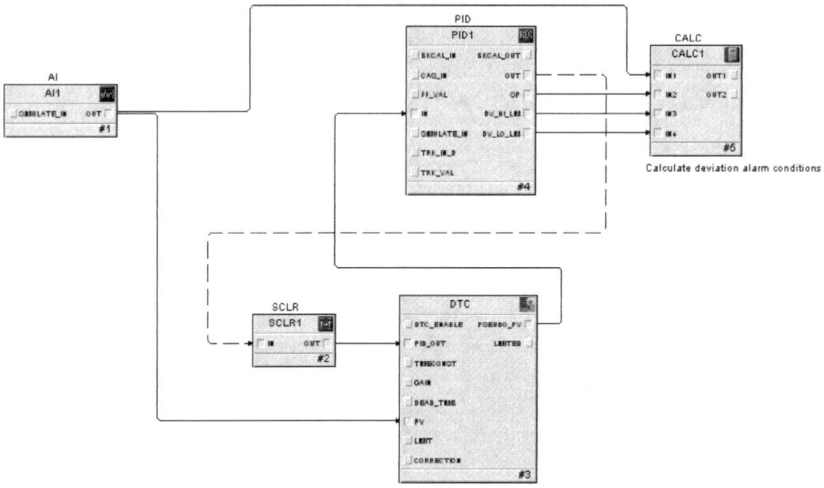

Smith-Predictor PID – New Technology Solution #2

- Embedded MPC Solution

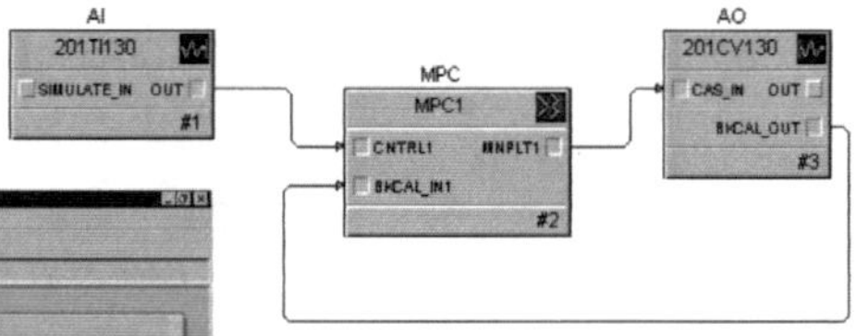

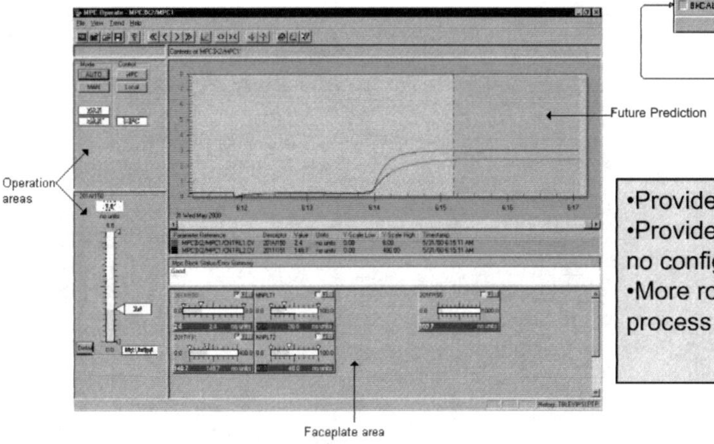

- Provides Future Prediction
- Provides MPC Faceplate – no configuration needed
- More robust under changing process dynamics

Smith-Predictor PID –
New Technology Solution #2

- Embedded MPC Solution

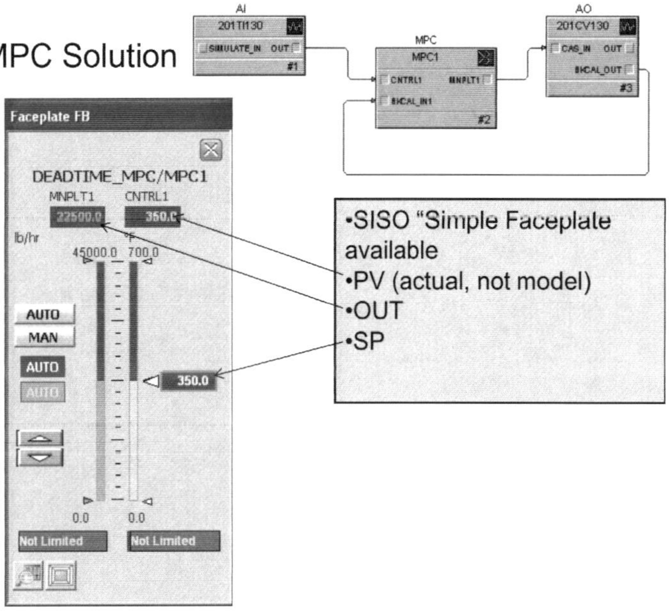

Summary Smith-Predictor PID

New Technology Solutions

- Standard Smith Predictor (#1) and Embedded MPC (#2)
- Both easy to implement and understand!

New Technology Solution #2 (Embedded MPC)

- More robust solution under a changing process model
- Provides visual display of the "future" PV
- Save money through better control performance

Agenda

- Introduction
- Examples – Best Practices & Lessons Learned
  - Split Range Duty
  - Smith Predictor
  - Gain Scheduling/Adaptive Control
    - Control Challenge
    - Legacy Solution
    - New Technology Solution
    - Summary
  - Signal Characterization
- Summary

Gain Scheduling – Control Challenge

- Legacy systems have limited capability to modify tuning based on process conditions
- Others had add-on application software, e.g. PROVOX "Gain Scheduler" system provided 3 sets of Gain, Reset, and Rate applied in 3 separate regions of a "state variable"
  - State variable = PV, OUT or "other"
  - Transition between tuning sets interpolated over "dead band"
- Beware: some suspend loop execution when tuning parameters change, so don't change the tuning every scan

Gain Scheduling – Legacy Solution

- Custom code
- Multiple points, registers, displays
- Restricted (slower) update rate

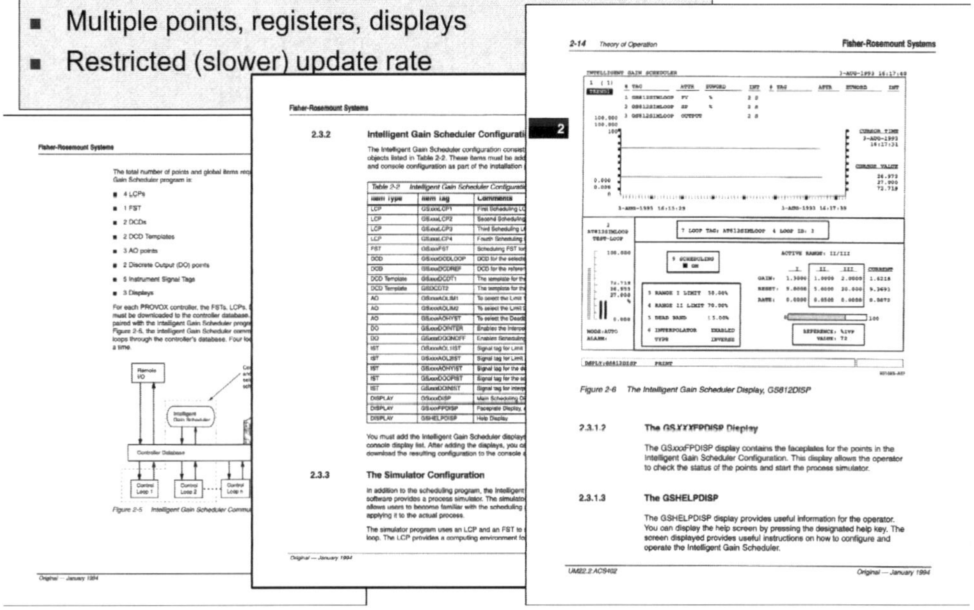

Gain Scheduling – New Technology Solution #1

Standard Modules

- Standard configuration
- Preconfigured faceplate & detailed display
- Easy to implement
- Tuning updates at loop execution

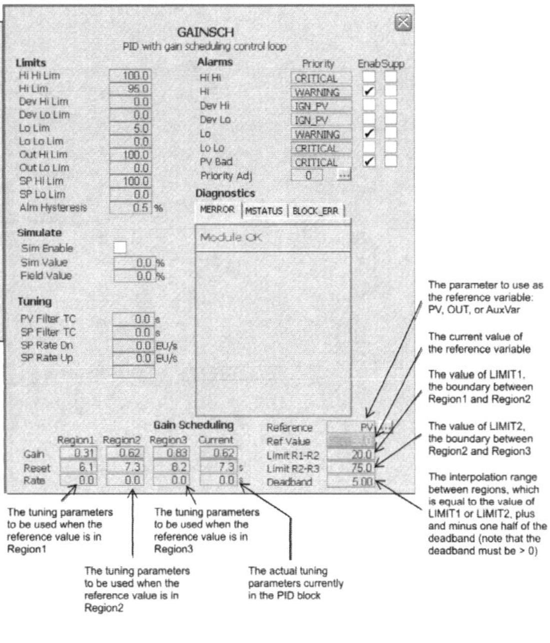

Gain Scheduling – New Technology Solution #2

New Technology "Adaptive Control"

- "Learns" process models from normal operating moves
- Allows up to 5 tuning regions of the "state variable"
- Partial Adaptive Control
  - Use "approved model" only & selected tuning rule in each region
  - Lambda PI or IMC PID; "Closed loop time constant" or "Lambda Factor".
- Optional Full Adaptive Control
  - continuous "Learning" of process model in each region
  - updates tuning while in each region per selected tuning rule
  - set allowable deviation (%) from approved model in each region

New Technology – Adaptive Control

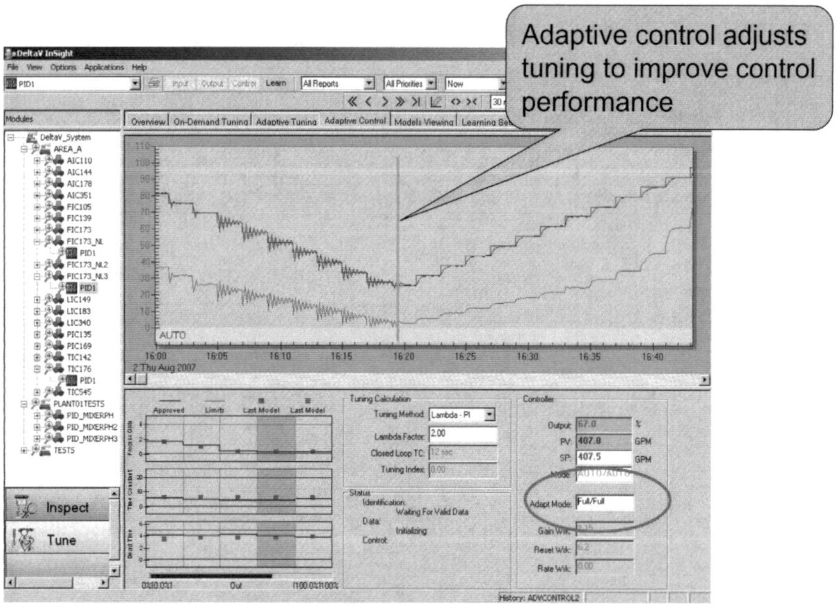

Summary Gain Scheduling

New Technology Solution(s) vs. Legacy Solution

- Gain Scheduling easy to implement and understand!
  - Standard configuration
  - Preconfigured faceplate & detailed display
  - Tuning updates at loop execution
- Adaptive Control provides additional benefit and ease of implementation
  - Learns models during normal operation – NO STEP TESTs!
  - More Regions – Up to 5
  - Automatic tuning updates when process model changes
- Improved control performance saves money!

Summary Gain Scheduling

New Technology Solution(s) vs. Legacy Solution

- Gain Scheduling easy to implement and understand!
  - Standard configuration
  - Preconfigured faceplate & detailed display
  - Tuning updates at loop execution
- Adaptive Control provides additional benefit and ease of implementation
  - Learns models during normal operation – NO STEP TESTs!
  - More Regions – Up to 5
  - Automatic tuning updates when process model changes
- Improved control performance saves money!

Agenda

- Introduction
- Examples – Best Practices & Lessons Learned
  - Split Range Duty
  - Smith Predictor
  - Gain Scheduling/Adaptive Control
  - Signal Characterization
    - Control Challenge
    - Legacy Solution
    - New Technology Solution
    - Summary
- Summary

Signal Characterization - Control Challenge

- Input Characterization (AI)
  - Combustion control – characterize fuel and air
  - Level – characterize pv (e.g. level to volume)
- Output characterization (AO)
  - modify installed flow characteristics of valves, dampers or other final control elements
  - Use to modify loop gain, e.g. output path to slave loop
- Legacy System Challenges
  - Lack of resolution (few points)
  - External black box characterizers may shift, lose documentation
  - Potential operator/maintenance confusion on signal going to final control element

Signal Characterization – Legacy Solution

- Legacy system using external analog characterizer

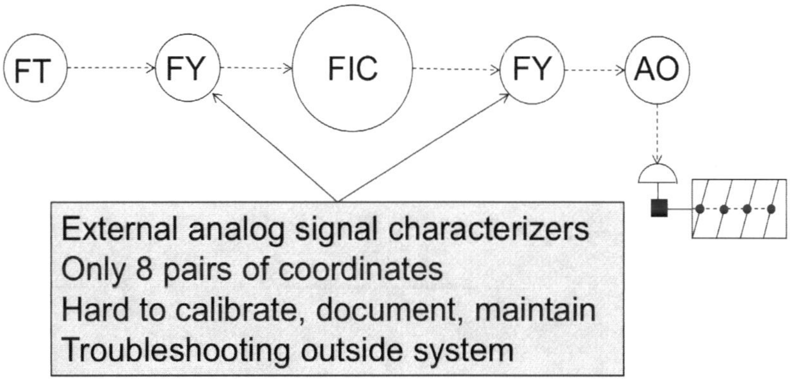

Signal Characterization – New Technology Solution #1

- Standard Module implementation provides 21 point signal characterizer function block (SGCR)

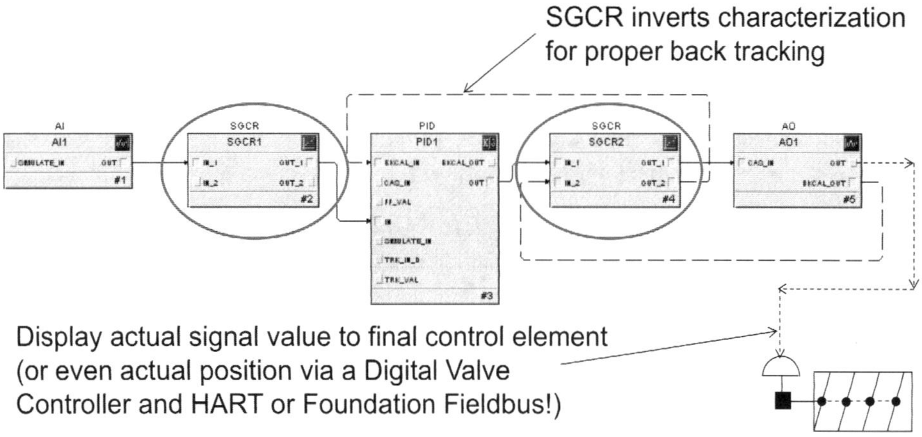

Signal Characterization – New Technology Solution #2

- Adaptive Control
  - Adjust tuning to compensate for final control element characteristics

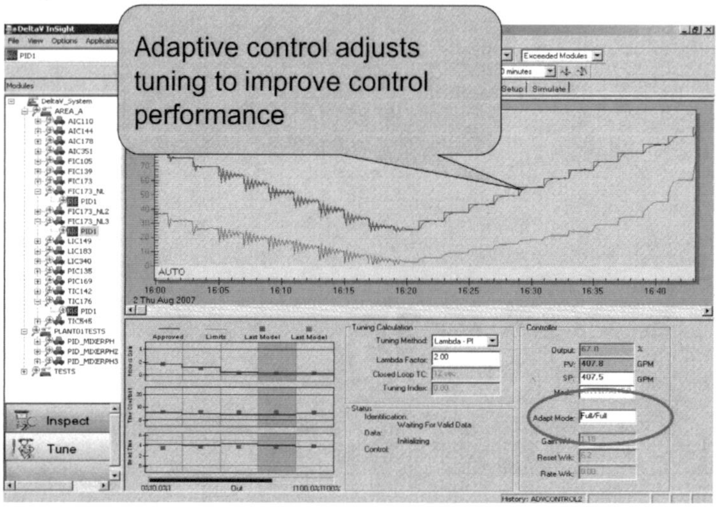

Summary Signal Characterization

- Newer Technology Signal Characterization function block
  - Better resolution (21 pairs of coordinates)
  - Self documenting (exists in control system)
  - No calibration required vs. 'external black box'
  - Provides "inverse" characterization for analog output characterizations
  - Reduce confusion: display signal being sent to final control element
- Applying Adaptive Tuning may provide a superior solution for AO characterization of final control element

Agenda

- Introduction
- Examples – Best Practices & Lessons Learned
  - Split Range Duty
  - Smith Predictor
  - Gain Scheduling/Adaptive Control
  - Controller Characterization
- Summary

Summary

- Practical application of control features that deal with common challenges
- Highlighted best practices & lessons learned
  - Split Range Duty
  - Smith Predictor
  - Gain Scheduling/Adaptive Control
  - Signal Characterization
- Do look at new functionality available when modernizing
  - Allows jump start to process optimization
  - Leads to improved process performance
  - Easier to implement and maintain over the life investment

Contacts & Questions?

James.Beall@Emerson.com
903-235-7935

Laurie.Ben@Emerson.com
512-834-7237

Paralleled Double-Effect Distillation: Simulative Case Studies

*Pei Wang, Leilei Dai, Ming Shi , Fan Wang, Hong Gao, Jinsheng Sun**
School of Chemical Engineering and Technology, Tianjin University, No. 92 Wei Jin Road, Tianjin 300072, P. R. China

Abstract

The increasing scarcity and growing demand for energy have led to an increasingly attention to the energy-efficient methods. Double-effect distillation is a common method to achieve remarkable energy conservation, which has three kinds of typical configurations including the feed splitting (FS), the light split/forward (LSF) and the light split/reverse (LSR). In spite of a good pathway for energy saving, better options are expected to cut down energy expenditure by a bigger margin than double-effect distillation does.

In this paper, case studies of paralleled or combined double-effect distillation schemes are proposed on the basis of current double-effect distillation. In order to verify the feasibility of this method, the improved methanol distillation scheme and organosilicon distillation scheme are optimized under this principle as examples. Exergy losses and the energy consumptions are measured as a tool to compare the two prevalent schemes with their corresponding optimized counterparts, respectively. The results of the total exergy loss and the energy consumption of these schemes show that the optimized schemes can cut the corner of the exergy loss by 22.8% and 40.41% and the energy consumption by 15.61% and 44.28%, respectively, in comparison to their popular cousins.

Keywords: double-effect distillation, paralleled double-effect distillation, methanol distillation, organosilicon monomer distillation, distillation process simulation, energy saving, exergy analysis

1. Introduction

Distillation, in chemical process and related industries, is by far the most widely applied for the separation of liquid mixture. In spite of its well-known benefit and widespread application, the major drawback is its high energy consumption. In classical distillation process, heat is transferred to the reboiler of distillation column. Along with the distillation column to separate the mixture, a large number of heat are directly discharge into the surrounding environment and cooling water. Araujo AB [1] has done some research to achieve that distillation has a relatively low thermodynamic efficiency and it requires the input of high quality energy in the reboiler to push the separation of liquid mixture. Therefore, the focus of energy conservation is to consider how to recover or reuse the discharged heat, or how to reduce the quantity of transferred heat to the column. In the early 20th century, many scholars have devoted to the study of saving energy in distillation, and now most of correlative techniques are widely used in distillation process [2,3]. These energy saving methods involve multi-effect distillation, heat pump distillation, heat integrated distillation and etc [3-8]. But without extra rotating machines by electricity or other power producers, multi-effect distillation has been more favored in research and application [9].

Many researches confirmed multi-effect distillation to be one of the effective methods to reduce the energy consumption [10-12]. The energy saving effect of those schemes used multi-effect distillation is associated with the number of the effects and the characteristics of separation system. Generally speaking, those schemes can offer energy savings around 30% with respect to the original or classical multicomponent distillation scheme [23, 27, 28]. Considering the interaction between investment of plants and energy saving effect, the double-effect distillation has been the most commonly used in not only research but also field application.

The use of the double-effect distillation technology for optimizing the conventional distillation schemes can cut a large amount of the energy requirement in comparison with the correspondingly prevalent schemes [13]. Although the system is very difficult to control in the sense that it is nonlinear, multivariable and interacting [14]. More and more published studies [15-19] have been focused on further questing the way of energy saving on the basis of the double-effect distillation schemes. However, the overhead vapor of one high pressured column only can integrate with the reboiler of one low pressured column in those published studies. In view of this, in this paper, the paralleled or combined double-effect distillation is proposed to broaden the thought for the application of the double-effect distillation, in which the overhead vapor from one of the high pressured column is divided into two or more streams to successively integrate with the reboiler of the other columns [9].

2. *Theory*

2.1 Double-effect or heat integrated distillation system

A double-effect distillation system is made up of two adjacent distillation columns running in different pressure, which can lead to a temperature difference between the top of the high pressured column and the bottom of the low pressured column. Through the thermodynamic connection, the reboiler of the low pressure column plays a role of the condenser for the high pressure column. According to the direction of the mass flow and the heat integration, the double-effect distillation can fall into three kinds of configuration: the feed splitting (FS), the light split/forward (LSF) and the light split/reverse (LSR), as showed in Figure. 1.

One of the configurations with a high thermodynamic efficiency is the feed splitting (FS), in which the feed is divided roughly equal between the two columns running in different pressure, and the products can be obtained from both of the top and bottom of the two columns, as illustrated in figure. 1 (a). The overhead vapor form the high pressured column is used to heat the reboiler from the low pressured column. There is only energy exchange between the two columns, so the heat duties of the feed splitting (FS) are approximately equal to the heat duty for single column with the same conditions divided by the number of effect.

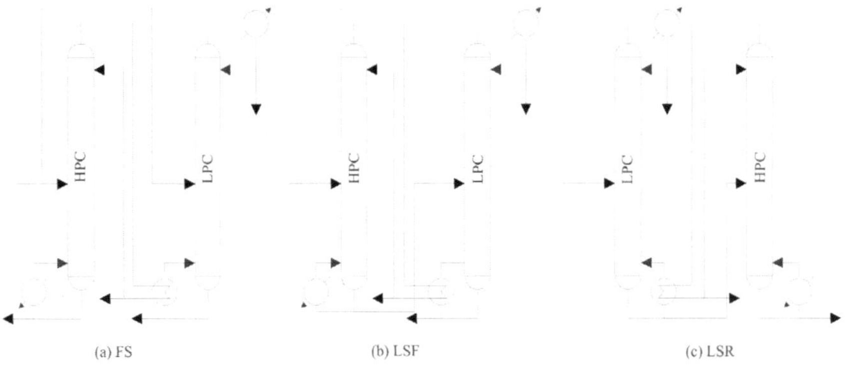

Fig. 1. Three configurations of double-effect distillation

Figure. 1 (b) and (c) respectively shows the configuration of the light split/forward heat integration (LSF) and the light split/reverse heat integration (LSR). Different from the feed splitting (FS), the total feed stream was sent to one column in the other two configurations. In the light split/forward (LSF), the total feed is pumped to the high-pressure column (HPC) and the bottom stream of HPC flows into the low-pressure column (LPC). In this way, all of the heavy components are from LPC and the light components from the top both of the two columns. Figure. 1 (b) indicates the direction of mass flow and heat integration is the same. The last configuration is the light split/reverse (LSR), as shown in figure. 1 (c), all of the feed is sent to the low-pressure column (LPC). It can remove the two kinds of light components from the top both of the two columns and one kind of heavy components from the bottom of HPC. Moreover, the direction of heat integration is in the reverse with the mass flow. These two kinds of configurations have not only energy exchange but also substantial exchange, therefore, when the operating parameters of one column are changed, the other column will also change.

Theoretically, the three kinds of configurations are empirically 30% less energy consumption than the single column distillation. As for choosing what kind of the configurations of the double-effect distillation, it depends on the properties of the multi-component mixture, the pressure of saturated steam, the temperature of cooling water, structure parameters of existing distillation column and so on.

2.2 Exergy analysis of the distillation process

According to the First Law of Thermodynamics, also called the Law of Energetic Balances, no matter what happens, the total amount of energy has not yet changed. However, when the various types of energy generate the mutual conversion, the quality of the energy is always decreased. The separation process in distillation is irreversible. In order to separate the mixture in distillation column, it needs the driving force which must consume a certain amount of external work, by this means, the decline of the quality of energy is inevitable. Therefore, we should choose a reasonable driving force in distillation process to avoid unnecessary loss of the quality of energy. It defined the concept of exergy (E_x) to compare the quality of the different types of energy. Exergy, which is a measure for evaluating the extent of available energy or the quality of energy, is based on the Second Law of Thermodynamics. Exergy analysis reveals the essence of the energy, and can be used as an important tool to evaluate the effect of energy saving. And it usually combines with process optimization studies, and as its guidance, to identify cost saving opportunities and amend the existing scheme.

Szargut [29] has summarized a comprehensive definition of the exergy, that is: 'Exergy is the amount of work obtainable when some matter is brought to a state thermodynamic equilibrium with the common components of the natural surroundings by means of reversible processes, involving interaction only with the above mentioned components of nature'. The state of one stream recorded as 1, according to the definition of exergy, the exergy of the stream can be expressed as:

$$E_x = (H_1 - H_0) - T_0(S_1 - S_0) \tag{2}$$

where, H_1 and S_1 is the enthalpy and the entropy of the stream 1, and T_0 is the temperature of the surrounding and H_0 and S_0 represent the enthalpy and the entropy of the surrounding.

The state of another stream recorded as 2, when the system stream change from state 1 (T_1, P_1) to state 2 (T_2, P_2), the variation of exergy (ΔE_x) is given by

$$\Delta E_x = E_{x,1} - E_{x,2} = \Delta H - T_0 \Delta S \tag{3}$$

In the conventional distillation process, it must input the heat or energy to drive the occurrence of separation process. The bottom of the column input the heat (Q_R) at this temperature (T_R), and the top of the column was sent to the heat (Q_C) at this temperature (T_C). The net work consumption (W) in distillation is

$$W = Q_R(1 - \frac{T_0}{T_R}) - Q_C(1 - \frac{T_0}{T_C}) \tag{4}$$

When the enthalpy difference between the product and the feed is very small, $Q_R=Q_C=Q$ can be approximately gotten, then Equation (4) is rewritten to be:

$$W = QT_0(\frac{1}{T_C} - \frac{1}{T_R}) \tag{5}$$

Then, the total exergy loss (Ex_{loss}) of the distillation column can be calculated by

$$Ex_{Loss} = W - \Delta E_x \tag{6}$$

3. *Prototype case: the prevalent distillation schemes*

3.1 The improved four-column methanol distillation scheme

In the methanol industry, it has gradually appeared two-column, three-column, four-column and five-column distillation scheme [20-23]. With the increase of the number of distillation column, the corresponding distillation scheme can not only increase the efficiency of the process, but also cut the energy consumption. At present, the four-column scheme has a dominant status to produce the purified methanol. However, the improved four-column distillation scheme can cut some more energy requirement with respect to the traditional four-column distillation scheme. The improved four-column distillation scheme is seen as the comparative reference in the paper.

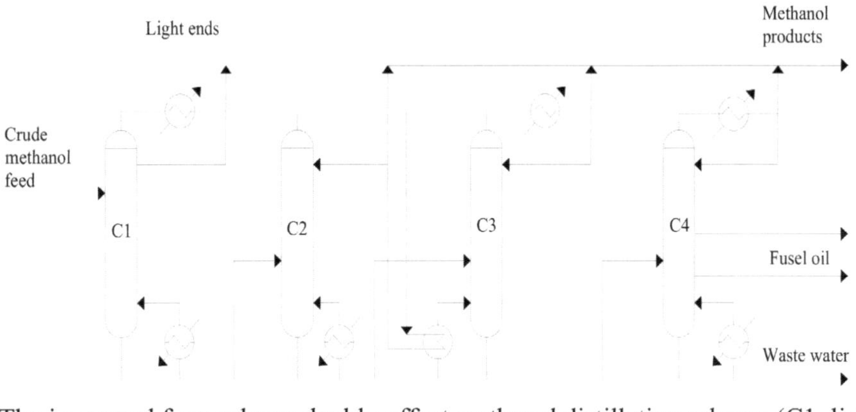

Fig. 2. The improved four-column double-effect methanol distillation scheme (C1, light ends column; C2, pressured column; C3, atmospheric column; C4, water column)

In the improved four-column distillation scheme, as shown in figure. 2 [24], the crude methanol flows into the light ends column (C1) in 87.5 ℃ for removing the light ends and the remnants of dissolved gas from the top of C1. Besides, in order to raise the methanol recovery rates, the light ends column (C1) designs a two-stage condenser system. The crude methanol was distillated successively from the pressured column (C2) and the atmospheric column (C3), and most of the methanol products can be received at the top both of C2 and C3. The bottom stream of C1, C2 and C3 is pumped respectively as the feed into C2, C3 and the water column (C4). Residual methanol, fusel oil and waste water are extracted from the corresponding position of the top, side stream and the bottom of the water column (C4).

In the improved scheme, the operation pressure of C2 is 0.9MPa, and the other three columns run under the atmospheric pressure. Due to the different of operating pressure, the overhead vapor of C2 can undertake thermal coupling connection with the reboiler of C3.

3.2 The current organosilicon monomer distillation scheme

According to the survey, there are several kinds of organosilicon distillation schemes. However, the 9-column organosilicon distillation scheme [9], in China, has an important role in the organosilicon industrial, as illustrated in Figure. 3, it is used as the base case.

Figure. 3 shows that the crude monomer is delivered to the heavies removal column (C1') and the bottom stream of C1' has a lot of the unsteady substances of high boiling points. These unsteady substances of high boiling points are pumped to the high boiling point column (C9') as the feed, and the products from the top of C9' continue to serve as the crude monomer by circumfluence. By this way, it can accomplish the separation between the unsteady substances of high boiling points and the useful substances at a relatively low temperature in the high boiling point column (C9') and avoid the unsteady substances of high boiling points from thermal cracking to generate many unknown substances which would greatly affect the subsequent separation of product. The low removal column (C2') provides trimethylchlorosilane (Me3), methyldichlorosilane (MeH), and other light components for lights removal column (C3'), MeH column (C4'), azeotrope removal column (C5') and Me3 column (C6'), which can finally achieve two products (MeH and Me3) and three wastes. It also provides methyltrichlorosilane (Me1) and dimethylchlorosilane (Me2) and other heavy components for Me1 column (C7') and Me2 column (C8'), which can finally get two products (Me1 and Me2).

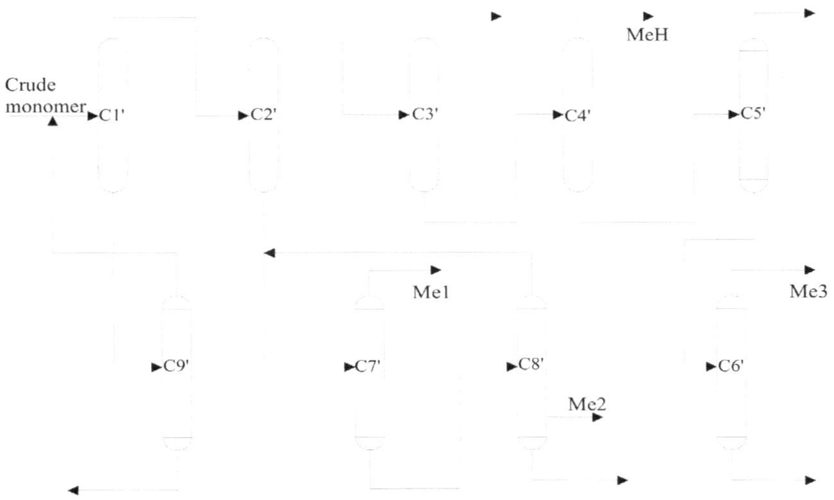

Fig. 3. The prevailing 9-column distillation scheme (C1', heavies removal column; C2', low removal column; C3', lights removal column; C4', MeH column; C5', azeotrope removal column; C6', Me3 column; C7', Me1 column; C8', Me2 column; C9', high boiling point column)

In this scheme, the operating pressure of all nine columns is relatively low, and the highest operating pressure is 260 kPa in the lights removal column (C3'). Although it has a different operating pressure between C3' and the other columns, but the temperature difference is not enough to make them use double-effect or heat integrated distillation.

4. Simulation and analysis both of the prevalent schemes and the optimized schemes

4.1 The novel five-column heat integrated methanol distillation scheme

Although the energy saving of the improved four-column methanol distillation scheme can reach 10% compared with the traditional four-column scheme, it still exist several shortages, which will be illustrated by the results of computer simulation to the improved four-column methanol distillation scheme. The flow rate of the crude methanol is 239230 kg/hr, and the temperature is 87.5 ℃, and the composition of the crude methanol is exhibited in table 1. The inlet and outlet temperature of the cooling water are 30 ℃ and 40 ℃.The methanol content is specified as 99.99 wt% in the methanol products and 10ppm in the wastewater of the water column. Table 2 shows the operating parameters of each column in the improved four-column scheme.

Table 1. Crude methanol feed data

Component	Mass fraction%	Component	Mass fraction%
H_2	0.0100	Pentanol	0.0338
N_2	0.0710	Formic acid	0.0030
Methane	0.0100	Acetic acid	0.0030
CO	0.0360	Methyl formate	0.0092
CO_2	0.6640	Acetone	0.0193
Methanol	95.0000	Methyl ethyl ketone	0.0030
Dimethyl ether	0.0309	Diisopropylether	0.0056
Trimethylamine	0.0010	Water	3.7100
Ethanol	0.0900	i-Pentane	0.0562
i-Propanol	0.0024	n-Pentane	0.0169
Propanol	0.0264	n-Hexane	0.0487
i-Butanol	0.0338	n-Heptane	0.0009
Butanol	0.0185	n-Octane	0.0050
3-Pentanol	0.1407	n-Nonane	0.0050

Table 2. The operating parameters of improved four-column scheme

The operating parameters	C1	C2	C3	C4
Top pressure (kPa)	150	900	110	110
Top temperature (℃)	81.0	132.6	67.7	67.7
Bottom temperature (℃)	83.8	139.1	96.5	104.5
Reflux ratio	0.58	2.44	1.68	11.57
Number of theoretical stage	33	52	58	61

The results of simulation indicate that the hot utility is still in a high level because the reboilers of C1, C2 and (C4) still need the saturated steam to transmit the heat. Table 2 shows the reflux ratio of C4 is 11.57 and this large reflux ratio can cause the increasing of the energy requirement of the water column (C4). Besides, according to the flowsheet of this scheme, the diameter of column is so large that the emergence of many urgent cases.

In view of above these defects in the improved four-column scheme, the novel five-column heat integrated methanol distillation scheme (shown as Fig. 4) is designed in this paper to reduce the impact of these factors as far as possible by using the paralleled or combined double-effect distillation. Juntao Zhang [23] and Jinsheng Sun [25] are also proposed a five-column methanol distillation scheme by using triple-effect distillation and double-effect distillation respectively, and the effect of energy saving of the two schemes is outstanding.

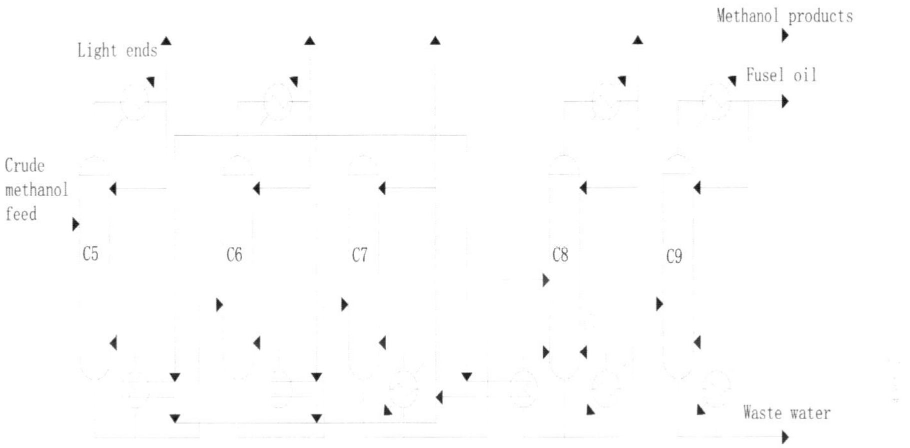

Fig. 4. The five-column heat integrated methanol distillation scheme (C5, light ends column; C6, prepositive atmospheric column; C7, pressured column; C8, postpositive atmospheric column; C9, water column)

According to the principle of the paralleled double-effect distillation, the four columns are unable to constitute the heat integration network in the improved four-column scheme. Through repeated verification, the prepositive atmospheric column (C6) is added after the light ends column (C5), meanwhile, the operating pressure of the pressured column (C7) is reduced to increase its throughput. In this way, the overhead vapor of C7 is split into three streams to respectively integrate with the light ends column (C5), the prepositive atmospheric column (C6) and the postpositive atmospheric column (C8). Due to use the paralleled double-effect distillation, the duties of the hot utility have decreased in a great margin. However, there are still part of heat from the background process is absorbed by the cooling water. So an inter-reboiler with a low temperature level is designed in the postpositive atmospheric column (C8) and it can share the heat with the background process streams, besides, we have proved the feasibility by using the column grand composite curve analysis of columns. Figure. 4 shows that the product of the top of the water column (C9) is fusel oil instead of purified methanol, which can not only reduce the reflux ratio of C9, but also make the diameter of column smaller to reduce the difficulty of operation.

Due to produce the fusel oil instead of purified methanol in the water column (C9) in the novel five-column distillation scheme, the methanol content of the feed for the water column have to limit at 0.016 wt%. In the five-column scheme, the operating pressure and the number of theoretical stage for each column are determined through the existing process, the onsite data and the requirement of the heat coupling. The proportion of the methanol products from the C6, C7 and C8 is determined by their own double-effect or heat integrated distillation system. The simulation results of the novel five-column scheme are listed in table 3.

Table 3. The operating parameters of five-column scheme

Column	C5	C6	C7	C8	C9
Top pressure (kPa)	160	110	400	110	110
Top temperature (℃)	83	66.6	104	66.6	86.5
Bottom temperature (℃)	85.6	84	118.8	104.5	103.2
Reflux ratio	1.46	1.49	2.23	1.87	4
Number of theoretical stage	33	54	54	65	29

4.2 Energy consumption and exergy loss analyses for the two methanol schemes

In the improved four-column scheme, the pressured column and the atmospheric column constitute the distillation system of heat coupling, there are three columns (C1, C2, C4) need the hot utility, and the entire duty of reboiler is 145.4×10^3 kw. As described above, the light ends column (C5) and the prepositive atmospheric column (C6) don't need the saturated steam in the five-column scheme and the total duty of reboiler is reduced to 122.7×10^3 kw. The detailed data of the reboiler duty as shown in table 4.

Table 4. The duty of reboiler in the two schemes

Column	C1	C2	C4	C7	C8	C9
Duty of reboiler (10^4 kw)	3.30	10.04	1.20	8.97	2.86	0.44

Obviously, the novel five-column scheme is much better than the improved four-column scheme in energy saving and the energy consumption of the novel scheme can be reduced by 15.61%. However, the recovery rate of methanol of the novel scheme is approximately equal with the improved four-column scheme.

Only the energy consumption analysis is not revealed the energy saving essence of the novel five-column scheme. Exergy analyses are used to further compare the two schemes, and the exergy flow rates are calculated through using the computer simulation data. The activity coefficient of the liquid is calculated using the Alcohol Package thermodynamic model. The calculation of the net work consumption of distillation column should pay more attention to the temperature of the cooling water. When the cooling water exchanges heat with other streams, the temperature of the cooling water is momentarily changed, so it is replaced the temperature of cooling water with the arithmetic average of inlet and outlet temperature of cooling water. Figure. 5 show the overall exergy loss of each column in the two methanol distillation schemes, in addition to the overall exergy loss of C6. Due to optimize the four-column scheme, the prepositive atmospheric column (C6) is added in the transformed scheme, and the overall exergy loss of C6 is 9157.6 kw. The exergy loss of each column, besides the pressured column (C7), in the novel five-column scheme is obviously less than their counterparts in the improved four-column scheme.

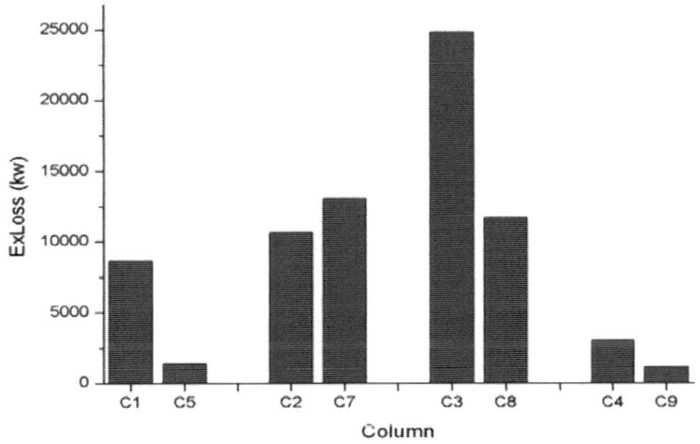

Fig. 5. The overall exergy loss analysis of each column in the two schemes

The total exergy loss of the improved four-column scheme and the transformed scheme is, respectively, 47100.1 kw and 36355.1 kw. The total exergy loss of the optimized scheme is dropped by 22.8%. Through the comparison between the exergy loss and energy consumption of the two schemes, it is indicated that the novel five-column scheme should be applied to produce the purified methanol in industry.

4.3 The novel heat coupling organosilicon distillation scheme

The primary source of the crude organosilicon monomers is the catalytic reaction products between chloromethane and silicon at high temperature and the table 5 displays the normal boiling points and mass percentage of the main components of the crude organosilicon monomers. Methyltrichlorosilane (Me1), dimethylchlorosilane (Me2), trimethylchlorosilane (Me3) and methyldichlorosilane (MeH) are the most valuable components in the crude organosilicon monomers. The purification requirements on organosilicon of these materials are shown in table 6.

Table 5. The crude organosilicon monomer feed data

Component	Normal boiling point at 1 atm/℃	Mass percentage %
LB		1
Methyldichlorosilane	41.8	4
Silicon tetrachloride	57.6	0.5
Trimethylchlorosilane	57.9	2.5
Methyltrichlorosilane	66.4	11
Dimethyldichlorosilane	70.2	80
HB		1
Temperature (℃)	40	
Pressure（kPa）	100	
Flow rate (kg/h)	26200	

Table 6. Purification demand on organosilicon

Monomer	Purification demand
MeH	99.5%
Me3	99.5%
Me1	99.5%
Me2	99.95%

The conventional nine-column scheme is simulated with the aid of existing process and onsite data to achieve the operating parameters of each column, as illustrated in table 7. The bottom temperature of the heavies removal column (C1') and the high boiling point column (C9') are obviously much bigger than that of the other seven columns in table 7. According to the principle of the double-effect distillation, the top temperature of the high pressured column is relatively high in comparison with the bottom temperature of the low pressured column, and the high pressured column can integrate with the low pressured column. Although the top temperature of C1' is less than the bottom temperature of the other seven columns, the overhead vapor of C1' still has a great opportunity to conduct heat with the other seven columns with the increasing of its top temperature.

Table 7. The simulation results of prevailing 9-column

Column	C1'	C2'	C3'	C4'	C5'	C6'	C7'	C8'	C9'
Top temperature (℃)	70.8	65.3	57.3	66.7	72.1	70.6	71.2	74.9	72.7
Bottom temperature (℃)	124.1	89.1	80.4	85.2	74.9	75.4	75.3	75.3	126.5
Reboiler duty (MW)	11.67	5.61	1.13	0.51	1.19	0.46	12.27	16.08	0.02
Condenser duty (MW)	-12.47	-5.53	-1.08	-0.53	-1.20	-0.47	-12.69	-16.12	-0.02
Top pressure (kPa)	110	180	260	220	170	150	110	110	110

According to the way of the paralleled double-effect distillation, the increasing operating pressure in the heavies removal column (C1') can make the top temperature of C1' ascending to integrate with the other seven columns. The idea is still in study to overcome some difficulties. However, we have got another feasible method, in which the operating pressure of C1' and the Me1 column (C7') is raised to 360 kPa and 280 kPa, respectively. In this way, the top temperature of C1' and C7' is increased to ensure they can integrate with the other columns, besides the high boiling point column, and the parallel double-effect or heat integrated organosilicon monomer distillation scheme is designed as shown in figure. 6 [26]. The feasibility of this heat coupling scheme has confirmed through three x-y diagrams including the heavies component and Me1, the heavies component and Me2, Me1 and Me2 [9].

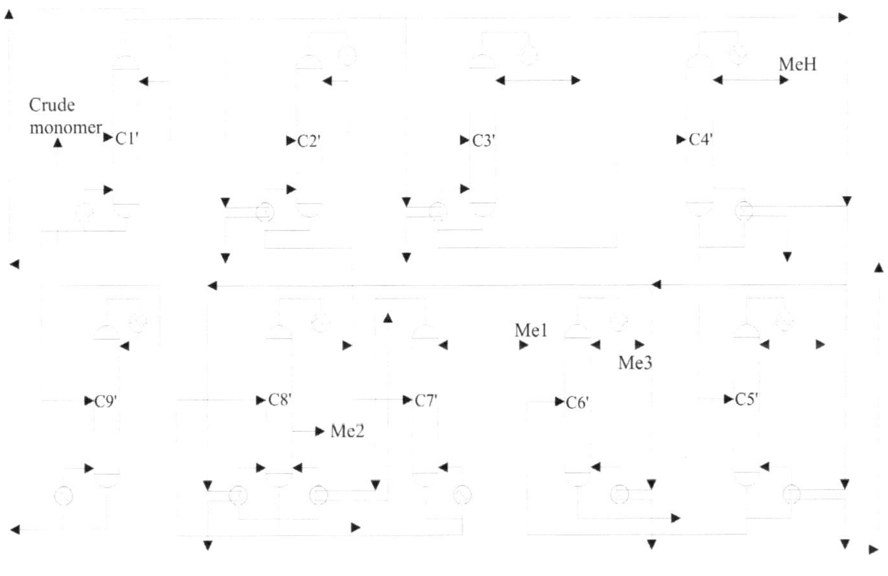

Fig. 6. The heat coupling 9-column scheme (C1', heavies removal column; C2', low removal column; C3', lights removal column; C4', MeH column; C5', azeotrope removal column; C6', Me3 column; C7', Me1 column; C8', Me2 column; C9', high boiling point column)

Figure. 6 shows the heavies removal column (C1') make the heat coupling with the low removal column (C2'), the lights removal column (C3'), the MeH column (C4'), the azeotrope removal column (C5'), the Me3 column (C6') and the Me2 column (C8'), successively. In order to gain a better energy saving effect, C8' is designed with two reboilers, the one integrated with C1' and the other integrated with the Me1 column (C7'). The operating parameters and the duty of reboiler and condenser of each column in the heat integrated 9-column are enumerated in table 8.

Table 8. The simulation results of heat integrated 9-column

Column	C1'	C2'	C3'	C4'	C5'	C6'	C7'	C8'	C9'
Top temperature (℃)	114.3	65.3	57.3	66.7	72.1	70.6	105.1	74.9	72.7
Bottom temperature (℃)	124.9	89.9	78.8	84.5	75.4	75.4	110.9	76.6	121.4
Reboiler duty (MW)	13.06	5.62	1.13	0.51	1.19	0.46	13.83	14.12	0.38
Condenser duty (MW)	-13.72	-5.54	-1.08	-0.53	-1.20	-0.47	-14.02	-14.43	-0.41
Top pressure (kPa)	360	180	260	220	170	150	280	110	110

4.4 Energy consumption and exergy loss analyses for the two organosilicon schemes

Figure. 4 indicate that each column of the prevalent nine-column scheme is not integrated with the other columns and the reboilers of all of these columns are required the saturated steam (600 kPa). According to the data from the table 7, the total duty of reboiler is 48.94 MW for the prevailing 9-column scheme. In the new nine-column scheme, C1' integrate with C2', C3', C4', C5', C6' and C8' successively and C7' integrate with C8', so C2', C3', C4', C5', C6' and C8' don't consume the saturated steam. The total duty of reboiler is 27.27 MW. Obviously, the heat integrated 9-column scheme has an absolute advantage over the prevailing 9-column scheme in energy saving, and the energy savings can reach 44.28%.

As the same with the methanol distillation scheme, the inlet and outlet temperature of cooling water are 30 ℃ and 40 ℃, and the total exergy loss is calculated through the results of computer simulation of the two schemes. The overall exergy loss comparison of each column in the two schemes is illustrated in figure. 7. In the normal scheme, the throughput of C9' is so small that the overall exergy loss of C9' is very near to zero, so it only have the exergy loss of C9' of the new scheme in figure. 7. The overall exergy loss of each column in the optimized scheme is declined in comparison with that of the corresponding column in the normal scheme except the high boiling point column (C9'). Because the crude organosilicon monomers contains some unstable components, in order to avoid these components happening thermal cracking in the heavies removal column (C1') in the novel scheme, more liquid from the bottom of C1' is pumped to C9' as the feed. That can cause the ascent of the throughput and the overall exergy loss of C9'.

The total exergy loss is 14035.5 kw and 8363.2 kw, correspondingly, in the conventional scheme and the novel scheme, and the exergy loss of the new scheme can be reduced by 40.41%. Through the above analysis and comparison, the heat integrated nine-column organosilicon distillation scheme should be a substitute for the prevalent nine-column scheme.

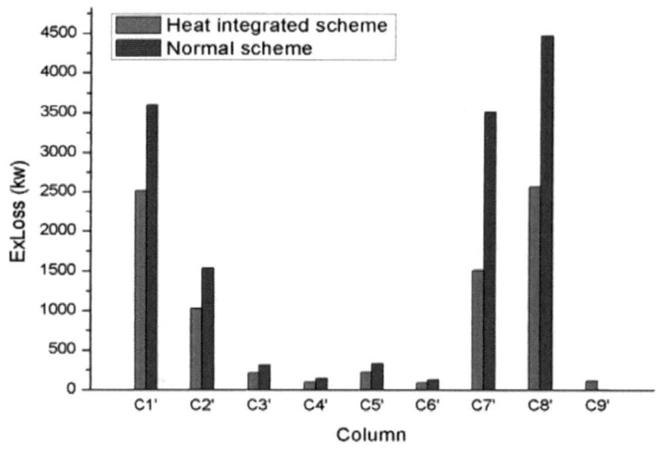

Fig. 7. The exergy loss comparison between the heat integrated scheme and the normal scheme

5. Conclusions

Paralleled double-effect or heat integrated distillation for energy saving is proposed, exampled by heat integrated five-column methanol distillation scheme and the heat coupling 9-column organosilicon monomer distillation scheme, with the aid of computer simulation. Besides, the optimized methanol distillation scheme and the transformed organosilicon scheme are comprehensively analyzed through the energy consumption and the exergy loss in comparison with the correspondingly prevalent scheme.

Adding a prepositive atmospheric column and decreasing the operating pressure of the pressured column in the novel five-column scheme can make the pressured column to simultaneously integrate with the light ends column, the prepositive atmospheric column and the postpositive atmospheric column. Besides, an inter-reboiler is added to integrate with the background process streams. The energy consumption and the exergy loss of the new scheme are 15.61% and 22.8% lower than the improved scheme.

Increasing the operating pressure of the heavies removal column and the Me1 column can make them to integrate with the other columns, in addition to the high boiling point column, to constitute a paralleled double-effect distillation system in the optimized nine-column scheme. The energy consumption and the exergy loss of the novel scheme are 44.28% and 40.41% less than the normal scheme.

The two cases expound that the key of paralleled double-effect distillation process lies in the reasonable adjustment of the energy distribution, the operating parameters and the configuration of some columns. It not only reduces the quantity of energy consumption, but also avoids the decrease of the quality of energy. For factories with a large numbers of distillation schemes of high energy requirement, the paralleled schemes lead to improve economic benefit.

References

[1] Araujo AB, Brito RP, Vasconcelos LS, "Exergetic analysis of distillation processes-a case study," Energy 2007;32:1185-93.

[2] Michele Mascia, Francesca Ferrara, Annalisa Vacca, Giuseppe Tola, Massimiliano Errico, "Design of heat integrated distillation systems for a light ends separation plant," Applied Thermal Engineering 2007;27:1205-1211.

[3] Amiya K. Jana, "Heat integrated distillation operation," Applied Energy 2010;87:1477-1494.

[4] Hilde K. Engelien, Sigurd Skogestad, "Multi-effect distillation applied to an industrial case study," Chemical Engineering and Processing 2005;44:819-826.

[5] James G. Gebbie, Michael K. Jensen, Piotr A. Domanski, "Experimental transient performance of a heat pump equipped with a distillation column," International Journal of Refrigeration 2007;30:499-505.

[6] Massimiliano Errico,Giuseppe Tola, Michele Mascia, "Energy saving in a crude distillation unit by a preflash implementation," Applied Thermal Engineering2009;29:1642-1647.

[7] Massimiliano Erricoa, Giuseppe Tola, Ben-Guang Rong, Daniele Demurtas, Ilkka Turunen, "Energy saving and capital cost evaluation in distillation column sequences with a divided wall column," chemical engineering research and design 2009;87:1649-1657.

[8] Ivar J. Halvorsen, Sigurd Skogestad, "Energy efficient distillation," Journal of Natural Gas Science and Engineering 2011;3:571-580.

[9] Jinsheng Sun, Fan Wang, Tingting Ma, Hong Gao, Yanzhen Liu, Fang Cai, "Exergy analysis of a parallel double-effect organosilicon monomer distillation scheme," Energy 2012;47:498-504.

[10] Hilde K. Engelien, Sigurd Skogestad, "Multi-effect distillation applied to an industrial case study," Chemical Engineering and Processing 2005;44:819-826.

[11] Kazuo Matsuda, Kenichi Kawazuishi, Yasuki Kansha, Chihiro Fushimi, Masaki Nagao, Hiroshi Kunikiyo, Fusao Masuda, Atsushi Tsutsumi, "Advanced energy saving in distillation process with self-heat recuperation technology," Energy 2011;36:4640-4645.

[12] M.A. Sharaf, A.S. Nafey, Lourdes García-Rodríguez, "Exergy and thermo-economic analyses of a combined solar organic cycle with multi-effect distillation (MED) desalination process," Desalination2011;272:135-147.

[13] Weiwen Wang, Wenjian Wang, "Energy Saving in Methylchlorosilane Distillation," International Conference on Computer Distributed Control and Intelligent Enviromental Monitoring 2012:290-293.

[14] Myungwan Han, Sunwon Park, "Multivariable control of double-effect distillation

configurations," J. Proc. Cont. 1996:247-253.

[15] Hilde K. Engelien, Sigurd Skogestad, "Multi-effect distillation applied to an industrial case study," Chemical Engineering and Processing2005;44:819-826.

[16] Debadrita Maiti, Amiya K. Jana, Amar Nath Samanta, "A novel heat integrated batch distillation scheme," Applied Energy2011;88:5221-5225.

[17] Anton A. Kiss, Servando J. Flores Landaeta, Carlos A. Infante Ferreira, "Towards energy efficient distillation technologies-Making the right choice," Energy2012;47:531-542.

[18] Diego C. Alarcón-Padilla, Lourdes García-Rodríguez, Julián Blanco-Gálvez, "Experimental assessment of connection of an absorption heat pump to a multi-effect distillation unit," Desalination2010;250:500-505.

[19] Diego C. Alarcón-Padilla, Lourdes García-Rodríguez, Julián Blanco-Gálvez, "Design recommendations for a multi-effect distillation plant connected to a double-effect absorption heat pump: A solar desalination case study," Desalination2010;262:11-14.

[20] Fiedler E, Grossman G. Methanol, "Ullmann's Encyclopedia of Industrial Chemistry," Weinheim : Wiley-VCH Verlag/Gmbh & Co; 2002.

[21] Meyers RA, "Handbook of synfuels Technology," New York : McGraw-Hill; 1984.

[22] Chu YZ, Qin LP, "Optimum design of methanol distillation process and columns," Chem Ind Eng Prog 2008;(27):1659-62.

[23] Juntao Zhang, Shengrong Liang, Xiao Feng, "A novel multi-effect methanol distillation process," Chem Eng Pro 2010;(49):1031-37.

[24] Jinsheng Sun, Fan Wang, Tingting Ma, Hong Gao, Peng Wu, Lili Liu, "Energy and exergy analysis of a five-column methanol distillation scheme," Energy2012;45:696-703.

[25] J. S. Sun, P. Wu, Y. H. Wang, H. Gao, L.L. Liu.CN Patent102039058 A. 2011.

[26] J. S. Sun, P. Wu, Y. H. Wang, Y. X. Su, H. Gao, J. Li, D. Yan. CN Patent 101798322 A. 2010.

[27] Michele Mascia, Francesca Ferrara, Annalisa Vacca, Giuseppe Tola, Massimiliano Errico, "Design of heat integrated distillation systems for a light ends separation plant," Applied Thermal Engineering 2007;27:1205-1211.

[28] Olga A. Flores, J. Carlos Ca´ rdenas, Salvador Herna´ ndez, Vicente Rico-Ramı´rez, "Thermodynamic Analysis of Thermally Coupled Distillation Sequences," Ind. Eng. Chem. Res. 2003;42:5940-5945.

[29] Szargut J, "Exergy Analysis of Thermal, Chemical and Metallurgical Processes," New York: Hemisphere Publishing Corporation;1988:55-152.

(45n) In Situ Transesterification of Jatropha Curcas Seed Particles with Alkaline Phase Transfer Catalyst

Subbarao Duvvuri[1], Shuhaimi Mahadzir[2] and Sintayehu Hailegiorgis[2], (1)Department of Chemical Engineering, Universiti Teknologi PETRONAS, Tronoh, Malaysia, (2)Chemical Engineering, Universiti Teknologi PETRONAS, Tronoh, Malaysia

Economic viability of biodiesel technology is at stake due to the high cost of edible oil feedstock and slow reaction rates due to limited solubility of oil in methanol. In situ transesterification of non edible oil seed particles can minimize the cost of feedstock. Microwave pretreatment of seeds can make oil molecules more reactive. Use of phase transfer catalysts can enhance the reaction rates. In the present work, these three concepts are utilized together to investigate alkaline in situ transesterification of microwave pretreated jatropha curcas seed particles in presence of benzyltrimethylammonium hydroxide (BTMAOH) as a phase transfer catalyst (PTC). It was observed that addition of PTC to alkaline catalyst has improved conversion and reduced reaction time. Microwave pretreatment of jatropha seed particles reduced reaction time drastically. The quality of biodiesel produced was within the limits of the international standard requirements of ATME D6751 and EN-14214. Thus, combined use of microwave pretreatment of seed particles, phase transfer catalysis in the presence of alkaline catalyst can increase the rate of in situ transesterification and product yield.

(45o) Catalyzed-Assisted Manufacture of Olefins (CAMOL): Updated for Use in Naphtha Service

Maximilian Walter, Petrochemicals, BASF SE, Ludwigshafen, Germany and **Suzanne Rech**, BASF Qtech, BASF Corporation, Iselin, NJ

Since the commercialization of the olefins plant, operators have searched for ways to be more competitive. One area always targeted includes keeping their furnaces in operation longer between decoking steps. Over the years, many anti-coking methods/technologies have become commercially available using different approaches to prevent coils from coking. One of the latest available technologies is the novel CAMOL^{TM} (Catalyzed-assisted Manufacture of Olefins) Technology from BASF Qtech. As reported in the introduction of the technology to the EPC in 2008 & 2010, the CAMOL^{TM}technology differs from other available anti-coking technologies in that the technology not only eliminates the catalytic (or filamentous) coke make through a physical barrier, but also gasifies the thermal (or amorphous) coke for an even greater coking resistance. The coating process has generated a coating with almost identical thermo-mechanical properties as the base metal substrate (alloy). This allows bonding to the base metal similar to being welded with a minimum risk of spallation when going through thermal cycling. Commercially, the CAMOL Technology has been available since 2006 and is now in operation in furnaces of various designs.

With CAMOL^{TM}, operators have experienced extended run lengths, reduced energy requirement and GHG emissions, lower average TMT operating temperatures, reduced DMDS consumption, and less carburization. In addition, the technology has exhibited high operating temperature stability (>1130°C/ 2065°F).

At BASF SE in Ludwigshafen, Germany, the technology has been in operation since early 2010 in mostly naphtha service. After overcoming some initial challenges associated with a catalytic surface technology, such as reactivating the catalytic species after decoking, the coil is now successfully operating with run lengths significantly higher than uncoated coils. This paper will report on the results and learnings from the experiences gathered over the past three years along with the advancements made since the last report in 2010.

(45p) Extending Furnace Run Length and Coil Life by Optimizing Decoking

Vinod Mishra, Ingenero, Mumbai, India and **James Brigman**, Ingenero, Houston, TX

Abstract:

In the Ethylene manufacturing process, coke deposited in the radiant coils and Transfer Line Exchangers (TLEs) during normal operation is removed periodically by decoking the furnaces using a mixture of steam and air. There is a significant variation in operating parameters adopted by different plants for decoking. These are often either too conservative or aggressive, leading to production loss from increased off-stream time, TLE limitation and deterioration in coil life. This paper describes how the decoking procedure was modified based on furnace and TLE operating conditions and the decoking steps changed to obtain the best possible Ethylene yield, without adversely impacting coil life and TLE performance. This optimization has now been done in several major ethylene facilities and furnace run length has more than doubled and production has significantly increased.

SOLVING OPERATIONAL PROBLEMS IN A NAPHTHA SPLITTER

Flávio Weissheimer, GE Water & Process Technologies do Brasil Ltda, São Paulo/Brasil
Giorgio Franceschetti, Braskem, Santo André/Brasil
Rafael Henrique Camparin, GE Water & Process Technologies do Brasil Ltda, São Paulo/Brasil
Patrick Lucas, GE Water & Process Technologies, The Woodlands/US

Abstract

This paper sets out how the polymerization process of Reboilers of the (A-100) Crude Naphtha Fractionation Unit of Braskem UNIB 3 ABC can be controlled by applying polymerization inhibitors developed by GE Water & Process Technologies Ltda and Braskem´s engineering solutions.

These heat exchangers have a high degree of polymerization because of the content of olefinic compounds present in the feed composition which, when in contact with high temperatures in the heat exchange areas of reboilers, tend to polymerize. The joint solution of GE technology and Braskem engineering solutions applied to this system ensures regular operation of such equipment for periods of time greater than those hitherto recorded without manifesting loss of heat transfer efficiency.

In the 2012 shutdown while the opening of these reboilers was being inspected, it was observed that pipes had not been clogged nor was surface fouling present, which is a very positive state when compared to the historical records of previous campaigns.

The strategy adopted by GE/Braskem to ensure a longer campaign for these pieces of equipment was based on the quality control of feeding by adjusting dosages and monitoring performance using statistical analysis tools.

The use of statistical models based on Multivariable Regression Analysis (MRA) to evaluate performance, which is validated by inspecting reboilers internally, has been an efficient way to determine the degree to which the system has been fouled and also to schedule shut-downs for cleaning the mechanical parts.

Description of the Process

The (A-100) Naphta Fractionation Unit includes two distillation columns, the objective of which is to separate the light (C_6), medium (C_6 - C_8) and heavy (C_{9+}) cuts, included in the feed supply of crude naphtha.

The light cut and a portion of the heavy cut are sent to pyrolysis furnaces so as to produce ethylene. The medium cut is sent to the Hydrotreating and Catalytic Reforming Unit, with the objective of increasing the aromaticity of the system so as to produce BTX.

The A-100 processes 100 tons/h of crude naphtha, and approximately 35% of this current represents the average cut. Figure 1 below shows the schematic flow diagram of this fractionation process.

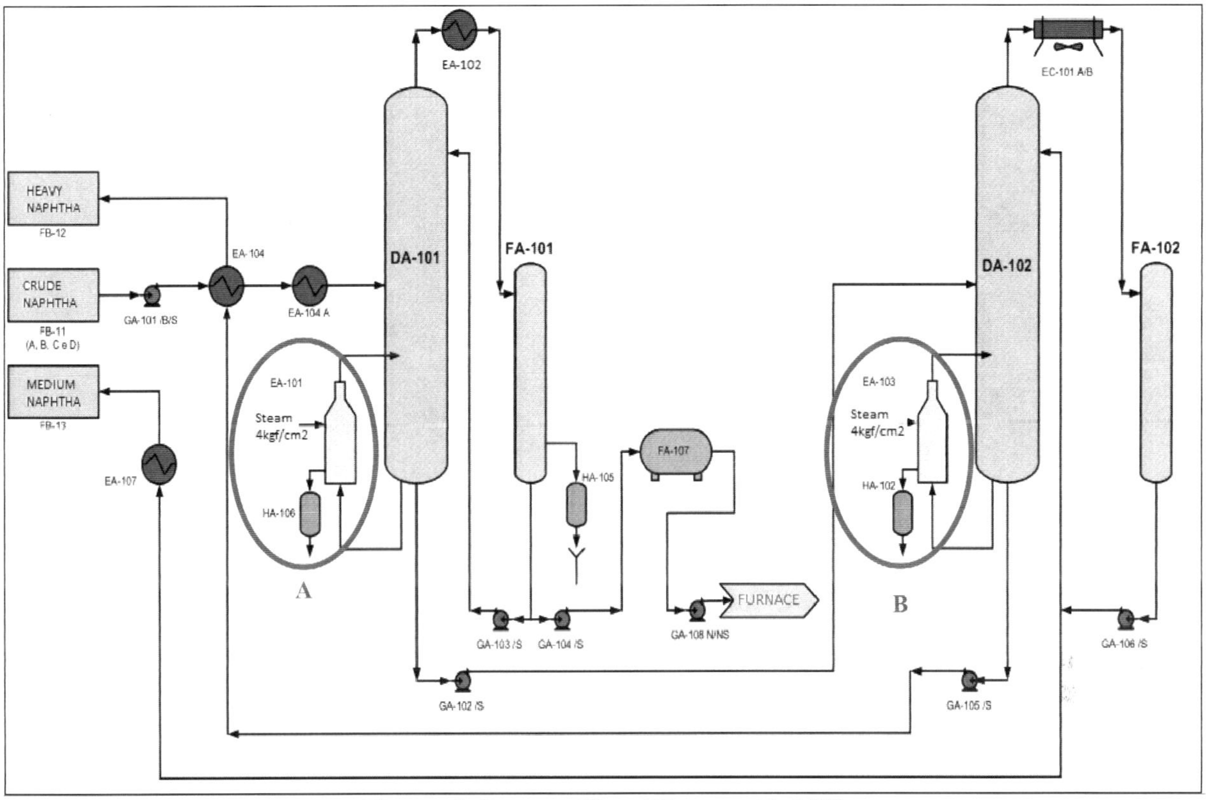

Figure 1. Process Flow Diagram (A-100)

The Reboilers, indicated by the letters A and B in Figure 1 above, operate with an injection of low steam (4kgf/cm2) to promote vaporization of the bottom stream of the A-100 distillation columns. These devices operate over a temperature range of about 150°C, which is sufficient to promote the polymerization of certain olefinic compounds present in the feed of crude naphtha.

Poor performance of these Reboilers can lead to production stoppages for maintenance and cleaning, which very often affects up to five days of production. The possibility of raising the times of campaigns and predicting the degree of fouling in downstream units helps to minimize production losses due to unscheduled stoppages.

This study will discuss the results of Reboiler A, which is the most critical equipment, with a view to synthesizing data, but gains in equipment B were also significant.

Evaluation System

Historical Record of the Campaign

Reboiler A was installed in 1972 but examination of the historical records did not reveal that the sheaf of tubes had ever been replaced. This exchanger has currently 11 tubes plugged into it.

The Campaign times of this equipment are shown in Table 1, with the last campaign being highlighted in blue so as to set it apart from the other ones. Table 1 gives a summary in the order of the timings of the campaigns, the average load processed in the period, the total load processed, the reason for operational stoppages etc.

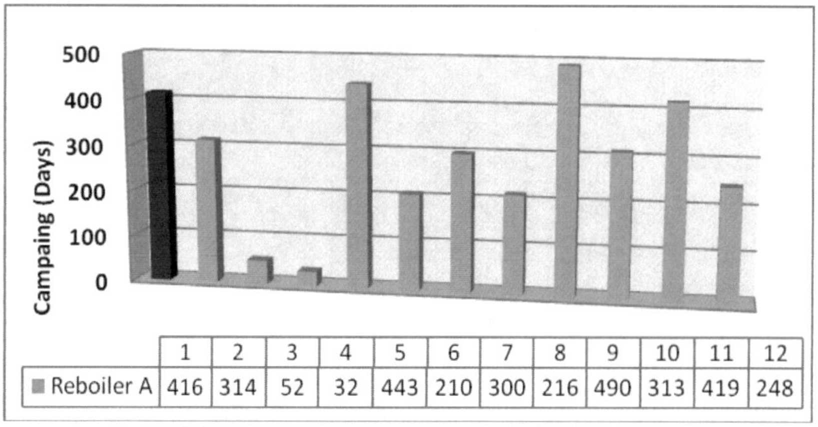

Figure 2. Campaign of the Reboiler A

Table 1. Historical Record of Reboiler A Campaigns

Date	Campaign *(days)*	Capacity *(ton/h)*	Capacity *(ton/campaign)*	Occurrence	Tubes Clogged *(%)*
13/06/12	416.0			Programmed Shutdown	0
17/03/11	314.0	93.1	701601.6	Heat exchange performance	30
03/05/10	52.0	90.8	113318.4	Stuck tube	-
11/03/10	32.0	71.1	54604.8	Heat exchange performance	50
09/02/10	443.0	72.0	765716.6	Heat exchange performance	20
16/11/08	210.0	98.1	494575.2	Heat exchange performance	-
16/04/08	300.0	104.0	748800.0	Programmed Shutdown	-
16/06/07	216.0	92.8	481075.2	Programmed Shutdown	-
10/11/06	490.0	-	-	Programmed Shutdown	-
30/06/05	313.0	-	-	Stuck tube	-
17/08/04	419.0	-	-	-	-
18/06/03	248.0	-	-	-	-
10/10/02	-	-	-	-	-

Conditions of Opening

In this review, a comparison is presented between two operational shut-downs of A-100, which occurred in March 2011 and May 2012, in order to observe the improvement in performance from the last campaign associated with the change in the GE Treatment strategy.

Campaign I: Production shutdown on 03/2011

Table 2 below summarizes the features observed during the inspection of the state of the inlet of this piece of this equipment during the shutdown of March 2011:

Table 2. Internal inspection of Reboiler A in 03/2011

Bottom Cover	Surface residue
Bottom Mirror	No deposits or internal blockage of the tubes were observed.
Top Mirror	Approximately 30% of the tubes were clogged, especially in the region of the vapor inlet, which has higher temperatures. Magnetic and non-magnetic deposits.

May/2011	**May/2011**

Figure 3. The mirror of Reboiler A before it was cleaned.	**Figure 4.** The mirror of Reboiler A at the vapor inlet before it was cleaned.

Note: It is difficult to carry out internal cleansing by Hydro jetting (pump pressure during cleaning is 20,000 psi). This reboiler has 11 tubes plugged into it, but none of them are plugged in during this shutdown.

Campaign II: Production shutdown on 03/2011

Table 3 below summarizes the features observed during the inspection of the state of the inlet of this piece of equipment during the shutdown in May 2012:

Table 3. Internal inspection of Reboiler A in 05/2012

Bottom Cover	Surface residue
Bottom Mirror	No deposits or internal blockage of the tubes were observed.
Top Mirror	Clogged tubes were not identified, not even in the region of the inlet vapor. There was only a thin layer of deposit that was easy to remove and which was evenly distributed over the upper surface of the mirror.
Spool	A thin deposit on the inner surface of the caisson.

May/2012

Figure 5. The mirror of Reboiler A before it was cleaned.

May/2012

Figure 6. The mirror of Reboiler A before it was cleaned.

Note: Hydro jetting was done quickly, without any obstructions due to deposits having been identified. This equipment has 11 tubes plugged into it, but none of them were during this shutdown.

Analysis of Deposits

The characterization of the deposits found in Reboiler A showed an 86.0% loss on ignition, thus indicating the strong presence of organic compounds in their composition, with the functional groups of carboxylic acid and ester standing out. The inorganic analysis of the sample indicated a high concentration of iron oxides and also sulfur oxides in the form of sulfides on the cover and spool of this reboiler. This is summarized in Table 4 below:

Table 4. Analysis of the Deposits of Reboiler A

Organic Compounds - LOI (%)	86.0
Water of hydration	Smaller
Sulfur compounds	Smaller
Carboxylic Acid and Ester	Greater
Dichloromethane soluble (%)	2.0
Elemental Sulfur	Greater
Dichloromethane insoluble (%)	98.0
Water of hydration	Smaller
Inorganics	Smaller
Carboxylic Acid and Ester	Greater
Composition of Dichloromethane insolubles (%)	
Carbon	56.0
Hydrogen	5.0
Nitrogen	3.0
Sulfur	8.0
Inorganic Compounds (%)	
Iron, $Fe_2O_3 + Fe_3O_4$	82.0
Sulfide, S^{2-}	16.0
Phosphate, P_2O_5	1.0
Sodium, Na_2O	1.0

Evaluation of Performance

The evaluation of performance of Reboiler A was prepared using the MRA tool, which is a GE good practice to predict fouling in critical systems. This statistical tool enables a dependent variable to be determined, in this case the percentage of opening the steam valve, by means of the numerical regression of the other operational variables of the column. This method generates a mathematical equation to predict the percentage of inlet steam when the equipment is clean, given that the equation obtained by regression includes data of this condition.

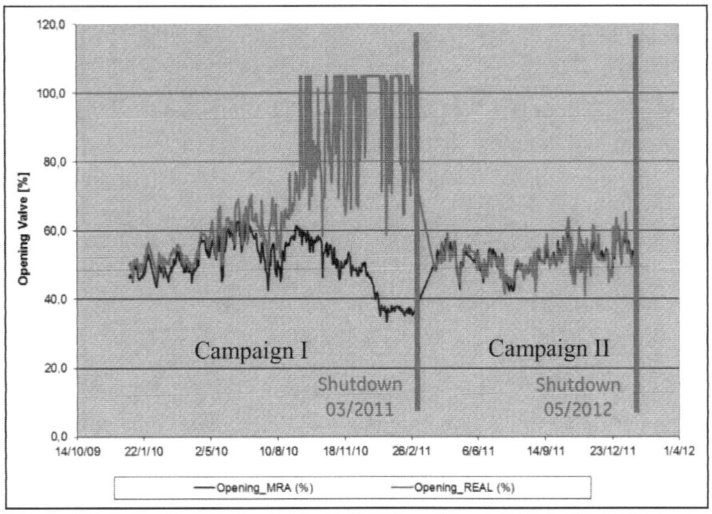

Figure 7. Performance Evaluation of Reboiler A

In Figure 7 the dark blue line represents the percentage of inlet steam when the equipment is clean, as per the equation defined by the MRA. The pink line represents the actual percentage of inlet steam. The sharp upward shift evidenced in the period between November 2010 and March 2011 shows the fouling of the system during Campaign I (prior to the 2011 operational shutdown).

After the 2011 shutdown, it can be seen that the two curves follow the same trend, without a shift, indicating that the unit has the equivalent performance of a clean piece of equipment. The change in the GE treatment strategy, taking the control of the dosages of the product Petroflo 20Y25 into account, was seen to have been effective in increasing the campaign time of this equipment.

Conclusion

The change of strategy in applying dispersant and controlling the dosages after the March 2011 shutdown, presented positive results, given the states of the inlet seen during the inspection of reboilers A and B in May 2012..

The low degree of fouling of these reboilers evidenced in the 2012 shutdown led to the production process being considered reliable, as the A-100 reboilers are no longer the pieces of equipment responsible for the loading limit of the unit. The downtime during the cleaning of the mechanical parts of the reboilers was about 10 times less than that observed in the previous shutdown, thus reducing Braskem´s maintenance costs.

In addition, the MRA statistical tool was able to produce a reliable representation of the degree of fouling of the system. This tool can be effective for scheduling shutdowns more than 03 months in advance.

References

1. Setal Lummus, "Manual de Operação da Unidade de Processamento de Nafta Bruta (Área 100)", Santo André, Brazil.
2. Ponce de Leon, Carolina; Weissheimer, Flavio and Camparin, Rafael H. (2011), "Proposta Técnica de Fornecimento de Especialidades Químicas, Equipamentos e Serviços para Tratamento Químico para a Unidade de Processamento de Nafta Bruta (Área 100)", Santo André, Brazil.
3. Kleinbaum, David G., Kupper, Lawrence L., Muller, Keith E. (2008) "Applied Regression Analysis and Other Multivariable Methods", Belmont, USA.

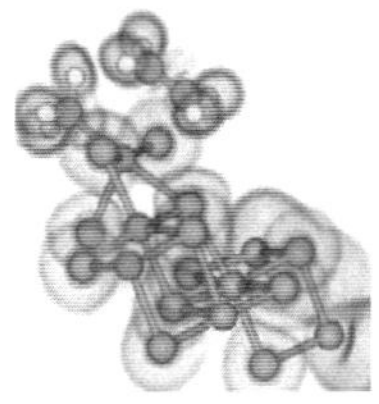

13AIChE
Spring Meeting, San Antonio, TX

SmartProcess® Application Strategy Achieves Faster and More Efficient Control of an Ethylene Complex

Patrick Truesdale
Principal Consultant
Industry Solution Group
Emerson Process Management

April 29, 2013
San Antonio, TX

EMERSON.
Process Management

Speaker Introduction

- Principal Consultant
  Emerson Process Systems & Solutions
- BS Electrical Eng, PE
- 40+ Years experience with Refinery & Petrochem Process Control
- Joined Emerson in 2007
- Previous employers: duPont, Bonner & Moore, SETPOINT, AspenTech, Comcept Technology

Patrick Truesdale

E-mail: patrick.truesdale@emerson.com

13AIChE
Spring Meeting, San Antonio, TX

EMERSON.
Process Management
Page 2

Agenda

- Performance Package Review
- SmartProcess Ethylene Overview
  - Furnaces
  - Distillation
- Benefits

SmartProcess Ethylene Packages

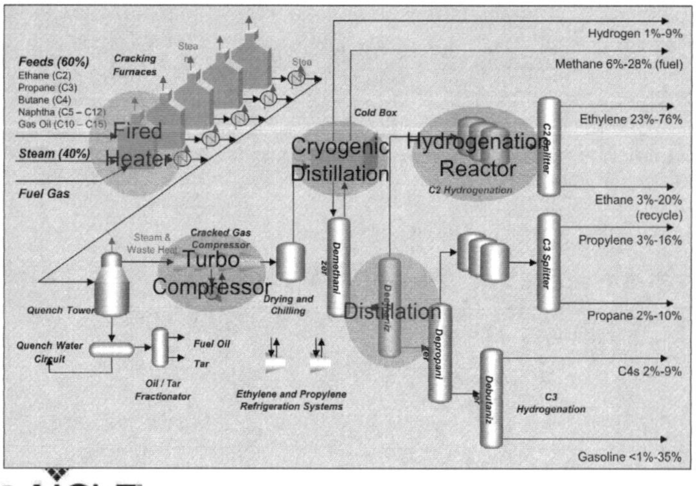

SmartProcess Scope

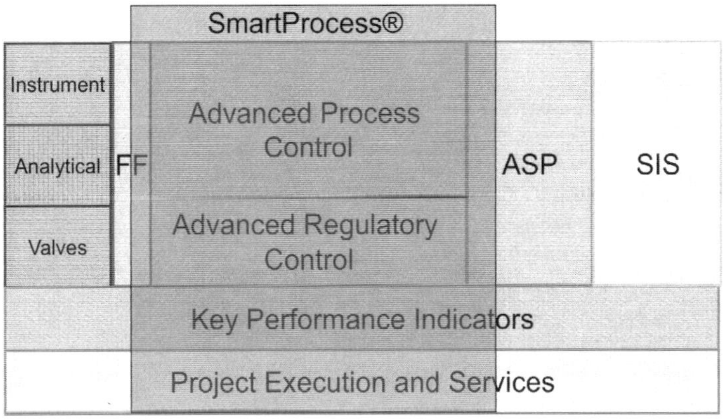

Basic Ethylene Furnace Variables

Key Operating Objectives:

- **Constant severity**
- **Maintain HC/Stm ratios**
- **Min excess air**
- **Maintain within constraints**
- **Max charge (if needed)**

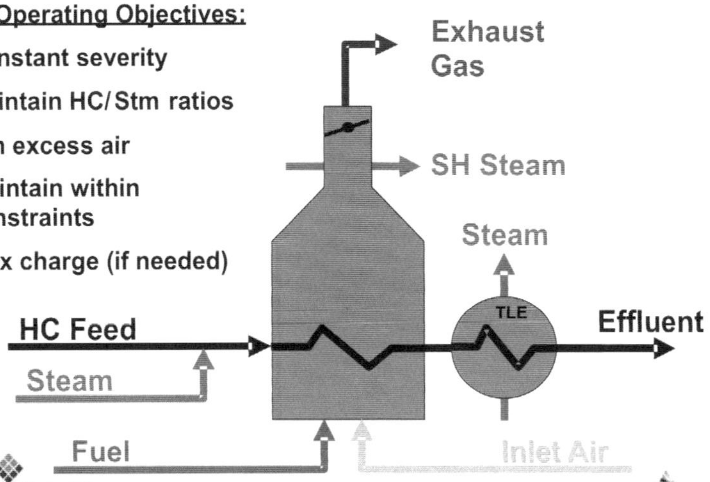

SmartProcess Furnace Module

- Design allows for different furnace configurations
  - Process Zones
  - Firing Zones
  - Multiple TLE analyzers
- Advanced Regulatory
  - Charge Control
  - Steam Control
  - Severity Calculation
  - Combustion Control
  - KPI's
- Advanced
  - Pass Balancing
  - Heat Balancing
  - Severity Control
  - O2 Control

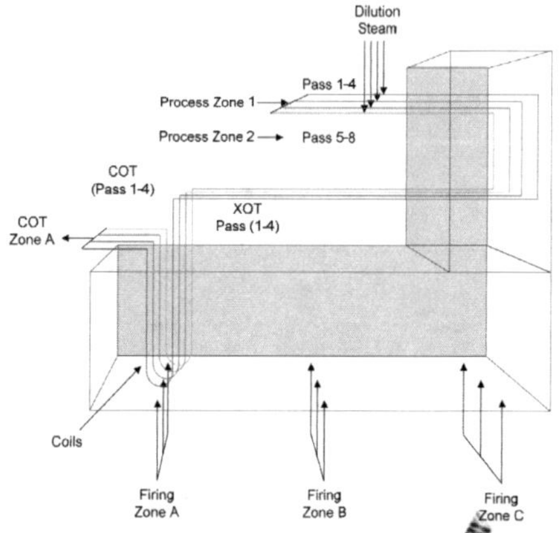

13AIChE
Spring Meeting, San Antonio, TX

EMERSON. Process Management

Standard Furnace Module Design

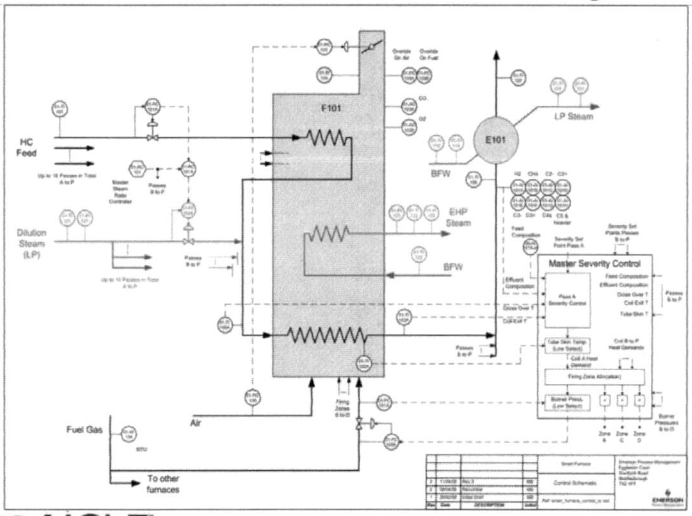

13AIChE
Spring Meeting, San Antonio, TX

EMERSON. Process Management

Charge Rate Control

Functions:
- Pass Balancing
- Constraints
- Feed Maximization

Functions:
- Ramping
- Limits & Alarms
- Multiple Feed Types
- Furnace Status
- Pass Offsets

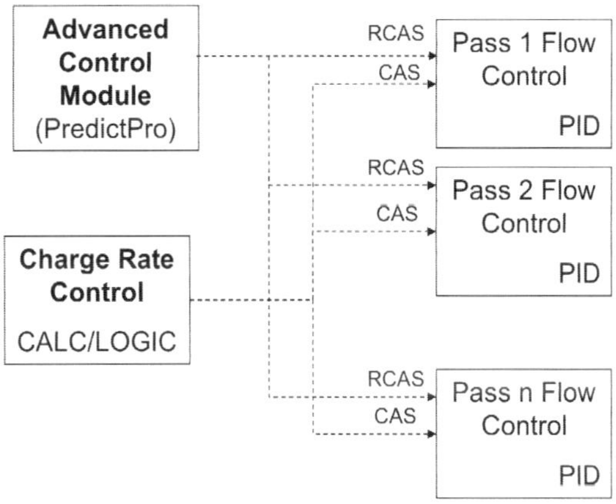

13 AIChE Spring Meeting, San Antonio, TX

EMERSON Process Management

Steam Ratio Control

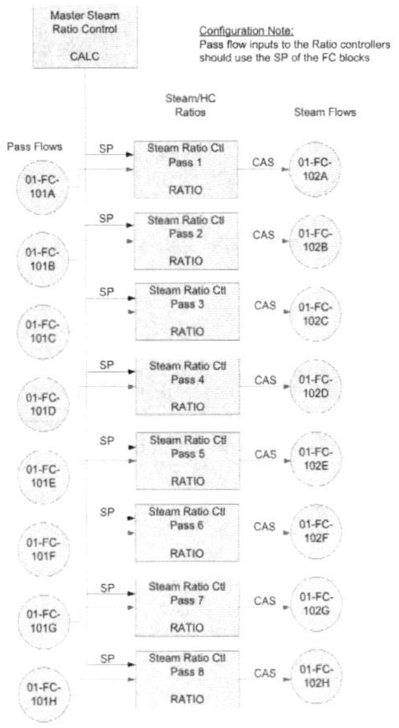

13 AIChE Spring Meeting, San Antonio, TX

EMERSON Process Management

Severity Control

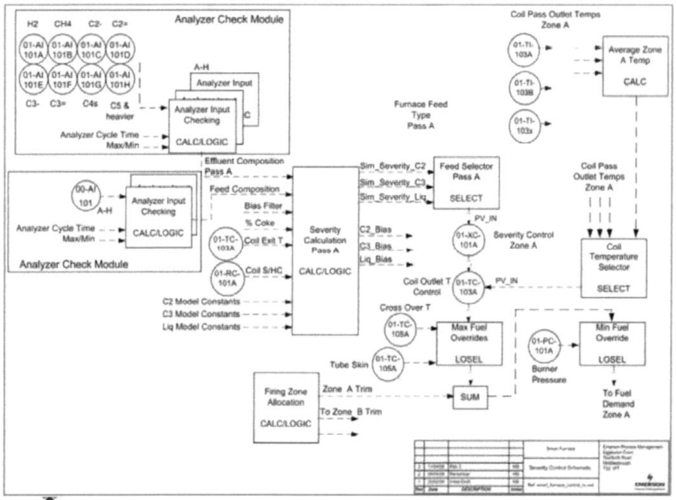

13 AIChE
Spring Meeting, San Antonio, TX

EMERSON. Process Management

Automated Furnace Sequences

- Standard module templates to be used for Furnace Startup, Shutdown and Decoking sequences.
- Utilizes preconfigured Sequence Function Charts (SFC) running in DeltaV controllers or App Station
- Standard sequence functions can be customized to match site requirements
- Manual operator steps with confirmation can be mixed with automated steps
- Sequence execution control graphics and faceplate templates

13 AIChE
Spring Meeting, San Antonio, TX

EMERSON. Process Management

Key Performance Calculations

- Furnace efficiency
  - Heat loss method
- Furnace Duty
- TLE Duty, Fouling factor
- Stack gas estimated composition
- Stack gas daily total tons exhaust for CO, CO2, S, N, etc.
- Stack gas NOx emissions (optional – based on CEMS)

13 AIChE
Spring Meeting, San Antonio, TX

EMERSON. Process Management

Distillation - Classic Multivariable Problem

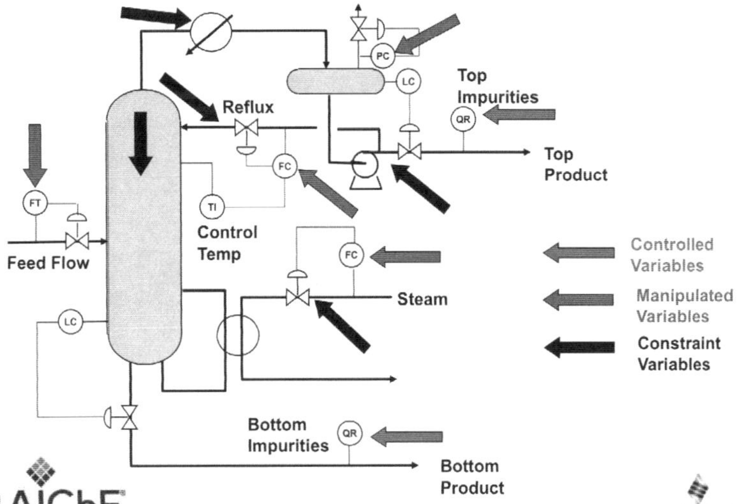

13 AIChE
Spring Meeting, San Antonio, TX

EMERSON. Process Management

Distillation Control Module

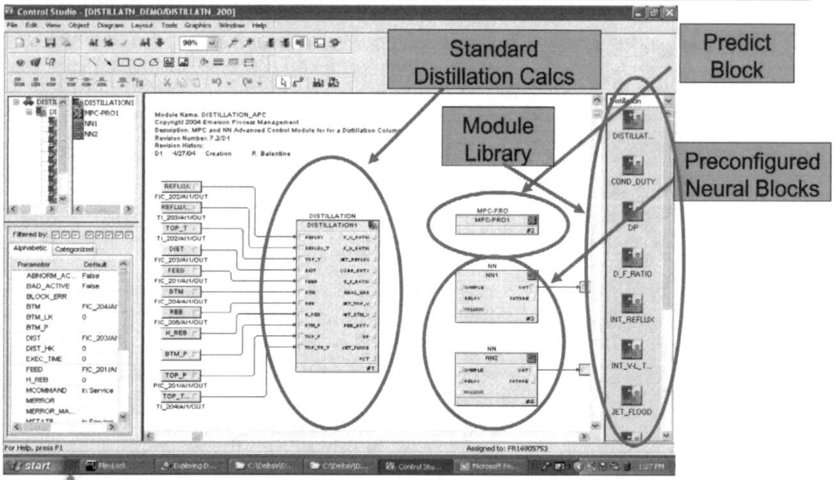

13 AIChE
Spring Meeting, San Antonio, TX

EMERSON. Process Management

Page 15

SmartProcess Distillation

- DeltaV PredictPro with up to 4 MV's
- DeltaV Neural for up to 2 product qualities
- Standard set of calculations & constraints
  - Internal reflux
  - Pressure compensated temperatures
  - Jet flooding
  - Reboiler & condenser duty
- Engineering Services to design, implement and commission

13 AIChE
Spring Meeting, San Antonio, TX

EMERSON. Process Management

Page 16

SmartProcess Applications

- Licensed industry applications for common processes that are:
  - Pre-engineered, tested, documented
  - Designed by APC Consultants with extensive practical experience
  - Supported & enhanced
- Bundled combination of:
  - DeltaV libraries: composite blocks, modules & templates
  - DeltaV simulation and demonstration example
  - APC tools (e.g. Predict, Neural)
  - Application code (e.g. VB, .Net)
  - Other databases and apps (e.g. SQL*Server, Excel, TankMaster)
  - Sample DeltaV Graphics
  - Documentation
- Templates to be used to configure for specific site requirements
- May be implemented by Emerson or Customer

Benefits

Ethylene Furnace High Temperature Imaging Cameras to Monitor Burner Flames ,Tube Temperature and Tube Failure Detection

Thomas M. Canty, P.E.
President
J.M. Canty Inc
Buffalo, NY USA

Abstract

The development of explosion proof, high temperature imaging cameras which operate in ethylene furnaces without cooling air now allows monitoring of ethylene furnaces to improve energy/operating efficiency along with providing important safety benefits. Imaging cameras work off the visual wavelength which provides 2 dimensional view and temperature measurement of the tube temperature which is more accurate than thermocouples and IR probes. In addition visual images of the furnace burners lets you detect flame detachment, smoking burners and flame shape in order to increase burner efficiency and save energy. The safety aspects of the Camera imaging systems is significant as it allows visual detection of pin hole leaks and tube failure in the furnaces.

Introduction

Ethylene furnaces, or any type of cracking unit, rely on thermal management in order to make product efficiently and to operate in a safe manner. Having 'eyes' in the furnace can help to identify problems in their early stages of development. High temperature camera technology is able to survive the environment and give a wide angle view of the tubes from which temperature and tube blockage may be determined. Achievable benefits from a camera system include:

1. Temperature analysis capacity for all areas of the image, not just single point. Multiple zone capability for user customization of analysis.
2. Flame detection and size measurement
3. Detection of tube movement and ability for software to automatically move measurement zones as tubes distorts.

Imaging Fundamentals

Basics of an imaging system include a high resolution sensor, proper illumination, appropriate hardware for the process and a robust software analysis package. For this particular application illumination is created by the process and need only be managed optimally for best results. A high density pixel array yields resolution of the image and measurement as well. The hardware and software are the more challenging aspects of the application and will be discussed in some detail.

Hardware

The design of the camera system is unique in that it's internally exposed lens assemblies are **designed to survive without cooling air**. When installed properly, the combination of system materials and heat transfer coefficients at the camera tip limit heat transfer to the system and allow it to run hot without chance of damage. This feature, when system is installed in a clean, gas burning environment, does not disrupt the temperature or air balances of the furnace because no cooling air is being introduced to the furnace. In addition, the design includes a disposable, quarts shield which protects the process exposed lenses from direct contact with the internal atmosphere. Should any particulate contact the system, the quartz shield is there to protect the optical lenses. Replacement of the protective quartz shield is inexpensive. Replacement of damaged lenses is very expensive and so the Canty design guards against that possibility.

Internally, the system is a two piece assembly consisting of the internal insertion lens and the external camera package. As mentioned earlier in this document, Canty employs a fused glass window as part of the electronic housing package. This has a dual purpose; Firstly, the camera housing assembly is a separate entity which carries WP and EXP ratings. Secondly, the camera housing bolts to the internal lens system and transmits the optical rays through the fused glass window to the CCD. This mounting also creates a hermetic barrier between the process and the electronic housing thereby preventing migration of hot, process gasses outside the pressure boundary of the system.

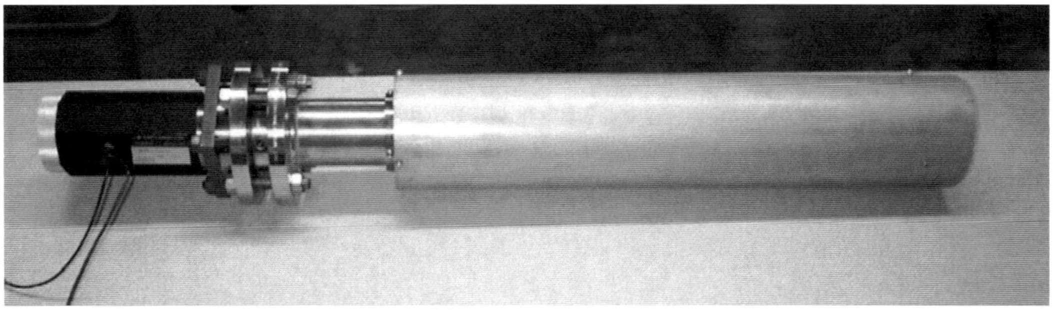

Figure 1. High Temp Camera Assembly

The camera electronics come with a full array of controls that allow it to be optimized to the software function to achieve accurate, consistent data output.

As previously mentioned, the installation is an important feature of the design. When installed correctly, the camera sits inside the refractory (see Figure 2) where its exposure to the process environment is minimized thus minimizing heat transfer into the internal lens.

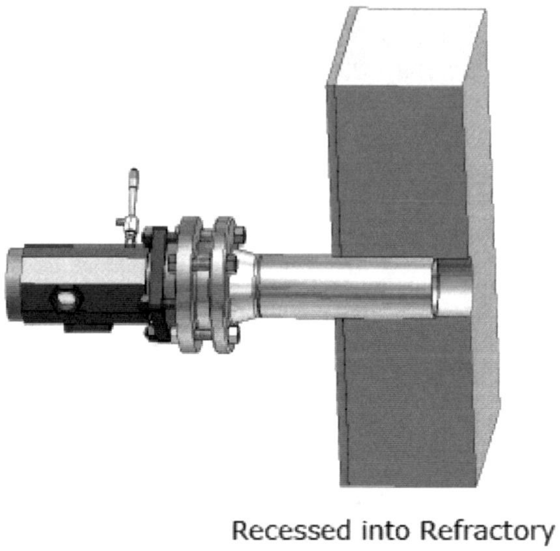

Figure 2. Camera Installation Schematic

Software

With respect to software design, the Canty software is fully flexible in analyzing any portion of the captured image. Multiple analysis zones can be drawn to monitor discreet areas of the view and each zone can run its own particular algorithm depending on need. The benefit here is that multiple tubes can be continually surveyed for changes and trends. The analysis data is also not single point, but is comprised of all the data points in the analysis zone. By choosing proper software filters various information can be obtained from the full array of data such as hot or cold spots. This data can also be archived in total or as snap shots at determined increments of time. The on screen data output can also be updated in the short or near term and can thus be used to judge what is happening in the furnace over time with just a glance at the screen. For instance, using the histogram feature with a 1x daily output a long term trend can be presented on screen in real time.

The basic theory of temperature measurement is based on the radiative emissions of a body at temperature – as the body temperature increases, the radiation curve shifts towards the shorter wavelengths and that is why as a body heats up the eye cannot detect it until it reaches 1000F which is the approximate visible point where emitted radiation is dark red in color. As the temperature increases, the color graduates to bright red, then orange then toward white. A pyrometer functions by measuring a narrow band of wavelength intensity and comparing it to the intensity of a black body to determine the temperature. From the graph below it can be seen that if a single wavelength is monitored the intensity of that measurement changes as the body temperature changes (curve shifts) which then corresponds to a different temperature on the calibration scale.

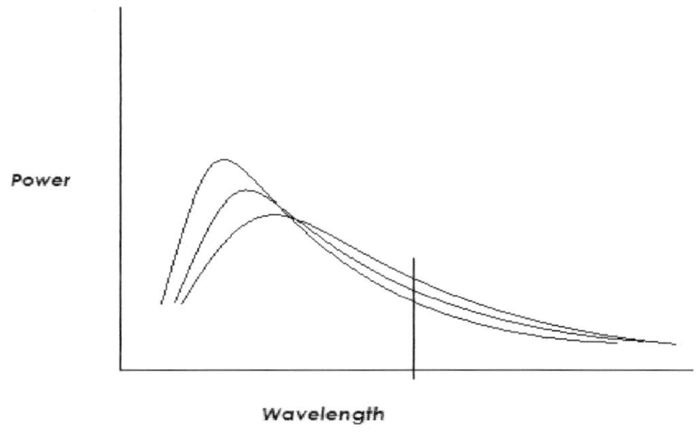

Figure 3. Power Curve at Varying Temperatures

The same is true of a camera system measuring the increasing intensity of an object as it radiates from approx 1000 F (dim, dark red color) on up the temperature scale to 3000+ F for instance which would be a bright white. From the graph above one can see that as color goes from red toward white the peak intensity is always increasing and it is the measure of this intensity which indicates the object temperature. Using an RGB camera the measurement can also be made by monitoring the individual color components of the signal impinging on the chip. For industrial uses, the measurement of intensity usually suffices.

Standard instrumentation used in industry include thermocouple type devices (contact) and thermal imaging and pyrometer systems (non contact). In severe temperatures and atmospheres, the contact devices deteriorate quickly and provide gradually less reliable information leaving the operator running blind. They also create a discontinuity on the surface they are welded to which, in the case of a furnace tube, can lead to premature tube failure.

Non contact systems have difficulties with calibration drift and emissivity adjustments to account for the fact that many items being measured are not black body emitters. In addition, these technologies are single point measurement systems and are best used as hand held devices which can be difficult to use and provide subjective measurement results based on operator preferences for targeting.

Vision technology offers several benefits these other technologies do not. The visual aspect of the system allows the operator to view the process and confirm the readings received. It also allows flexibility to set up measurement zones on screen at any time to assess any or all areas of the process in view should a change be required. Since vision systems are generally fixed installations, an added benefit is that the information is taken from the same process points and features within a given measurement zone at all times which means there is no error in sighting of the instrument.

The vision system is also unique in that it does not suffer calibration drift as almost all instruments do. Once the calibration is set the readings repeat themselves within a couple degrees F. The calibration is based on thermocouple or pyrometer readings initially, however after that calibration is finished it does not have to be redone in the future as long as the system remains in its location.

Process history and performance can also be best documented by a vision system. First off, the camera can analyze any area of process that is within the systems view. This affords the operator the capability of viewing many points of interest instead of just single point temperature readings. As an example, if the system is monitoring a tube bank it can assess the temperature of the tubes for the full length of view. If there is a local flow problem in a tube it will become overheated in one section. The camera can locate that section and indicate its temperature rise over the rest of the tube or adjacent tubes.
This information allows the tube location to be identified as well as the position in the tube that has encountered the upset condition. Over time this historical data, recorded in video or still image format by the control software, may be significant in improving the process. In addition to the ability to monitor the full area of the view, the full area of view can also be segmented into multiple zones of measurement, i.e. each tube having its own measurement zone for automated process control capabilities. This capability allows the tube to be operated at optimum efficiency since the addition of excess heat can be avoided thereby reducing energy draw to operate the furnace.

Thermal Analysis

Tube Temperature

The following figures illustrate the temperature analysis capability for tubes in a live test situation. Measurement zones are placed over the tubes (figure 4) as the furnace is heated and cooled. The temperature plots are shown in figures 5 and 6. The results are as expected and uniform from tube to tube, although a small number are included in this demonstration.

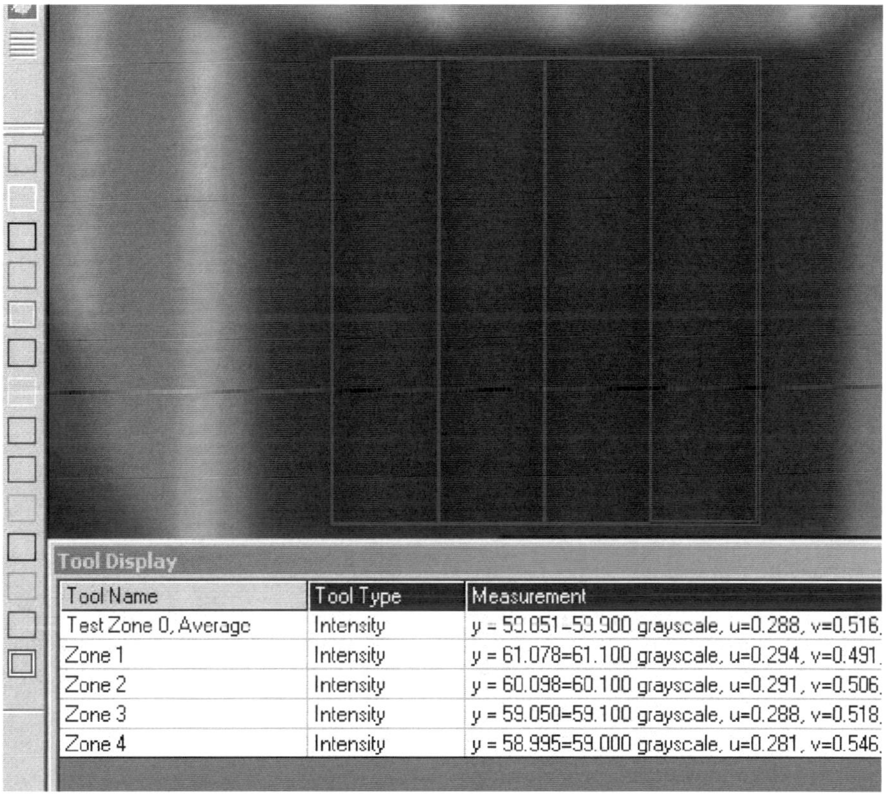

Figure 4. Software Screen Shot

As thermal loads are fluctuated up and down the following graphs track those changes. The cooling cycles are left to natural convection and thus are not linear in nature.

For the following curves, the vertical scale is gray scale intensity which is measured 0 – 255 (pitch black to bright white). The hotter the tube, the higher, or more intense, the gray scale reading. This value in process would be calibrated to the temperature. The horizontal scale is the image number, or time.

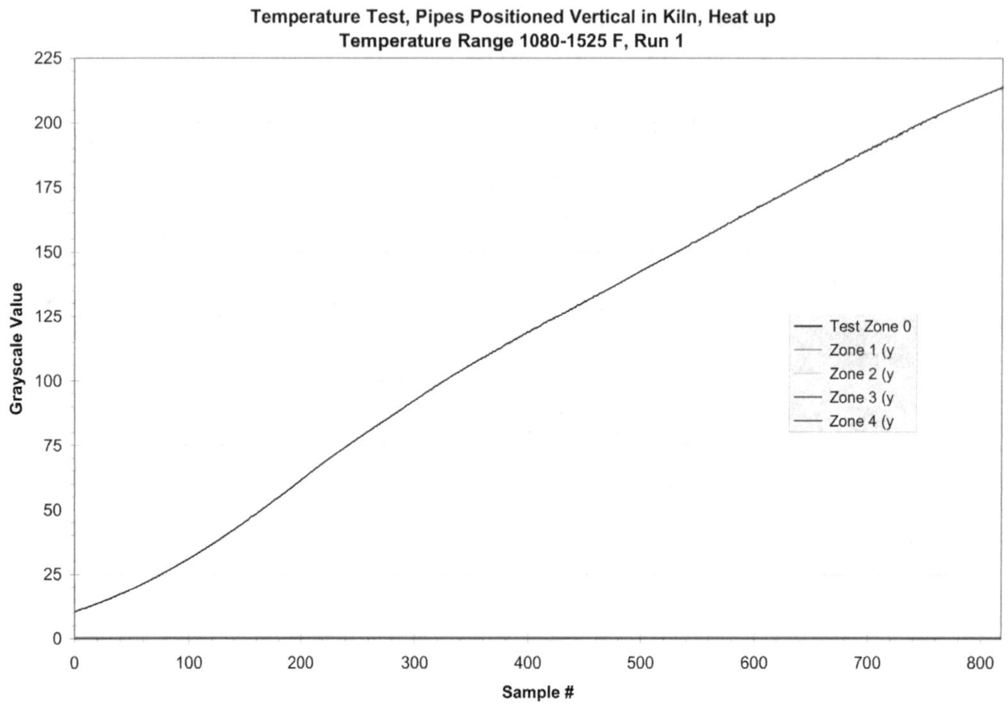

Figure 5. Heating Cycle

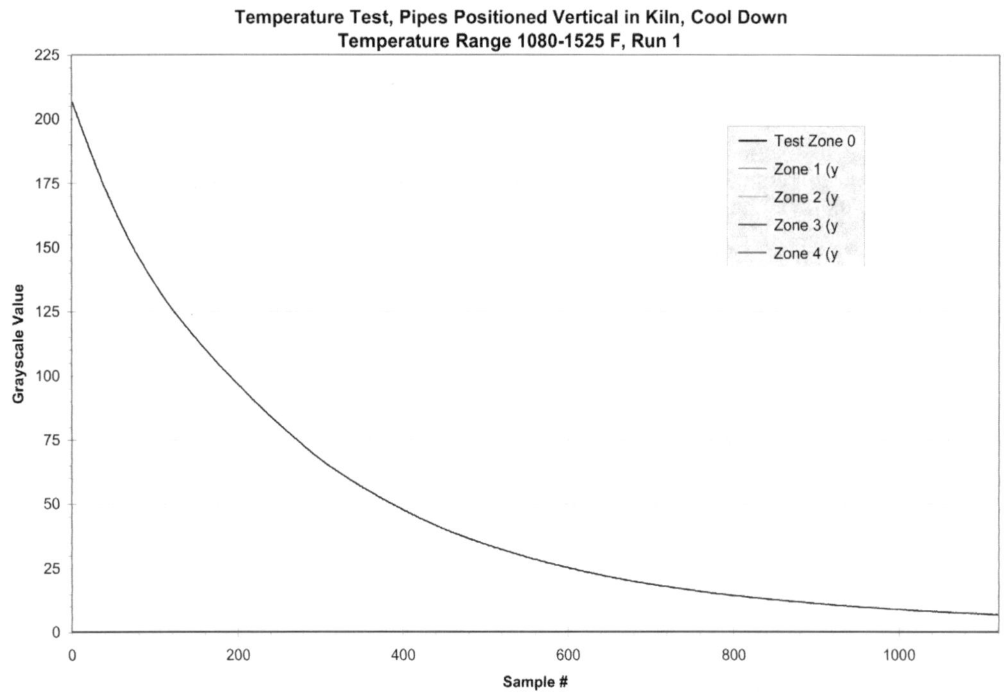

Figure 6. Cooling Cycle

Flame Measurement
Flame temperature can be monitored in the same way as described previously, however the flame size, or presence, can also be monitored. Figure 7 shows a live flame and Figure 8 shows a digital capture. Computing the flame area, or shape, can be done to track its changes over time.

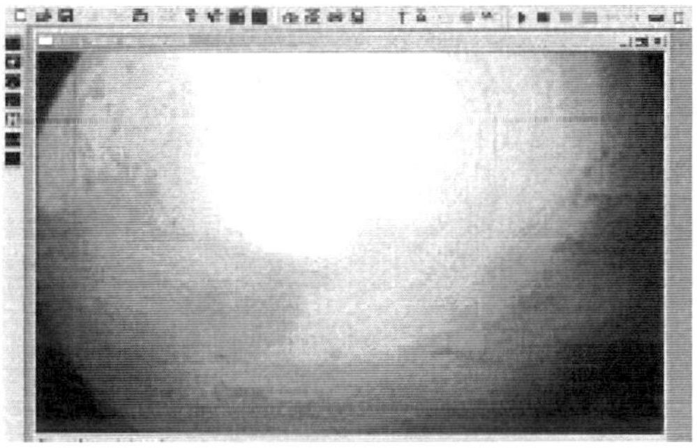

Figure 7. Kiln Flame (Live)

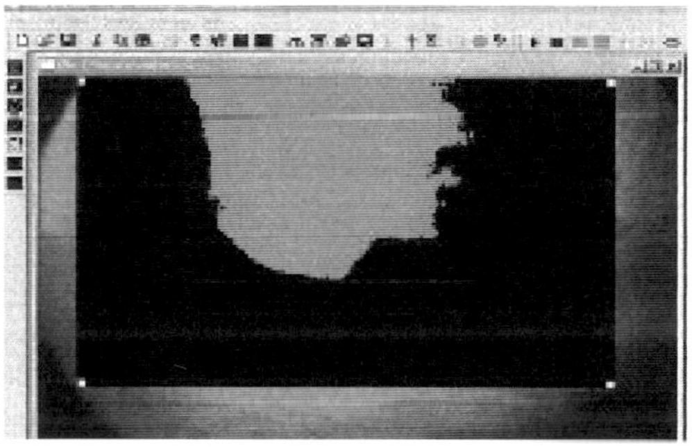

Figure 8. Kiln Flame (Digital)

Conclusion

Vision Technology presents an opportunity to increase effectiveness of furnace monitoring through hardware and software capabilities. The thermal design at the core of the hardware enables the system to provide the operator a view into the furnace and feed back on its thermal performance over time. This data can assist in detecting building upset conditions, but also provide a record for study to assist in improving the operational method of the furnace.

(45v) Straighten up! Non-Linear Compensation Using Feed Forward Control

Barbara Hamilton, Process Systems and Solutions, Emerson Process Management, Hampden , ME

Non Linear processes occur throughout the petrochemical plant. While the PID loop is surprisingly robust for most situations, there are cases where compensation of these non-linearity's is critical for stable control. Similarly, disturbances can cause major disruptions in control stability.

A method of feed forward control, using data based models, can provide robust control throughout the process range. Historical data are used to build models to drive the output toward the solution with PID control providing the final drive to zero error.

The model is also used to develop dynamic tuning. The result is a simple to configure, robust solution to manage non-linear loop and/or disturbance variations.

(45w) Adsorption of CO2 by Alkaline Activated Ca-Al2O3-SiO2 Material

Naim Faqir, King Fahd University of Petroleum & Minerals, Dhahran, Saudi Arabia and Reyad Shawabkeh, Chemical Engineering, King Fahd University of Petroleum & Minerals, Dhahran, Saudi Arabia

Natural gas contains a slight level of carbon dioxide which leads to several industrial and domestic heating problems. In addition, it is one of main greenhouse gases that contribute for global warming, where its pollution is a wide-spread problem arising from a wide range of industries, leading to complex contaminant in atmosphere. Given the increasing pressures worldwide to minimize the emission of greenhouse gases into atmosphere, CO_2 must be remediated efficiently and cost effectively.

In this short abstract a chemically treated limestone and alumino-silicate material were utilized for removal of CO_2 from natural gas using a combined adsorption-chemical reaction processes. This functionalized Ca-Al_2O_3-SiO_2 material was characterized according to its crystalline phase, surface area, pore size distribution and surface morphology. Then it was tested against different concentrations of CO_2 in CO_2-N_2 gas streams.

Pressure Gauges: Hidden Danger or Visible Safety

by Jeff Placek, WIKA Instrument, LP, Lawrenceville, GA

Introduction

In recent years, refining plants, petrochemical plants and chemical plants have been required to increase capacity, but have been provided with very few upgrades to existing equipment to accommodate the increased capacity requirements. This has caused plants to have increased stress and strain on their systems and operating units. For example, higher pulsating pressures and heavy mechanically-induced vibrations are translating to all areas of the processing equipment, especially pressure instrumentation.

Pressure gauges are one of the most important, yet most overlooked or ignored keys to ensuring the continued safety, productivity and profitability of process plants. Process pressure gauges are used to read pressure on transfer lines, discharge pumps, and other places where it is important to be able to read pressures.

Problems Associated with Misapplied, Faulty or Ignored Gauges

Incorrect, misapplied, damaged, faulty or ignored gauges can lead to a number of issues, including safety problems, downtime, and costly regulatory agency fines and other penalties.

Process Safety:

One group of problems relates to process safety issues, such as loss of containment that leads to fugitive emissions, fires and explosions. Bourdon tubes in malfunctioning gauges can rupture, causing the release of media, such as gasoline, oil or gas. Because this media is often volatile and corrosive, it can cause fires, explosions, or other serious safety issues, such as a toxic gas releases or acids spraying onto workers.

Media releases are common results of misapplied, faulty or ignored gauges. Fires and explosions are less common but they do occur. When reviewing plant instrumentation, findings have revealed instances where the wrong gauges have been installed in locations on pumps. In one instance, a gauge vibrated so severely when the pump came out of alignment that it caused the Bourdon tube in the gauge to crack. The media coming out of the tube leaked onto the pump, which was excessively hot because of age and misalignment, causing a fire.

Expensive Downtime:

A second group of problems relates to expensive downtime, which can result from the breakdown of process safety just mentioned, or simply from having to shut down a process in order to replace incorrect, misapplied, broken or otherwise damaged gauges.

1 - Safety-Related Downtime:

As a result of the problems identified above (emissions, fires, explosions, etc.), there can be any number of causes of subsequent downtime, including fire suppression activities, released media containment, line shutdown, personnel evacuation, and then eventually having to rebuild or repair the lines, replace equipment, etc. In addition, if there is a media release with no injuries or fatalities, the plant is required to notify the EPA. However, if there is a media release that does cause injuries or deaths, the plant is required to notify OSHA. These can lead to very long process shutdowns, anywhere from a day to a month or more. (More on OSHA and the EPA below.)

2 - Non-Safety-Related Downtime:

When a line fails, causing a gauge to fail, even if there is no release or safety incident, downtime can still occur as a result of the process needing to be shut down in order to replace the gauge. Gauge failure rarely causes other parts of the process to fail, although gauge failure is usually an alert that there is a problem. In most cases, a plant worker will simply replace the pressure gauge and move on. When another failure occurs, the worker will do the same, until the root cause of the failure, such as a misaligned pump, causes the plant to shut down the line, resulting in more costly, unplanned downtime and a major maintenance time.

Regulatory Agency Fines and Penalties:

A third group of problems relates to the potential for significant fines and other penalties from federal regulatory agencies (primarily OSHA and the EPA), as well as from state and local jurisdictions, that result from compromised process safety.

1 - OSHA:

OSHA addresses many process-related safety issues in 1910 Subpart H, "Hazardous Materials." More specific is 1910.119, "Process safety management of highly hazardous chemicals." In its Process Safety Management section, OSHA notes that, "Unexpected releases of toxic, reactive, or flammable liquids and gases in processes involving highly hazardous chemicals have been reported for many years in various industries that use chemicals with such properties. Regardless of the industry that uses these hazardous chemicals, there is a potential for an accidental release at times they are not properly controlled, creating the possibility of disaster."

OSHA goes on to note that: "To help ensure safe and healthful workforces, OSHA has issued the Process Safety Management of Highly Hazardous Chemicals standard (29 CFR 1910.119), which contains requirements for the management of hazards associated with processes using highly hazardous chemicals."

OSHA's standard emphasizes the management of hazards associated with highly hazardous chemicals and establishes a comprehensive management program that integrates technologies, procedures, and management practices.

While OSHA does not monitor gauges in specifically, they will enforce guidelines for worker safety. OSHA investigates injuries and deaths, and may levy fines or other penalties.

2 - EPA:

According to Greg Neneman, director of business development solutions for Valin Corporation (San Jose, Calif.), fugitive emission regulations from the Federal level have tightened up over the past several years. In many states, such as California, Oregon and Washington, the regulations are even tighter.

"One challenge companies face is ensuring compliance with emission regulations when they upgrade or expand their processes," points out Neneman. "They are not allowed to produce additional emissions beyond the original limit set by regulations."

From the standpoint of federal EPA regulations, each company that produces or emits volatile organic compounds (VOCs) is required to come up with a leak detection and repair program. "In some areas, county and local requirements are even tighter than federal requirements," continues Neneman.

In certain areas, companies that produce or emit VOCs are required by regulation to do audits on valves, flanges, gauges, and all other process connections on a quarterly basis. "And in some jurisdictions, the regulatory agencies can give plants 24 hours notice before they descend on that plant to check all of their records and conduct other investigations," adds Neneman.

In May 2012, the U.S. Environmental Protection Agency (EPA) levied a $1 million civil penalty on a plastics company for alleged violations of the Clean Air Act at two of its chemical manufacturing plants. In addition to paying the penalty, the company agreed to implement a comprehensive program to reduce fugitive emissions of hazardous air pollutants (HAPs) from leaking equipment such as valves and pumps. The settlement also required the company to implement enhanced work practices, including more frequent leak monitoring, better repair practices, and new efforts designed to prevent leaks. After being fined by the EPA, the company realized that instrumentation was one of the areas where they having fugitive emissions. Until that time, workers had been replacing gauges left and right, because they weren't aware of the problems that could occur. The gauges weren't a big expense to them, so they considered the replacement activity to be more of a nuisance than anything. It's when they had the rupture that they realized there was a real problem.

An audit of the plant, which involved reviewing at all of the instrumentation, found that in many applications, they were using the wrong instrumentation. To correct the issues, the right instrumentation was specified for them, and the correct gauges were ordered. The plant now has its instrumentation under control.

While conforming to EPA regulations has always been a challenge for process facilities, it is going to become even more challenging in the future as a result of EPA's recent introduction of new requirements related to fugitive emissions and leak detection and repair (LDAR) strategies. On April 17, 2012, the EPA Administrator signed a suite of final air quality regulations that affect both upstream and midstream oil and gas sectors - the production/processing and transmission/storage sectors of the oil and natural gas industry.

For the first time, EPA will regulate air emissions from natural gas wells that are hydraulically fractured, as well as other emission sources associated with exploration, production, processing, and transportation of oil and natural gas.

Promulgation of the final rules will result in new control equipment requirements, as well as increased requirements for monitoring, recordkeeping, and reporting for owners, operators, contractors and vendors.

The final rules require the use of emission controls, as well as work practices, through two different air regulatory programs. One is the New Source Performance Standards (NSPS). The other is the National Emissions Standards for Hazardous Air Pollutants (NESHAP).

EPA finalized a new NSPS at 40 CFR Part 60, Subpart OOOO, which regulates emissions of volatile organic compounds (VOC) and sulfur dioxide (SO2) from a variety of oil and gas exploration, production, processing, and transportation facilities, including natural gas wells, certain compressors and pneumatic controllers, storage vessels, and sweetening units at onshore natural gas processing plants. Subpart OOOO also includes startup, shutdown and malfunction provisions.

The EPA has also made it clear that oil and gas production and processing sources will be the focus of its National Enforcement Initiative (NEI) for fiscal years 2012 and 2013. With the new NSPS and revised NESHAPs in place, EPA now has the tools it needs to enforce in a comprehensive manner.

As a result of these new regulations, plants should look at everything more closely, including gauges.

How Problems Can Occur With Gauges

Technical problems can occur with process pressure gauges in a number of ways, which can be conveniently categorized into seven broad categories with specialties or subsets. The most common reason for gauge problems is that workers often install gauges without knowing how they should be installed or accessorized. As a result, the gauges can fail because they were never designed to function in those specific situations. They can also fail because they were installed incorrectly, or because they were not properly maintained.

One way to eliminate or minimize these problems is to consider specific pressure gauge accessories. These can prove to be the extra step needed to ensure a successful, proactive preventive maintenance program, and for extending the life of the gauges. Accessories can provide additional protection for the gauges during their lifetimes.

1 - Mechanical Vibration.

This is oscillating, reciprocating or other periodic motion of a rigid or elastic body or medium forced from a position or state of equilibrium. Typically, vibration is the main cause of pressure gauge failure in process plants. Pump applications produce most of this vibration.

Gauges that aren't designed to withstand vibration may be installed in applications where vibration occurs, causing the gauges to come out of accuracy. Even more serious, a gauge can vibrate so much that the Bourdon tube inside it can crack, releasing media."

One way to prevent vibration-related problems is to install liquid-filled gauges, which help to dampen the vibration and also protect the internal components. The liquid case filling also lubricates the internals and prevents corrosion.

2 - Pulsation.

When media rapidly cycles through the pressure system of the gauge, pulsation occurs. If the pointer pulsation increments are greater than 5 percent of the full scale value, you must intervene to prevent damage to the gauge.

A restrictor or throttle screw can solve most pulsation problems. It will slow down the media by reducing the size of the orifice and averaging out pressure fluctuations. For extreme pulsation, such as spiking and surging, a snubber works best. It allows more control over the amount of snubbing and the flow of media into the pressure system.

3 - Temperature:

If a gauge is exposed to a media with very high temperature, the metal alloy in the gauge may not be able to handle the temperature, causing the Bourdon tube to fail and release media.

Long pipes that separate the pressure gauge from the process, or small cooling elements with radiator fins, protect the pressure instrument from high process temperatures by allowing ambient air to flow over the surface area. In environments with steam, using a siphon with the pressure gauge cools the steam down and drives it back to its liquid state before it gets to the gauge. With very high-temperature media, a diaphragm seal with a cooling element will isolate and prevent the hot media from reaching the gauge.

4 - Spikes or Overpressure.

In some cases, process media may be under extreme pressure. Pumps also can cause spikes and over-pressure, both of which shorten gauge life. Gauges can fail as a result of these pressure spikes. When workers turn pumps on or off, or open or close valves, a surge of media or oil goes through the pipe and hits the pressure gauge, which causes a huge spike. For example, if a line is pumping 100 psi of oil, someone may install a 100 psi gauge, not realizing that this leaves no allowance for any overpressure, surging, or spiking. When pressure does increase, it causes overpressure on the gauge, sometimes to the point of where the Bourdon tube ruptures and leaks media.

One way to prevent this problem is to use a different type of gauge, such as 200 psi gauge for a 100 psi line. Another option is to accessorize the gauge with an over-pressure protector, which would shut off the pressure if there were a surge or spike. Overpressure protectors are designed to avert spikes and the resulting overpressure damage. These protectors provide protection for pressure gauges against the effects of pressures exceeding their maximum pressure ratings.

5 - Corrosion.

Corrosion problems can occur with gauges if the process media is not compatible with the pressure gauges' wetted parts material. Media can sometimes be so corrosive that it corrodes through the Bourdon tube and leaks onto the ground.

When the media is not compatible with the wetted parts material of the pressure gauge, a diaphragm seal is required to isolate the gauge from the media. The benefit of a diaphragm seal is that it can be made of special alloys that are designed to withstand corrosion. It is essential to select a diaphragm material that is compatible to the process media.

6 - Clogging.

In some processes, the media may eventually crystallize or may otherwise harden. If a standard gauge is used in such an application, it will eventually get clogged and will not be able to read the pressure accurately. For example, if a gauge gets clogged and ends up reading zero pressure, when in fact it is under tremendous pressure, there will be an unexpected and dangerous release of media if and when the piping system is opened. Diaphragm seals protect the pressure gauge from solids, particles and viscous media by creating a barrier between the process and the pressure inlet of the gauge. There is also an option to equip the diaphragm seal with flushing ports to constantly flush material out from the diaphragm surface.

7 - Mishandling and Misapplication.

Gauges can also fail prematurely if they are misapplied. For example, a gauge may have been ideal for the application when it was installed. However, if there are equipment modifications or more components subsequently added, this can change process parameters, resulting in the gauge not longer being right for that application. Misapplication can lead to serious problems associated with failure and subsequent media release.

It is also possible to accidentally damage gauges during installation. Gauges should be installed using the proper sized wrench on the flats. Gauges should never be installed by turning the case. After installation, gauges can also fail if they are mishandled or used improperly.

Why Problems Can Occur With Gauges

In a typical process plant, about 25 percent of pressure gauges have failed or are about to fail.[1]
A lack of management focus and worker education and experience are key reasons plants experience issues with gauges.

Management Focus.

Gauges are often viewed as inexpensive commodities, with management failing to realize that they are very critical parts of the processes. Often, plants will replace failed gauges with the same replacement gauges, which will also fail, only to have to be replaced again. Eventually, their luck may run out and a serious failure may occur because the incorrect gauge has been installed.

Why do gauges tend to receive such little attention from management? Management tends to focus on the expensive equipment - $100,000 or million dollars items. They often don't understand the importance of focusing attention on $100 gauges. This is not the fault of management. The problem is that no one has explained to them just how important gauges are.

As noted earlier, though, ignoring instrumentation can cause serious safety and financial problems. Plant process gauges can be thought of in the same way as vehicle gauges. If the temperature gauge on an automobile is defective, it is possible to destroy the engine as a result of overheating that is not registered by the defective temperature gauge. Similarly, if the oil pressure gauge on an automobile is defective, it is again possible to destroy the engine as a result of seized components resulting from loss of oil that is not indicated by the defective oil pressure gauge.

Worker Education and Experience.

A lack of worker education and experience also contributes to gauge issues. In many instances, when maintenance technicians look at gauges, they don't pay much attention to them unless they leaking or look as if they are going to explode. They don't realize all of the information that that gauge can provide. In addition, technicians may not realize that a gauge has failed or is about to fail. Again, as long as it is not leaking media, they will often just keep walking. This is rarely the fault of the technicians. Many have not been properly trained to understand process pressure gauge performance.

Along the same lines, the industry has been experiencing a "brain drain" for several years, with large numbers of experienced technical workers and engineers retiring and being replaced by younger people who lack the education and experience to understand what gauges are telling them, or, worse, not being replaced at all.

According to research conducted by Cambridge Energy Research Associates in 2007, more than half of all oil-field professionals will reach retirement age during the next decade. The Interstate Oil & Gas Compact Commission found that the average age of production and service company employees is 46 to 49 years, and the average retirement age in the oil and gas industry is 55 years. [2]

Historically, plants had specialists whose jobs focused exclusively on instrumentation. They spec'd the gauges, maintained them, and made sure they were operating correctly. However, as plants got leaner in recent years, they started letting these people go. Plants first shifted gauge responsibility to the pipefitter whose main responsibility was to make sure the pipes were put together correctly and not leaking.

[1]WIKA data aggregated from more than 250 instrument audits.
[2]Peter Peek, Jocelyn Fenard, Patrick Gantes, Christine Theiler. December 2008, "Skills Shortage in the Global Oil and Gas Industry: How to close the gap" Part I. Pages 14-15.

Then, as plants began letting pipefitters go, they shifted the responsibility to the unit operators or technicians, who now install the gauges. These people usually don't have any experience or knowledge of how to install instrumentation. They usually just replace old gauges with similar gauges. The problem is that the gauges are usually the wrong ones to begin with. The result is that, these days, few plants have people who have much if any education and experience with instrumentation.

Taking Stock of Process Pressure Gauges

When it comes to improving the safety, productivity and profitability of process plants through an effective gauge management program, companies can take a couple of steps. First, an understanding of the vital importance of gauges must be established. Second, workers must understand what kind of gauges will be effective.

Focus on Gauges:

World-class process plants maintain their assets proactively. In such plants, pressure gauges receive as much attention as pumps and control valves. Management understands the importance of proactively maintaining their assets, and they realize that one of the best ways to maintain their most expensive assets is to properly source and maintain the gauges that protect these assets. That is, they understand the value of spending $100 for a gauge that can subsequently prevent damage or destruction to a $100,000 or $500,000 piece of equipment and the resulting negative safety, downtime and regulatory implications.

There is an interest in predictive maintenance these days in process plants, which is a good thing. A lot of this, though, involves expensive software. In addition, when something is out of kilter, the system may be set up such that technicians may have to wait for it to get far enough out of the normal range before it will generate an alert on the computer. However, properly-maintained pressure gauges, studied by people who know what to look for, can be very inexpensive but reliable and useful mechanical predictive maintenance tools.

Gauge Procurement Considerations:

Over time, as noted earlier, the status of pressure gauges has unfortunately diminished to that of commodities in many process plants. In order to ensure process plant safety, productivity and profitability, it is essential that plant professionals understand the importance of selecting the right gauges for applications.

It is also important that these be high-quality gauges, not "commodity" gauges. That is, not all process gauges are created equal. The so-called "lowest price" gauges appear attractive at first, but are rarely the "lowest cost" in the long run, because they usually have to be replaced more often, leading to excessive downtime and inconvenience. Worse, they may fail unexpectedly, leading to catastrophic results.

Consider these 10 steps to select the right gauges that will help to ensure safe, productive and profitable plant processes.

1 - Know what is in the pipe. This way, you can select a gauge alloy that will be compatible with the process media.

2 - Next, know the temperature of the media. If it is very hot, for example, it will be unsafe to use a standard pressure gauge. A long pipe, siphon or diaphragm seal assembly may be needed to cool the media sufficiently before it reaches the gauge internals.

3 – Third, know the ambient conditions surrounding the gauge. For example, if it is corrosive around the gauge, you need to make sure that the external components of the gauge such as the case and window can handle the environmental conditions.

4 - Next, know the pressure in the piping that the gauge is meant to measure. A best practice is using a gauge that is designed for two times the normal operating pressure. For example, if normal pressure is 100 psi, use a 200 psi gauge.

5 – Know the conditions that affect gauge wear. These could include mechanical vibration, pulsation, pressure spikes, etc. For example, if vibration exists, a gauge with a liquid-filled case is recommended. If there is pulsation, use a throttle plug or restrictor accessory.

6 - Select gauges that are designed for a particular service. Process type gauges are designed with strong Bourdon tubes and other heavy duty components. If a gauge is misapplied, the internals of the gauge may not be rugged enough to resist failure and subsequent media release in these applications.

7 - The gauge case must be rugged and have the correct safety features. In some case designs, if there is a rupture of the Bourdon tube, media will be released out the front of the gauge, which can spray workers. Select a gauge where media would release out from the back of the gauge.

8 - Accuracy is important. In most process plants, +/-0.5% accuracy is the norm, and +/-1% is acceptable.

9 - Size is also important. The gauge dial needs to be large enough that operators can read it from a distance – four inches or larger.

10 - Finally, the gauge should be mounted properly. A gauge is usually mounted by its stem, but it can also be mounted in a panel or against a flat surface.

Looking Outside for Assistance

As noted earlier, it can be difficult for plant management to devote enough time and attention to an effective process pressure gauge program. Similarly, it can be difficult for technical personnel to get up-to-speed in terms of education and experience related to process pressure gauges.

There has been a growing trend over the last 30 years in virtually all industries for companies to outsource services that are not considered "core competencies." That is, companies have come to realize that it makes sense to focus their attention on their missions and what they do best, and leave secondary and peripheral activities to third parties that specialize in these other activities and services. Becoming experts in pressure gauges is certainly not a core competency for refining plants, petrochemical, and chemical plants, so this is an area that can easily be outsourced to an experienced third-party audit team.

If a plant elects not to gain the experience that it needs on its own, free assistance is available. To ensure that gauges are working properly and are able to protect the valuable assets that they were designed to protect, a five-phase instrument review is recommended. A best practices review identifies gauges with unwanted conditions and recommends corrective actions to ensure optimum performance. The review identifies issues that affect reliability, longevity and safety. Such a review can uncover evidence of pulsation, vibration, pressure spikes, excess process or ambient temperature, installation damage, clogging, chemical incompatibility, and other problems that can lead to serious problems.

Phase 1 - Interview. This involves discussions between the third-party audit team and plant personnel. During the interview, plant personnel learn the purpose of the instrument audit, how it will be conducted, and how they will benefit from the audit. The interview also helps determine the best way to match the audit to the needs of that specific plant.

Phase 2 - Pre-Meeting. This involves determining the units or sections of the plant to audit, as well as outline roles and responsibilities. It also identifies prohibited and exempt areas of the plant. The third-party team reviews problematic instrument installations and discusses safety procedures with plant personnel.

Phase 3 - Instrument Application Audit. This involves having key plant personnel lead the third-party team on an actual plant walk-through. The audit team reviews the target gauge installations, identifies gauges that demonstrate unwanted conditions, as well as those that demonstrate installation best practices.

Phase 4 - Post-Meeting. This involves explaining the instrument audit results to key plant personnel, specifically noting gauge installations that pose immediate safety concerns. Together, the audit team and plant personnel determine which installations to investigate immediately and which to block out. The audit team and plant personnel then set schedules for the follow-up meeting.

Phase 5 - Follow-up Meeting and Presentation. The third-party team presents its general findings and the executive summary resulting from the instrument audit. The team also presents a spreadsheet that lists details about the gauges from the instrument review, color-coded by severity. The team also provides recommendations for improvements based on experience and best practices. Together, the audit team and plant personnel develop a strategy for moving forward.

The plant should also consider leveraging the audit team in other ways. For example, over time a plant's storeroom inventory can get out of hand. A qualified third-party audit team can help plant personnel rationalize, consolidate and manage their inventory of pressure gauges. For example, if a plant stocks spares for 10 or more ranges, the audit team may be able to help reduce that number to five or six ranges.

Essentially, when used properly, gauges play a significant role in plant safety, productivity and profitability. They act like silent alarms, indicating when something is seriously wrong or helping to diagnose deeper issues within systems and processes that can eventually lead to serious safety issues. To ensure safety and reliability at your plant, make certain the proper gauges are in place by reviewing your processes, gauges and your inventory. Everything is at stake.

REFERENCES

1. WIKA data aggregated from more than 250 instrument audits.

2. Peter Peek, Jocelyn Fenard, Patrick Gantes, Christine Theiler. December 2008, "Skills Shortage in the Global Oil and Gas Industry: How to close the gap" Part I. Pages 14-15.

(56a) Selection of Optimized Gas Monetization Option with FLNG or Fixed FLNG

Joseph H. Cho, SK E&C, Houston, TX

In general, as the gas E&P technology advances, gas fields development costs with the same conditions decreases, results in allowing development of gas fields in remote areas or deep water field. Upon adoption of advanced technologies and widen their practical applications to gas monetization, those of stranded gas field developments are feasible and economical.

Even several FLNG project are under construction stage with overcoming technical limitations and commercial challenges; however, it may be said that project developers prefer to follow the conventional development options, which have been proven with commercial operation by others. It is mainly because the conventional development option is relatively easier to identify potential risks with required costs associated with mitigation methodologies than newly developed options, such as FLNG and Fixed FLNG options.

Main parameters that affect selection of development options are 1) proven technology, 2) potential risks and mitigation, 3) economic returns, and 4) gas reserve (location, size, reservoir characteristics). It is clear if the economic returns of a development option cannot absorb potential risks, a project developer would not take risks in the selected gas monetization option.

This paper discusses the cost analysis, cost savings, potential risks in gas monetization with FLNG concept compared to the green field development option with conventional technology. This paper also addresses potential benefits and risks of FLNG and Fixed-FLNG concept with a virtual project development case as a reference. Advantages and limitations of the Fixed-FLNG concept have also been presented. This includes cost benefit, application flexibility, permit issue, environmental impact, construction requirements, etc.

(56b) Smart Cargo Containment System Advanced (SCA) for LNG Carrier

Dai Gil Lee[1], Munkuen Ha[2] and Haeki Jang[2], (1)KAIST (Korea Advanced Institute of Science and Technology), Daecheon Si, South Korea, (2)Samsung Heavy Industries Co., Ltd, Gyeongsangnam Do, South Korea

Membrane type cargo containment system has been evolved in terms of sloshing resistance and low boil off aspect. Membrane type cargo containment system, so called Smart containment system Advanced, has been developed by Samsung Heavy Industries to achieve less barred range of filling and lower boil off rate.

Smart containment system Advanced can achieve less barred range of filling 15% to 70% of tank height and lower boil off rate, 0.12 % /day. It has been verified and approved by LR and ABS and now ready for real LNG carrier application. This paper discusses features of Smart containment system Advanced.

LNG FSRU and FPSO Pump Considerations for Ship Motion

Gregory P. Wood, P.E.
Engineering Manager
Ebara International Corporation
Sparks, Nevada USA

Scope

LNG Floating Storage and Regasification (FSRU) and Floating Processing and Send Out (FPSO) vessels are a growing component of the LNG infrastructure. These marine vessels use in-tank mounted vertical centrifugal submerged motor cryogenic pumps. Unlike land based pumps, marine pumps installed in FSRU and FPSO applications are exposed to three degrees of ship movement as a result of wave motion. There is concern that adding ship motion to a traditional stationary pump technology can cause issues such as unseating, free spinning, and bearing damage during non-operation. This paper reviews the technical impact of the ship motion and the effectiveness of current and proposed solutions.

Pump and Installation Description

The in-tank mounted vertical centrifugal submerged motor cryogenic pump used in the FSRU/FPSO service consists of a pump and motor assembly, suction valve, column, retraction system, head plate, and power system (see Figure 1). The vertical centrifugal pump uses a Thrust Equalizing Mechanism™ (TEM) (see Figure 3) to balance the axial thrust loads and weight forces during pump operation, eliminating all axial loads on the main ball bearing. During non-operation, the main ball bearing carries the axial weight of the rotating components. The pump is installed into a spring loaded suction valve that seals the column from the tank when the pump is not installed. The pump fluid discharge goes directly into the column and is separated from the tank inlet by a static face type seal between the pump and the suction valve. During non-operation, gravity secures the pump onto the suction valve seal and the rotating elements are allowed to rotate freely. During pump operation, the column fluid pressure applies additional force to the pump weight that is applied to the suction valve seal.

Ship Motion Conditions

The FSRU and FPSO vessels are built in shipyards, transferred to their final location and moored during operation. They may also be relocated to a new location periodically. FSRU and FPSO vessels sometimes contain their own propulsion system, but often rely on auxiliary vessels to provide propulsion. During transit the non operating FSRU and FPSO vessels are exposed to sea conditions. For this analysis, ship motion is described in linear and rotational terms, and is based on 50 to 80 foot sinusoidal wave motion over typical 8 to 20 seconds producing 0.13 to 0.05 Hz frequencies.

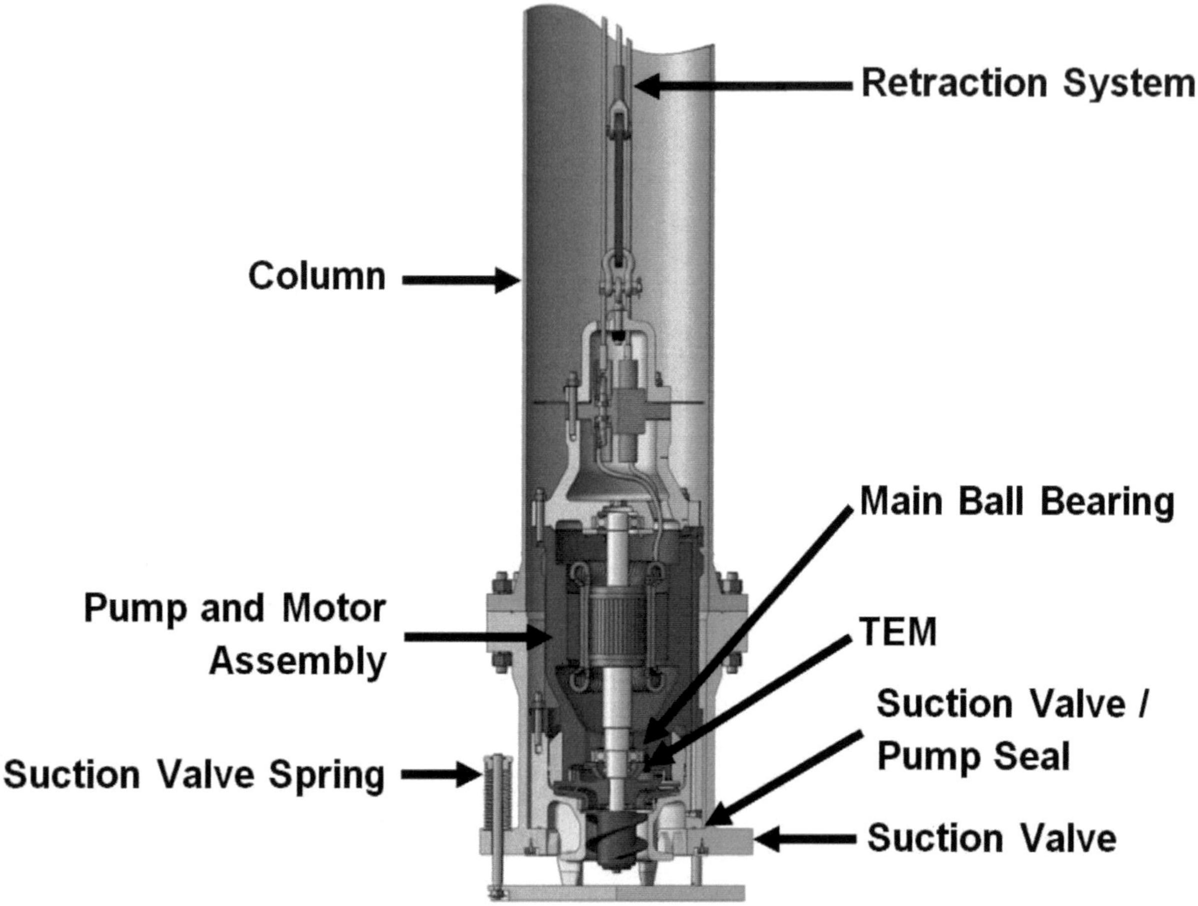

Figure 1: In-tank mounted vertical centrifugal submerged motor cryogenic pump installation

Ship Motion Description

Ship motion is described in linear and rotational terms (see Figure 2). The linear motions are heave (up/down), sway (side-to-side), and surge (front/back). The rotational motions are roll, the rotation of a vessel about its longitudinal (front/back) axis, pitch, the rotation of a vessel about its transverse (side-to-side) axis, and yaw, the rotation of a vessel about its vertical axis. Each type of ship motion introduces its own force and direction impact on the vertical in-tank pumps. The location of the pump in the vessel will also result in different force amplitudes and direction. The analysis going forward will focus on the worst case combination of ship motion and pump location.

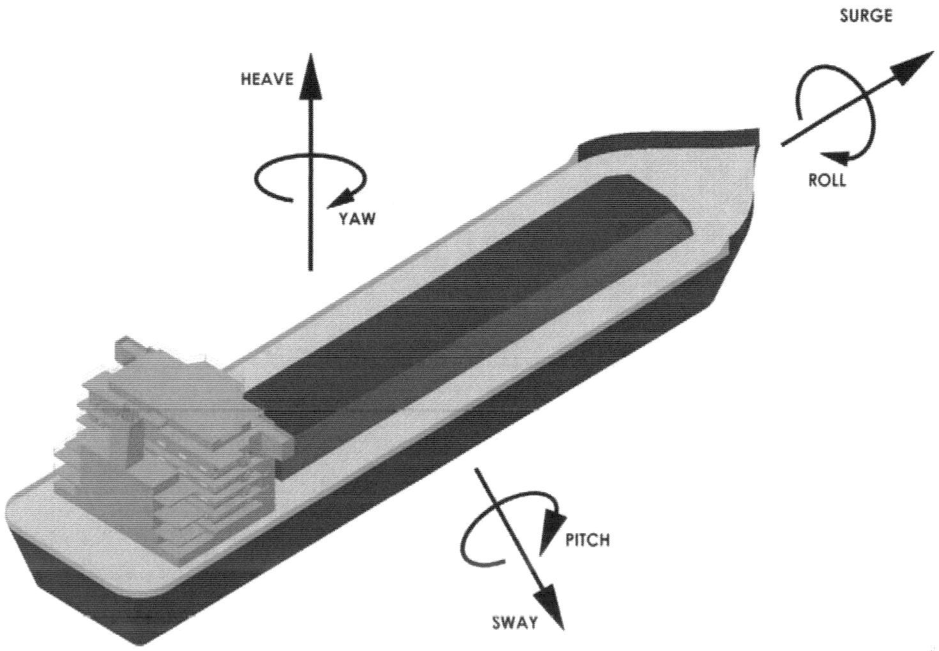

Figure 2: Linear and rotational ship motion

Ship motion will be based on sinusoidal waves. Although a ship will typically dampen the effects of sea wave motion, this analysis will use the actual sinusoidal wave motion as a worst case scenario to be conservative. The sinusoidal wave condition is based on 50 to 80 foot wave motion over typical 8 to 20 seconds producing 0.13 to 0.05 Hz frequencies (see Figure 4). Longer wave frequencies are not considered since they lessen the impact on the pump.

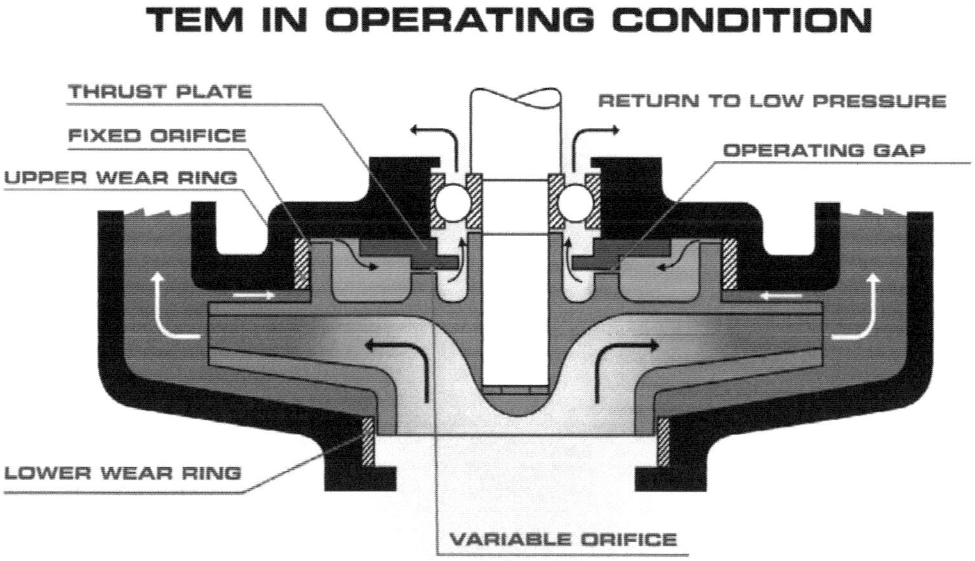

Figure 3: Thrust Equalizing Mechanism™

Period (Seconds)	8	12	16	20
Frequency (Hz)	0.13	0.08	0.06	0.05
Minimum Wave Height (Feet)	5	5	5	5
Maximum Wave Height (Feet)	50	80	80	80

Figure 4: Wave period (frequency) and height range

Ship Motion Impact on Non Operating Pumps

Operating pumps are not subject to ship motion damage. During pump operation there is a large amount of force on the suction valve seal from the pump discharge fluid pressure. This force is many times the pump weight, overcoming any tendency for the pump to unseat or tilt with regards to the suction valve seal. The Thrust Equalizing Mechanism™ balancing forces during pump operation are many times greater than the weight of the rotating element, eliminating any possibility of loading the main ball bearing as a result of ship motion.

During non-operation the pump is seated on the suction valve seal plate using pump weight and the rotating elements are resting on the main ball bearing. The concern is that adding ship motion to this traditional land based pump technology can cause issues such as unseating and/or tilting from the suction valve, free spinning of the rotating elements, and bearing damage due to excessive axial loading. Each of these potential impacts on the in-tank mounted vertical centrifugal submerged motor cryogenic pumps will be analyzed individually.

Pump Unseating Potential

Pump unseating from the suction valve would be the result of ship heave motion. The pump rests on the suction valve seal with a force equal to the weight of the pump minus the upward force of the suction valve springs. The typical spring force is equal to 25% of the pump weight. Therefore, the vertical acceleration required to unseat the pump from the suction valve is:

Equation 1: $a = (1 - f_s)g$
a = acceleration required to unseat pump
f_s = spring force as a percent of pump weight
g = gravitational constant (32.2 ft/sec^2)
$a = (1 - 0.25)\ X\ 32.2 = 24.2$ ft/sec^2

The peak acceleration for sinusoidal wave motion based on sea conditions must exceed 24.2 ft/sec^2 to unseat the pump from the suction valve. The sinusoidal wave characteristics are shown in Figure 5.

Equation 2: $a_{max} = 2\pi^2 F^2 \text{D}$

a_{max}= maximum acceleration based on sinusoidal wave

F = wave frequency (Hz)

D = wave height (feet)

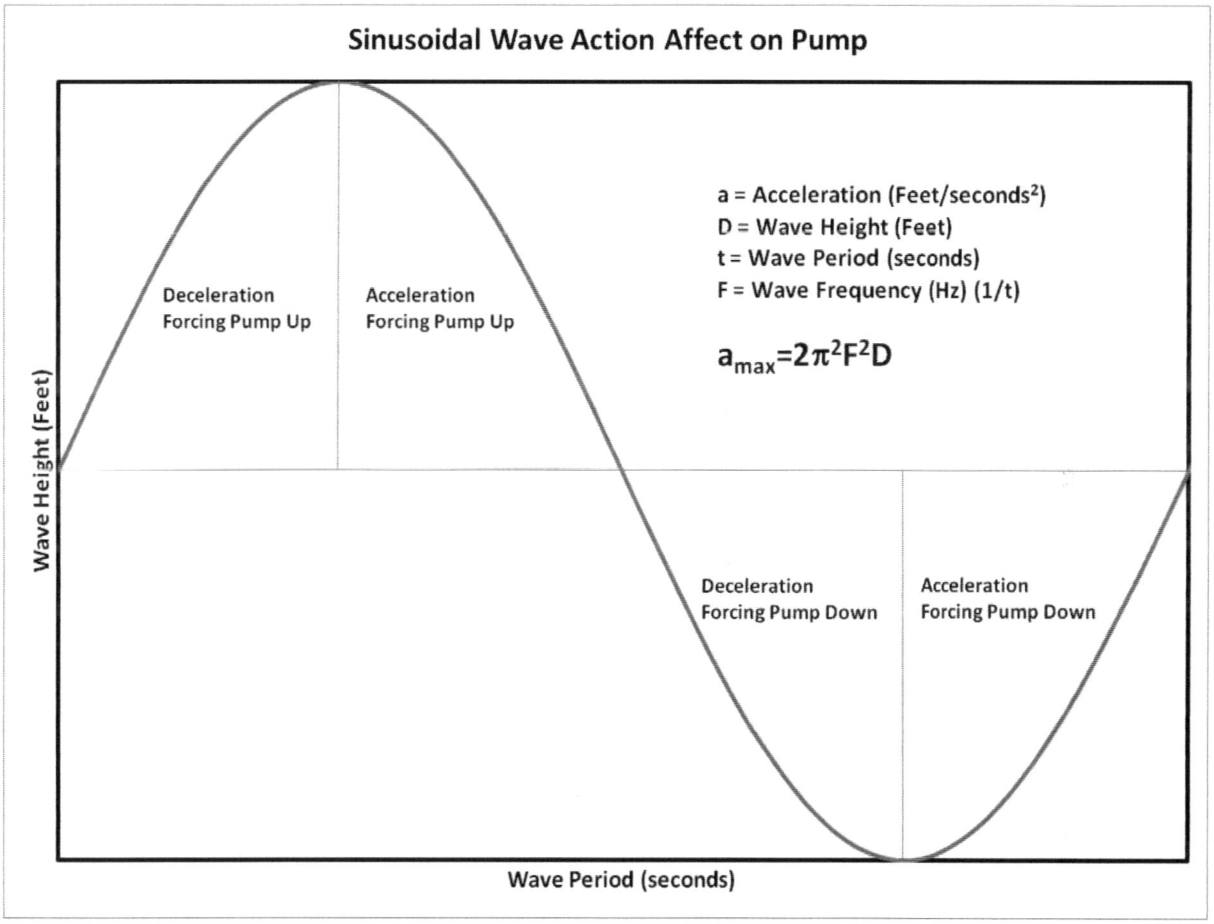

Figure 5: Sinusoidal wave motion

The results of calculating the maximum vertical acceleration derived from a sinusoidal wave form derived from sea conditions is 15.4 ft/sec^2, only 64% of the required 24.2 ft/sec^2 to unseat the pump from the suction valve (see Figure 6). Therefore, there is no potential to unseat the pump in the suction valve.

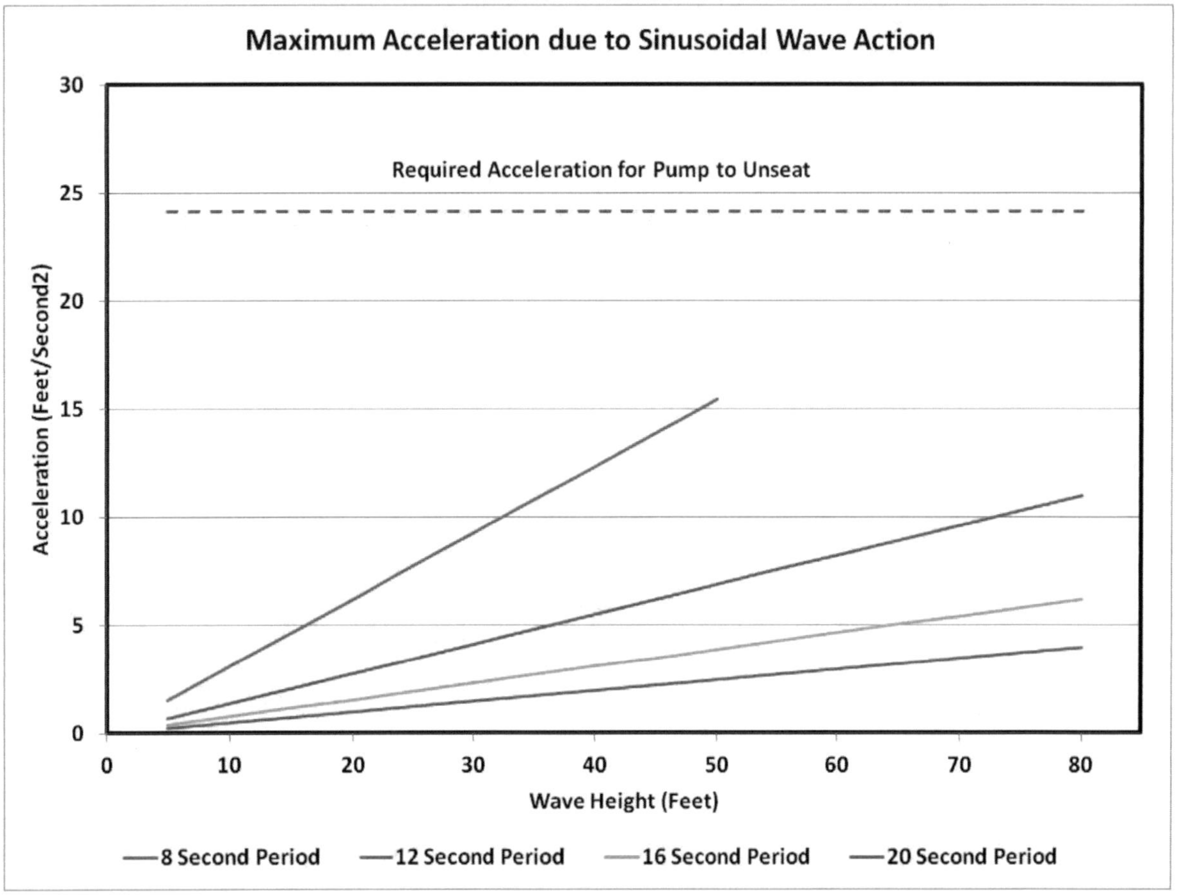

Figure 6: Sinusoidal wave maximum acceleration

Pump Tilting Potential

Pump tilting is based on ship roll movement. The pump is loose in a column and the pivot point is the suction valve seal. The center of gravity (CG) for typical in-tank vertical pumps is about 40% of the total pump length away from the seal plate pivot point (see Figure 7). Based on this CG location, when the ship rolls to some angle and combined with the angular momentum, the pump will tilt. The tilt stops when the guide pads on the side of the pump make contact with the column wall, approximately 0.25 inches. Although this is a small movement, it would cause the pump to suction valve seal to lose contact. Losing seal contact does not cause any immediate issues since the column is not sealed from the tank while the pump is installed. The seal would have to be damaged, however unlikely, to compromise pump performance to a noticeable degree. Even though this does not represent a high risk for premature pump failure, it is not a desirable condition.

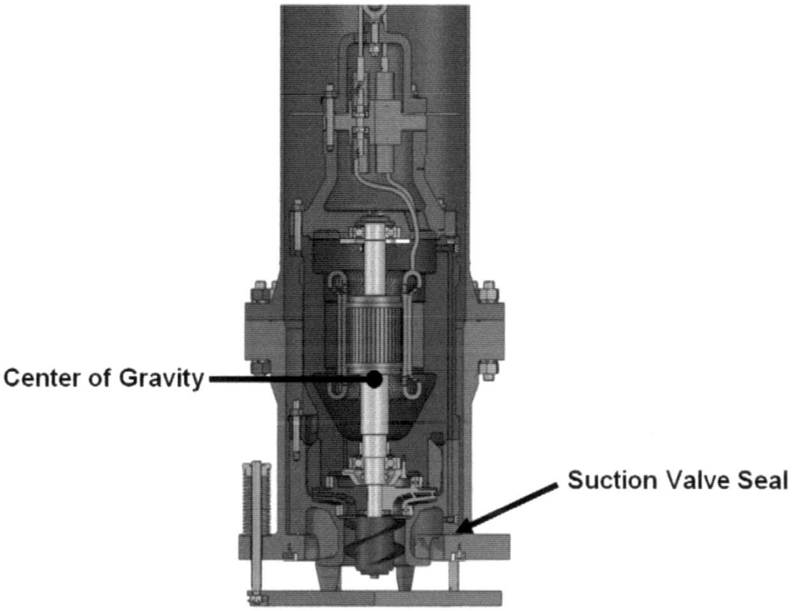

Figure 7: Typical Pump Center of Gravity in relationship to the suction valve seal

Main Bearing Damage Potential

Pump main bearing axial loading that could cause brinelling is based on ship heave motion. The vertical acceleration forces from the ship movement must be greater than one gravitational force to allow the rotating elements to rise above the main bearing. The rotating elements must then impact the bearing with enough force to cause brinelling. The typical main bearing is a deep groove ball bearing size 6320 constructed of 440C stainless steel. The axial force required to brinell the races of this 6320 bearing is 23,000 pounds. Based on this, a typical in-tank vertical centrifugal submerged motor cryogenic pump rotating element requires more than 12 G's to cause bearing brinelling. The maximum vertical force the pump would be exposed to (Figure 6) is less than 1 G. There is no possibility of brinelling the main bearing of an in-tank vertical centrifugal submerged motor cryogenic pump. Another bearing failure mode to consider is fretting wear false brinelling. This is caused by minute high frequency oscillations of the bearing balls displacing grease resulting in surface oxidation that creates indentations over time that look like brinelling. The pumps use stainless steel bearings, are exposed to low frequency vibration and typically are in an oxygen free environment eliminating this possibility.

Free Spin Potential

There is some thought that ship motion can potentially cause a pump to free spin. This spin has to occur around a vertical axis or yaw rotation motion, the only ship motion that rotates around a vertical axis. For a pump to see the full effects from ship yaw motion, it would have to be located close enough to the ship pivot axis to absorb most of the yaw motion. The yaw motion must produce acceleration large enough to overcome the calculated starting torque of the rotating elements that accounts for the ball bearing friction (see Equation 3). The friction coefficient for a deep groove ball bearing is 0.0010 to

0.0015. The typical yaw movement for large marine vessels is small with low acceleration. The yaw motion is considered to be a sine wave type motion.

Equation 3: $M = \mu Pd/2$

$M =$ friction moment (start torque)

$\mu =$ friction coefficient

$P =$ load on bearing (pounds)

$d =$ bearing bore diameter (inches)

Minimum Pump Rotation Start Torque: $M = .0010\ X\ 1244\ X \frac{3.937}{2} = 2.45$ in-lb

Equation 4: $T = I\alpha$

$T =$ torque on pump rotating element

$I =$ mass moment of inertia of pump rotating element

$\alpha =$ angular acceleration from ship yaw motion at the pivot point

Assuming the yaw rate correlates with the wave period as a worst case, the maximum angular acceleration is relatively low. To ensure the pump rotating element does not spin, $M \geq T$ must be true (see Figure 8). Based on this analysis, the pump rotating elements will not begin to rotate on the main bearing unless the pump is located at the yaw pivot point and the yaw motion exceeds 6 degrees in 8 seconds minimum, beyond any expected yaw motion. There is no free spin in a non operating pump.

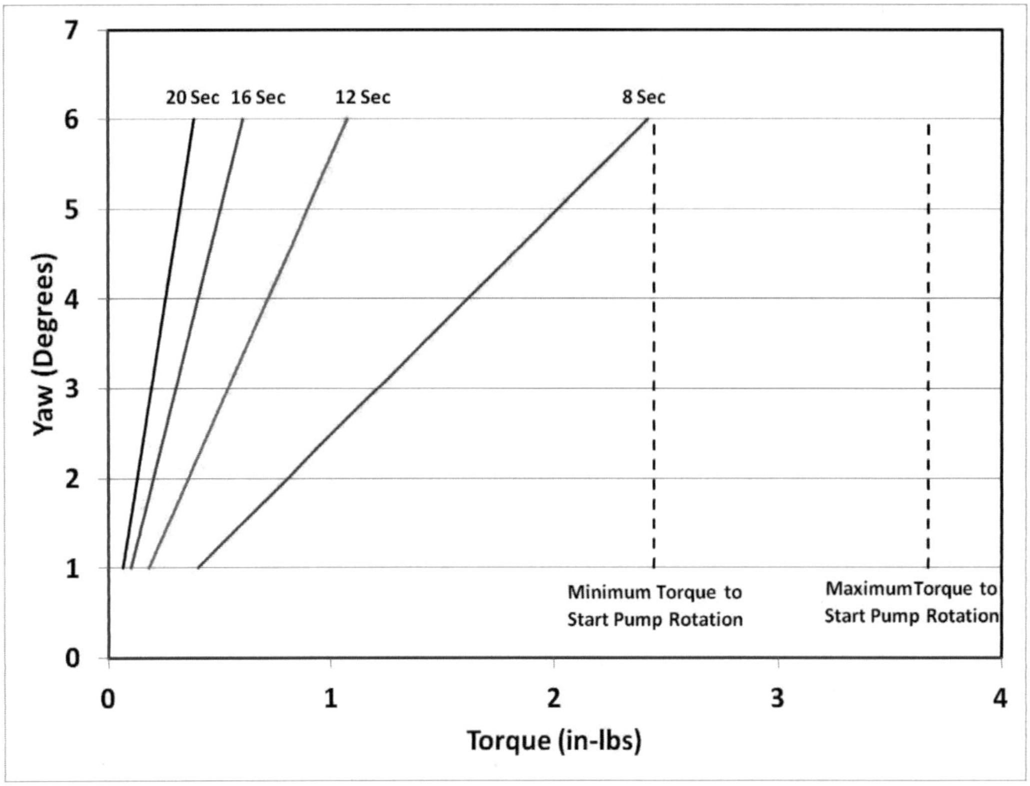

Figure 8: Correlation of torque from ship yaw motion to actual torque required to rotate pump based on 8 to 20 second wave period

Experience with Installed Marine Cargo Pumps

There are thousands of LNG marine cargo pumps in active service with some exceeding 30 years of active service. These pumps are used specifically for offloading LNG from cargo ships. Therefore, they do not operate during sea transit between ports. The pumps are continuously exposed to wave motion conditions during the transit times.

The marine cargo pump designs are essentially the same as a land based type pump (see Figure 9). They are rigidly mounted on a mast eliminating the suction valve which eliminates any potential pump unseating possibilities. There are no additional protection systems. There have been no reported bearing failures or degradations observed during periodic dry dock service. No cargo pump has failed to operate normally between dry dock services. There are thousands of marine cargo pumps in active service with no bearing failures or degradation. The service data and supporting calculations leads to the conclusion that marine vertical centrifugal submerged motor cryogenic pump bearings are not subject to damage during non-operation exposure to wave motion.

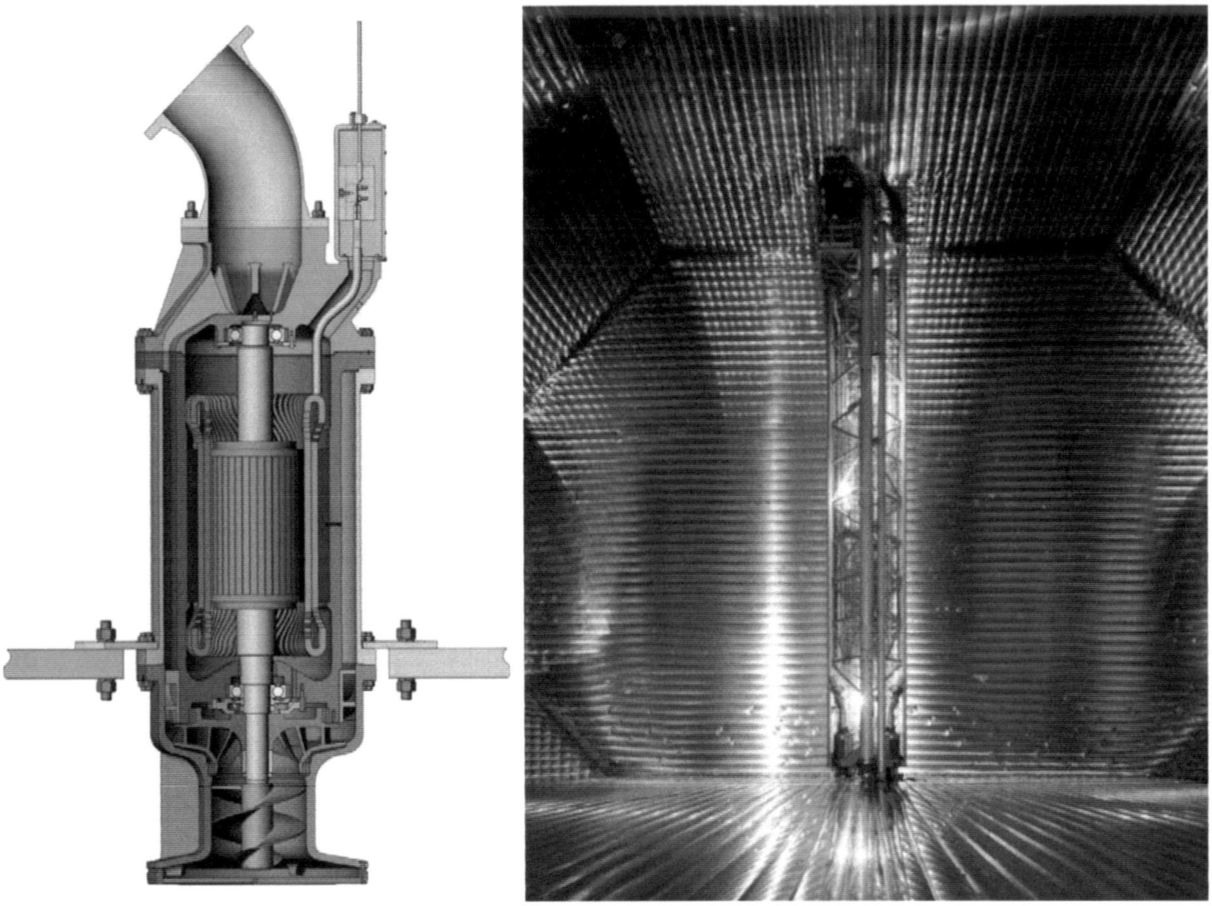

Figure 9: Marine Cargo Pump Cutaway (left) and Cargo Pump Ship Mast Installation

Pump Protection

Before any pump protection design is implemented, it would be prudent to determine if protection is necessary for the targeted failure mode. Ship motion analysis concludes that there is no pump unseating or bearing damage potential. A protection system can add weight, cost and complexity. If the decision is made to add protection to a pump from ship motion, several options may be considered.

Rigid Retraction System

A Rigid Retraction System is designed to eliminate any axial pump movement such as unseating or tilting. It is a retraction system constructed of rigid pipe sections instead of cables (see Figure 10). The rigid pipe sections transcend from the pump to the head plate. A constant down force is applied to the rigid pipe from springs located on the head plate eliminating any need for external control. In addition to a down force, the springs allow for coefficient of thermal expansion differences between the aluminum pump and 304 stainless steel piping and column. The coefficient of thermal expansion generates 0.008 inch per foot of pump length change at -160 degree C temperature (see Figure 11). The down force is recommended to be approximately equal to the pump weight. This will provide 1.75 G total down force, including the counter effect of the suction valve springs.

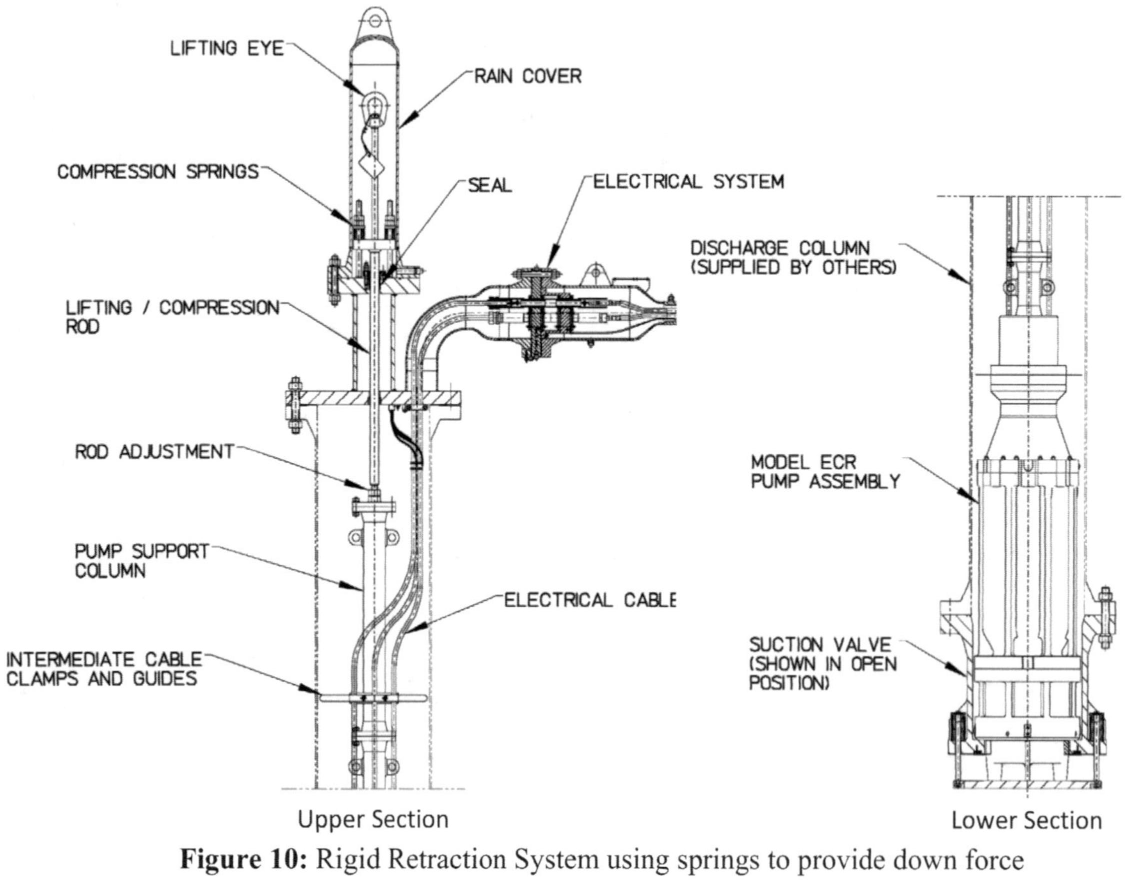

Figure 10: Rigid Retraction System using springs to provide down force

The spring type Rigid Retraction System provides continuous down force on the in-tank pump that ensures the pump will remain seated and in the upright position. However, the calculations show that wave motion will not cause an unseated condition. Another solution is to lower the pump Center of Gravity (CG) with respect to the suction valve eliminating the Rigid Retraction System while preventing pump tilting. This low CG system is presented in the next section.

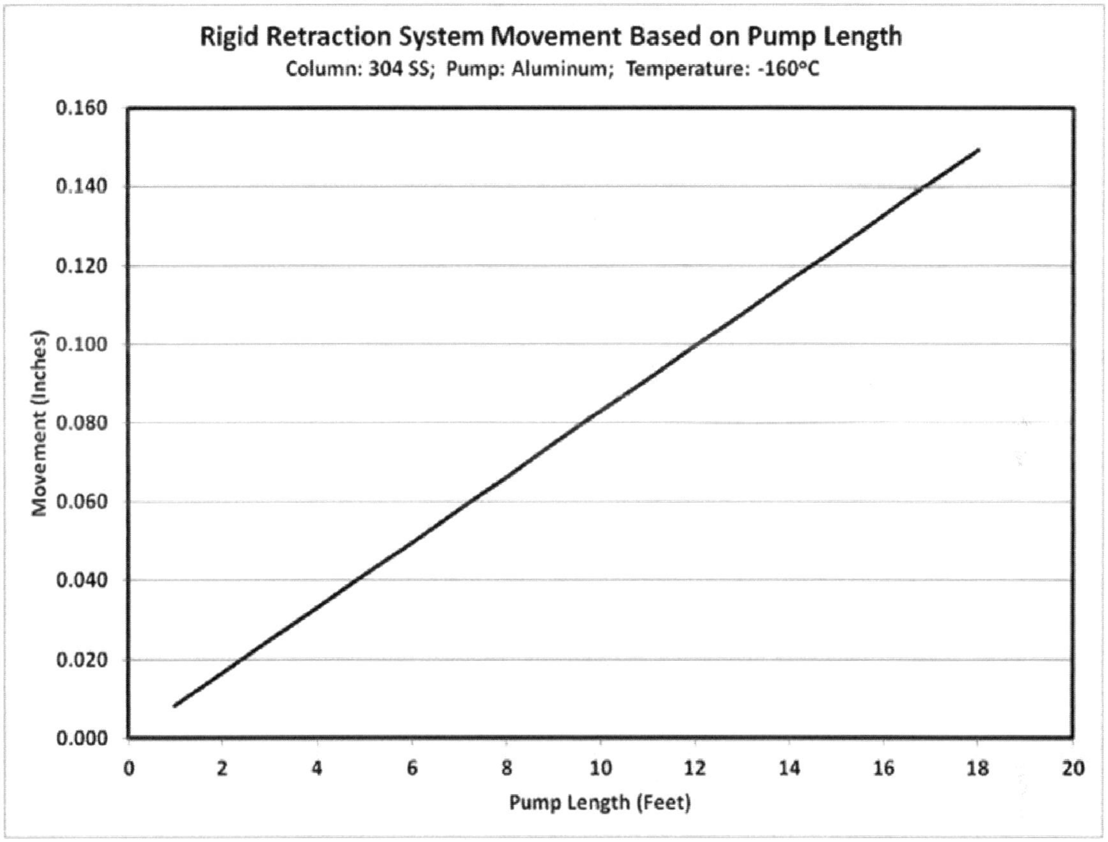

Figure 11: Rigid Retraction System differential length change at LNG temperature conditions

Low Center of Gravity Pump Seating

Since vertical pump unseating is eliminated as a possibility (see Pump Unseating Potential section), lowering the center of gravity (CG) relative to the suction valve sealing plate can now restrain the pump from tilting as a result of ship pitch and roll. The CG is lowered by lengthening the suction valve to reduce the pump sealing plate to pump CG distance (see Figure 12). The low CG pump seating effectively combines the pump weight and friction factors to ensure the pump remains seated. The suction valve seal is PTFE with a low coefficient of friction. If the ship were to tilt enough, the pump will slide until the bumpers make contact with the column and the pump makes contact with the suction valve. The column is 304 stainless steel and the pump, suction valve and bumpers are aluminum. The coefficient of friction between 304 stainless steel and aluminum is 0.45 and aluminum to aluminum is greater than 1.0. The pump weight, suction valve spring force, roll angle, and coefficient of friction are used to calculate the unseating. Since the seal is constructed of PTFE with a friction coefficient of 0.2, the pump will slide on the seal plate before it will tilt starting at 12 degrees. Once the pump bumpers make contact with the column, the friction forces between the column and bumpers keep the pump from unseating through 90 degrees of roll. The forces used for the calculations include pump weight, friction, and suction valve spring. The pump holding force is calculated based on the roll angle and the resultant friction forces, weight on springs, and spring force. The friction forces between the column and bumpers are enough to keep the pump from unseating all the way to a 90 degree angle (see Figure 13). The low center of gravity pump seating system ensures positive pump seating during ship motion eliminating the need for a more complex rigid retraction type system.

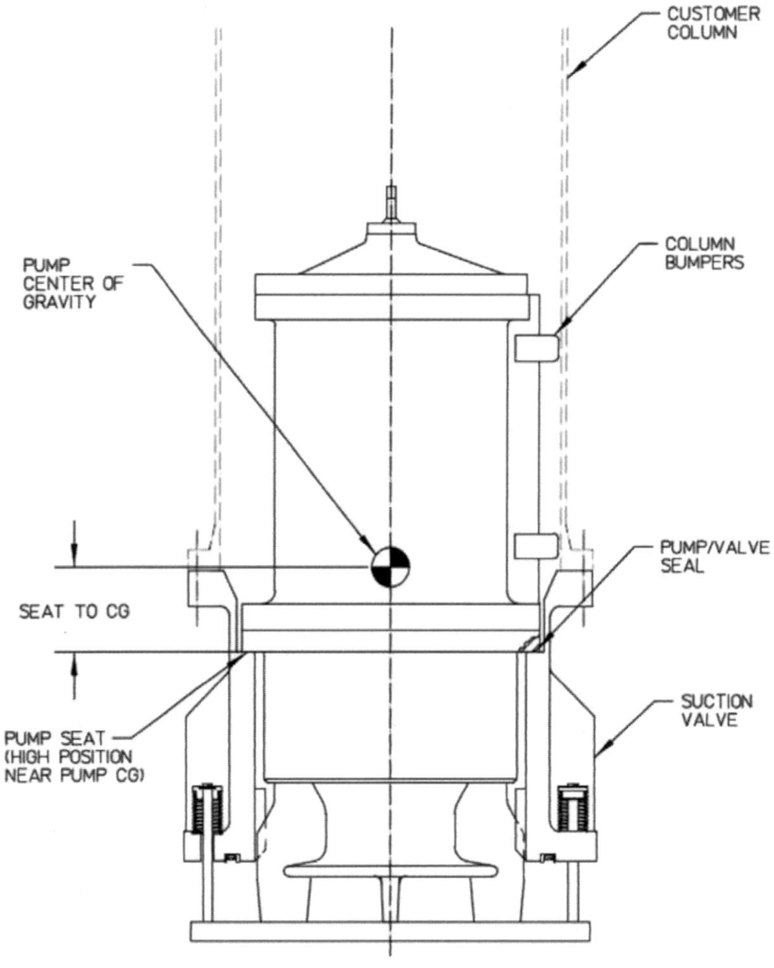

Figure 12: Low Center of Gravity System

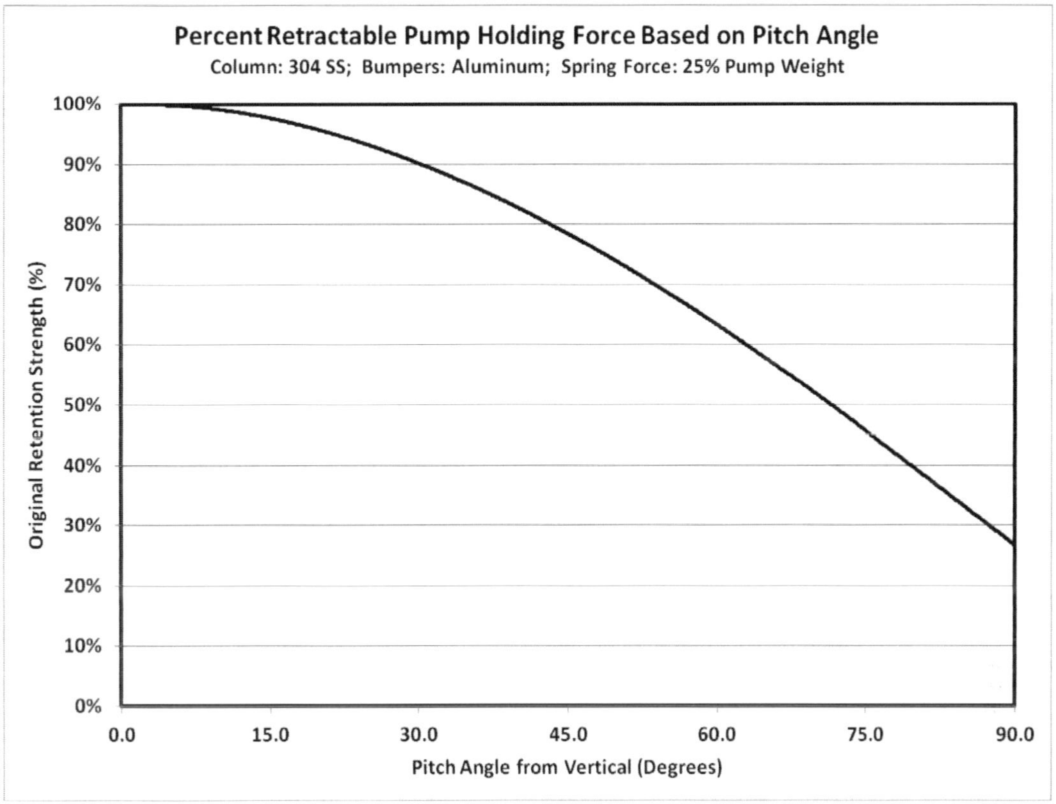

Figure 13: Retractable pump holding force compared to roll angle

Shaft Brake System

There is some thought that ship yaw and heave motions can potentially cause the rotating elements to spin, and load the main bearing during non operating conditions. A brake system can prevent shaft rotation and axial bearing loading. There are two types of shaft brake systems. One system presses a pad against the shaft to keep it from rotating (see Figure 14). The other lifts the shaft off the main bearing and presses against the shaft to keep the shaft from rotating and moving vertically (see Figure 15). Each system uses nitrogen actuated pistons which are actuated after the pump comes to a complete stop.

There is solid information that supports the idea that a brake system is not required to protect pumps against ship motion during non operating conditions. A typical in-tank retractable and rigid mounted cargo pump uses a 6320 type ball bearing that requires 23,000 pound axial load to cause brinelling. This would require more than 12 G's to cause brinelling in a typical pump. There are several thousand rigid mounted marine cargo pumps in service, some with over 30 years marine service. These pumps are non-operational during ocean transit conditions in both loaded and unloaded LNG cargo conditions. There have been no bearing failures or degradation.

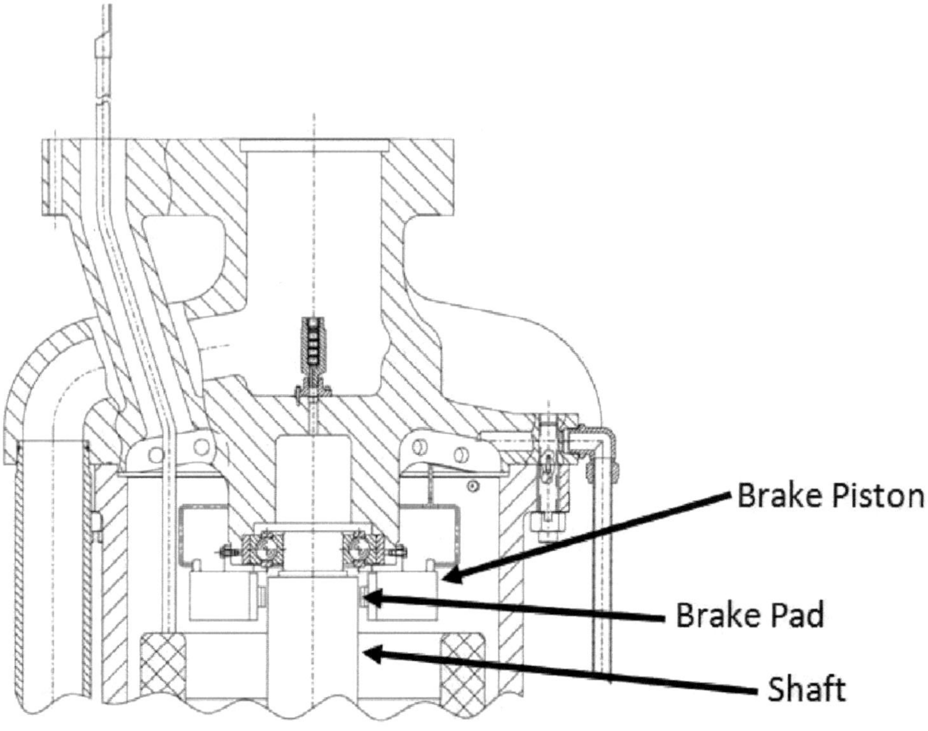

Figure 14: Shaft brake system that protects against shaft rotation movement

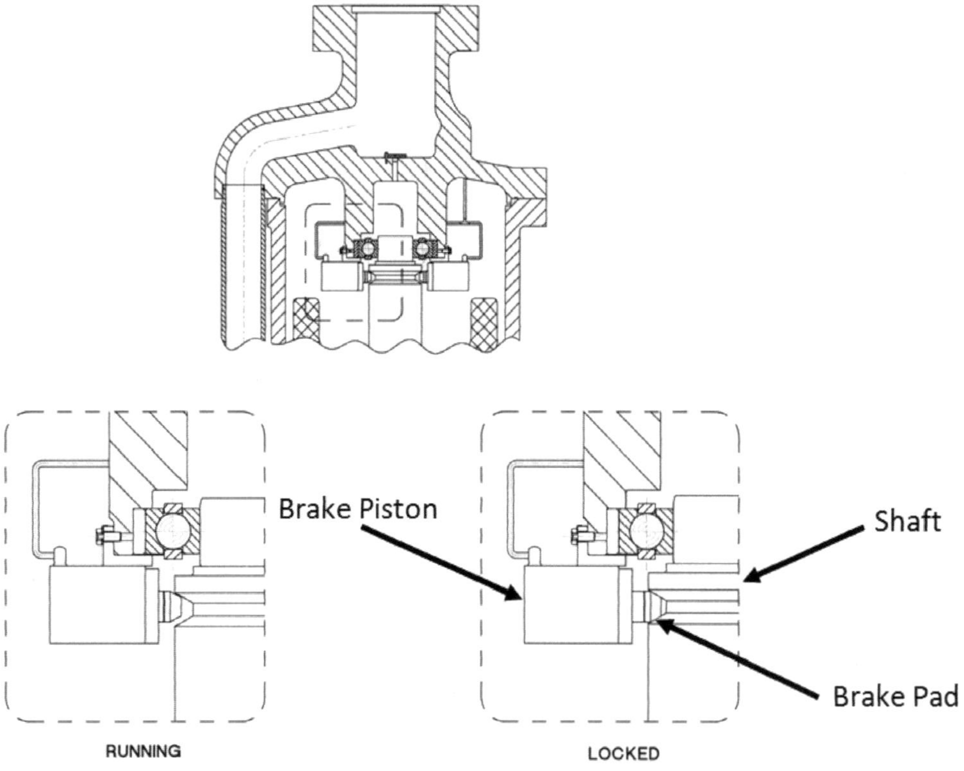

Figure 15: Dual shaft brake system that protects against bearing axial load and shaft rotation movement

Conclusion

LNG FSRU and FPSO vessels use in-tank mounted vertical centrifugal submerged motor cryogenic pumps that are exposed to three degrees of ship movement as a result of wave motion. The only recommended change to pumps used on LNG FSRU and FPSO vessels to provide protection against ship motion is a low CG suction valve. Ship motion does not cause free spinning and bearing damage during pump non-operation, which is confirmed by calculations and thousands of marine cargo pumps in actual service. Therefore, the pump life expectancy is not impacted. Adding additional equipment such as a rigid retraction and brake system to protect the in-tank pumps from free spinning and axial load bearing damage only adds complexity, cost, and weight. Ship roll motion may have the potential to cause a pump to tilt on the seal plate, which could potentially damage the seal. The least complex and most economical solution is to use a low CG suction valve design, moving the CG close to the pump/suction valve seal eliminating the possibility of pump tilting. The low CG suction valve design allows the ship to roll 90 degrees while maintaining the pump seated condition.

Study on determination of optimal operating conditions of single mixed refrigerant process for LNG FPSO

Yeonmi Suh, Taeyeong Lee, Gyeongho Han, Kiil Nam, and T.J. Kim, Hyundai Heavy Industries, Ulsan, Korea

Introduction

The demand and consumption of natural gas alike have been increasing all over the world. It is estimated that the global needs of natural gas will grow up to 345 million tons between 2010 and 2020, accordingly an average of 4.7% increase every year. Because 20% of the world gas reserves are in the offshore fields, LNG FPSO (Liquefied Natural Gas-Floating, Production, Storage and Offloading unit) has become to be economical for exploring natural gas well in this area and has been recently spotlighted.

The LNG FPSO is largely composed of two parts, i.e. hull and topside. While the LNG FPSO is moored steadily above the well, the unprocessed natural gas is pumped from well to the LNG FPSO's topside via riser. When the gas reaches the topside, it is processed to separate natural gas, LPG and natural gas condensate. The processed feed gas is treated to remove impurities and liquefied through the liquefaction process prior to being stored in the hull. When the liquefaction process turns the pretreated natural gas into LNG, the volume of natural gas shrinks to six hundredth. It is known that the liquefaction process accounts for 30~40% of overall capital cost as shown in Figure 1 [1]. Therefore, it should be regarded as top priority among all topside systems on the LNG FPSO.

The LNG liquefaction processes can be categorized according to the number of cycles. The number of cycles means the number of working fluids when the natural gas is pre-cooled, liquefied and sub-cooled. If the number of cycles increases, the LNG production efficiency will be enhanced accordingly. But, as the process gets more complex, consequently the capital cost increases. The single mixed refrigerant (SMR) process uses only one working fluid for pre-cooling, liquefaction and sub-cooling of the natural gas and is known for having the advantage that the system is relatively simple. In case of the large scale LNG FPSO and onshore plant, the DMR (dual mixed refrigerant) process, which uses two working fluids, can be applied because the space constraint is not a big problem. But in case of the small and medium scale LNG FPSO, the compact and light LNG liquefaction system is essentially required due to the limited space. For these reasons, the SMR process is generally suggested to an alternative of the LNG liquefaction process that catches up with the modern trend of LNG FPSO.

To build more economical liquefaction process, it is vital to optimize the liquefaction system. So, there have been a lot of research studies about the optimization of LNG liquefaction processes. Lee et al. [2] proposed a novel method for the selection of refrigerant composition using the nonlinear programming (NLP) techniques. Jensen [3] modeled and optimized the PRICO LNG from the perspective of process operation and control. Chang et al. [4], Wang et al. [5, 6] and Lee et al. [7] performed the optimization based on in-depth thermodynamic analysis. Nogal et al. [8], Alabdulkarem et al. [9] and Hwang et al. [10] used the stochastic optimization (genetic algorithm, GA) to overcome the local optima. Most of studies above dealt with onshore processes which were relatively large scale but didn't correspond to the offshore process.

In this study, the method was proposed to determine the optimal operating conditions of the SMR process on the LNG FPSO. The complicated configuration of real application process sufficiently reflected on this model by coupling the rigorous process simulation program (Aspen HYSYS®) with GA in MATLAB.

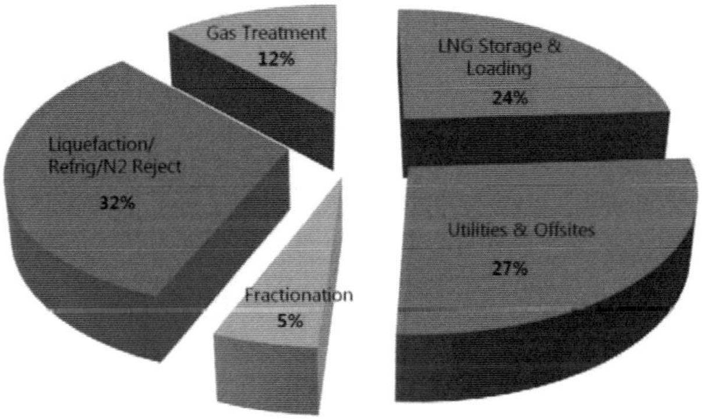

Figure 1 Ratio of capital cost for LNG FPSO

Optimization of SMR Process

Description of studied SMR Process

The SMR process consists of two compressors, two coolers, two phase separators, two mixers, three valves and three heat exchangers as shown in Figure 2. The mixed refrigerant is separated into two streams by means of molecular weight at phase separator 1. After entering into heat exchanger 1, respectively, the relatively heavy refrigerant is used for pre-cooling the natural gas, but the lightweight one is cooled by the heavier refrigerant. And then the cooled lighter refrigerant is separated into two fractions again at phase separator 2. After that, the heavier one is used for liquefaction in heat exchanger 2 and the lighter one for sub-cooling of the natural gas in heat exchanger 3. The mixed refrigerant is composed of methane, ethane, propane, n-butane, i-butane, n-pentane, i-pentane and nitrogen.

Modeling Conditions of SMR Process

The centrifugal type compressors are used for the SMR process. It is assumed that the compressors operate at 75% of adiabatic efficiency. If the inlet stream of the compressor is liquid phase, the compressor will be damaged. Therefore the inlet temperature of the compressor must be higher than the saturated temperature of the inlet stream. In cooler 1 and 2 located after compressors, seawater is used as cooling medium and the pressure drop effect is not considered. The temperature of the refrigerant has to be above seawater temperature of 25°C through the seawater cooler. The refrigerant is separated into vapor and liquid in the phase separator after compression. The pressure drop and the heat loss at each phase separator can be neglected because they are so small. Thus, the pressure and temperature of the outlet stream remain the same as those of the inlet stream in the phase separator. In heat exchangers 1, 2 and 3, the refrigerant exchanges the thermal energy with the natural gas. The pressure drop through the heat exchanger is ignored for simplicity. It is assumed that the outlet temperatures of the hot streams in the heat exchangers have the same value with the effectiveness of 100% for the heat exchangers [11]. All expansion valves are operated at an isenthalpic condition. To prevent back flow effects in the mixer, all inlet streams have to be the same pressure values.

With the above-mentioned conditions, the rigorous process simulation for the SMR process is implemented by Aspen HYSYS®, a commercial process simulation program. In this simulation, the Peng-Robinson equation of state is selected for modeling the property of substances. As the result of the simulation, the sixteen independent variables of the SMR process are defined and listed in Table 1.

Table 1 List of independent variables for SMR process

Variables	Baseline values
Heat exchanger 1 outlet temperature of natural gas (℃), T_{23}	-35.22
Heat exchanger 2 outlet temperature of natural gas (℃), T_{24}	-109.3
Cooler 1 outlet temperature (℃), T_2	27.1
Cooler 2 outlet temperature (℃), T_8	29.6
Compressor 1 outlet pressure (bar), P_1	26.8
Compressor 2 outlet pressure (bar), P_7	48
Valve 1 outlet pressure (bar), P_5	6.5
Mole fraction of mixed refrigerant composition (N2/C1/C2/C3/nC4/iC4/nC5/iC5)	0.0671/0.3115/0.3612/0.0619/0.1176/ 0.0580/0.0056/0.0170
Molar flow of mixed refrigerant (kgmole/s), m	14.05

Objective Function

In general, the required total power of the LNG liquefaction is summation of the required compressor power and the produced expander power. Because there is no expander in the SMR process, the objective function of this model is to minimize the total compressor shaft power required to liquefy natural gas.

Optimization Framework

Most of previous studies formulated their own mathematical equations for the thermodynamic properties, mass and heat balances [8, 10] to simulate and optimize their liquefaction processes. These equations are so complicated that it takes a lot of efforts to construct these equations and solve them. There are some assumptions to simplify the problems in the previous studies. It is difficult to utilize these approaches in the industrial company which do not have enough time and resources to construct such complicated models. To avoid this problem, we use Aspen HYSYS® which is a widely used flowsheet program. By using Aspen HYSYS®, we can easily construct the rigorous process models considering complicated mass and energy balances on all of the process units. Aspen HYSYS® has its own optimization program, HYSYS Optimizer. However, it is not powerful enough to solve the optimization problems of the SMR process which include the various compositions of streams as independent variables. To solve these problems, the SMR process is simulated with Aspen HYSYS® and then optimized with GA code in MATLAB Global Optimization Toolbox. In addition, a MATLAB code is created to make the objective function and to connect the process flowsheet in Aspen HYSYS® and GA code in MATLAB Global Optimization Toolbox. The optimization framework used to solve the optimization problem of the SMR process is shown in Figure 3.

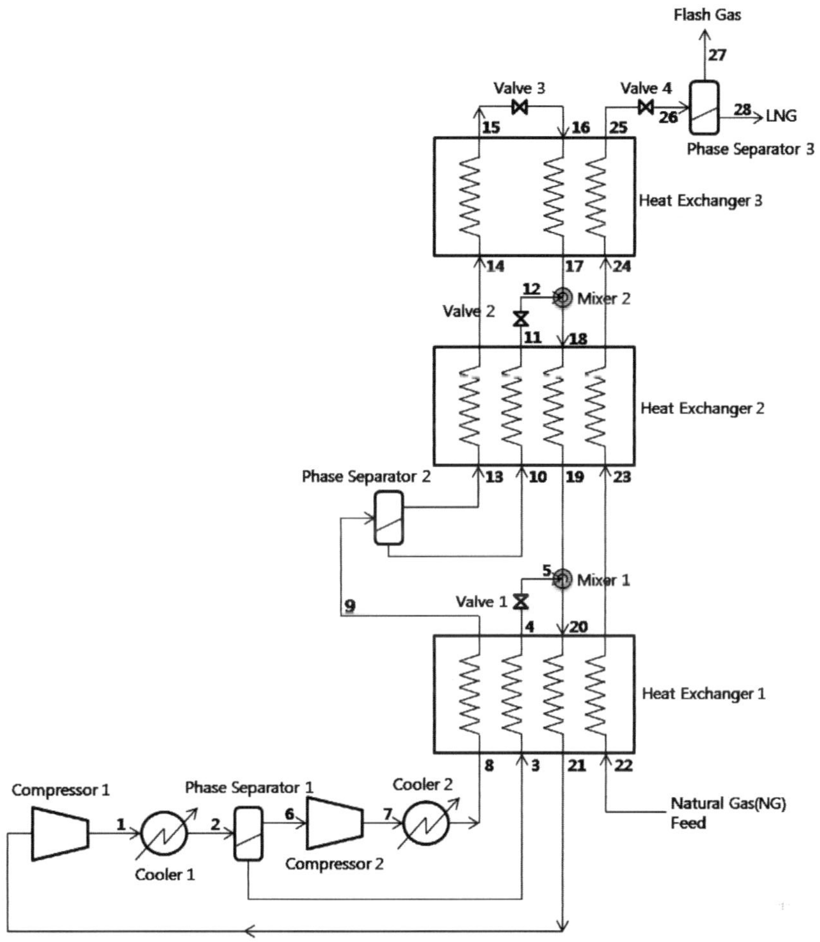

Figure 2 Schematic flow diagram of SMR process

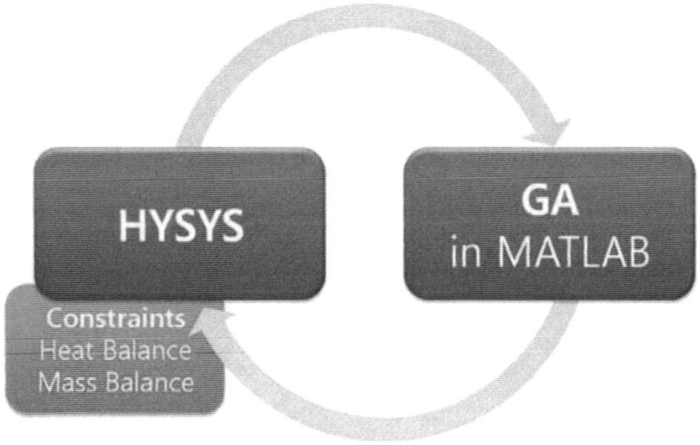

Figure 3 Optimization framework

Results and Discussions

As displayed in Figure 4, the SMR process is simulated by using Aspen HYSYS®. The composition of feed gas is listed in Table 2. The design specification of the SMR process is based on an engineering research project data and assumptions mentioned previously. The simulation result is only regarded as the baseline case. The baseline case is optimized by the proposed optimization framework.

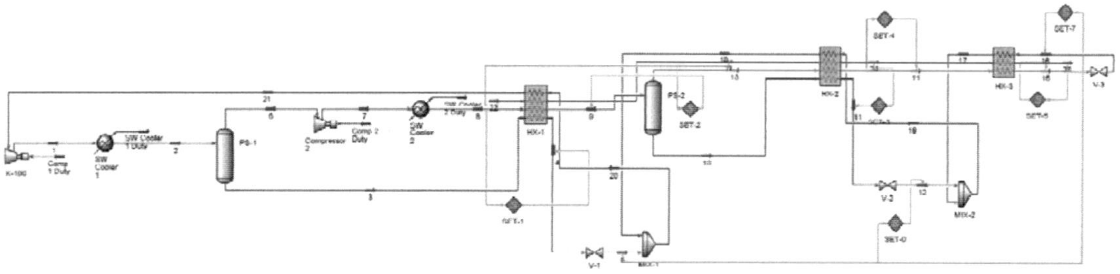

Figure 4 SMR process used in Aspen HYSYS®

Table 2 Typical composition of feed gas

Component	Mole fraction (%)
Nitrogen	1.5679
Methane	84.9954
Ethane	7.1809
Propane	3.5942
n-Butane	0.8727
i-Butane	0.7752
n-Pentane	0.2884
i-Pentane	0.3861
Hexane	0.3392

The model of the SMR process has sixteen independent variables as listed in Table 1. The objective function is to minimize the total power consumption of the SMR process, which is the sum of power consumption of all compressors. The optimal values of independent variables and the objective function values of each case are listed in Table 3. In baseline case of Table 3, the data are the simulation results to which no optimization tool is applied. The optimization results using HYSYS Optimizer program are shown as the HYSYS Optimizer case. The GA case means the results from optimizing the SMR process by utilizing GA in MATLAB Global Optimization. The GA tuning parameters used in the optimization are listed in Table 4. As shown in Table 3, the GA case has less molar flow of mixed refrigerant and lower overall compression consumption rate than any other cases. The mole fraction of methane, propane and i-pentane increase while the mole fraction of ethane and isobutene decrease by the optimization of the mixed refrigerant's composition. Thus the power consumption of the GA case is 15.9% less than that of the baseline case and is 6.52% less than that of the HYSYS Optimizer case.

The temperature profiles of the streams in the heat exchangers of the SMR process are shown in Figure 5. The efficiency of the heat exchanger can be maximized if the temperature of the hot stream (e.g. feed gas) approaches that of the cold stream (e.g. mixed refrigerant) along its entire length because the heat exchanger approaches to reversible operating condition under such circumstances [12]. The increased efficiency of the heat exchangers reduces the power consumption of the SMR process. The heat exchangers 1 and 2 in the GA case show small gaps between the temperature profile of the hot and cold components than any other case, while the heat exchanger 3 indicates larger gap. On the whole, the heat exchangers in the GA case show better efficiency than any other cases.

Table 3 Results of optimized SMR process

Case	Variables								
	T23	T24	T2	T8	P1	P7	P16	m	
Baseline	-35.22	-109.30	27.10	29.60	26.80	48.00	6.50	14.05	
HYSYS Optimizer	-10.00	-106.60	29.63	26.94	23.50	39.57	6.89	14.05	
GA	-25.06	-93.65	25.00	25.01	18.09	39.79	6.46	13.00	
Case	Variables								Objective function
	N2	C1	C2	C3	nC4	iC4	nC5	iC5	Power consumption (MW)
Baseline	0.0671	0.3115	0.3612	0.0619	0.1176	0.0580	0.0056	0.0170	82.27
HYSYS Optimizer	0.0671	0.3115	0.3612	0.0619	0.1176	0.0580	0.0056	0.0170	74.03 (Reduced by 10.02%)
GA	0.0714	0.3298	0.3148	0.0936	0.1191	0.0243	0.0045	0.0425	69.20 (Reduced by 15.88%)

Table 4 Typical GA tuning parameters

Tuning parameters	Values
Population size	100
Reproduction count	2
Maximum number of generations	500
Crossover fraction	0.8
Selection method	Stochastic uniform
Fitness scaling method	Rank
Mutation method	Adaptive feasible

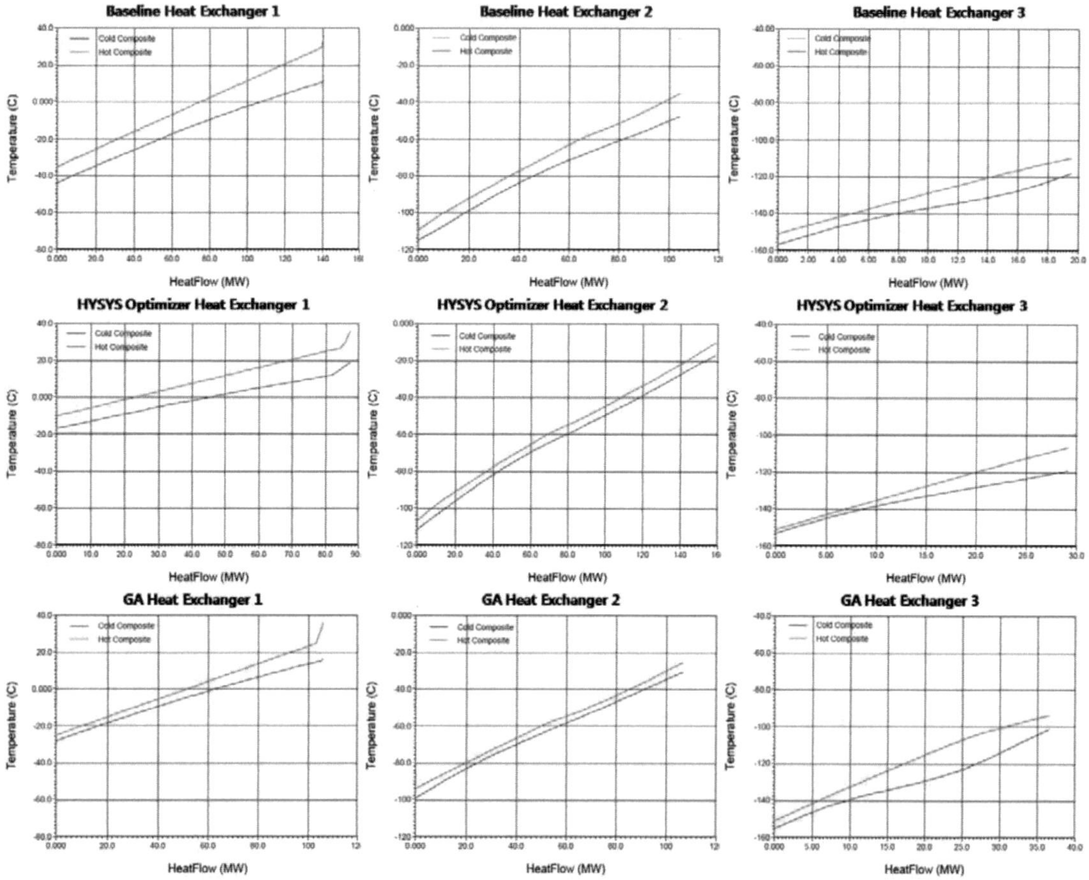

Figure 5 Temperature profiles of baseline, HYSYS Optimizer and GA cases

Usually, the heat exchanger of LNG liquefaction process is designed for minimum approach temperature to be within 1~3 °C [11]. To examine the effect of the minimum approach temperature, the additional cases, GA 1, 2 and 3, are optimized with each minimum approach temperature of 1, 2 and 3°C. As shown in Figure 6, the power consumption of the process lessens with the decrease in the minimum approach temperature. The saving in the power consumption from the heat exchangers with 1°C minimum approach temperature is 8.83% lower than that of the heat exchangers with 3°C minimum approach temperature.

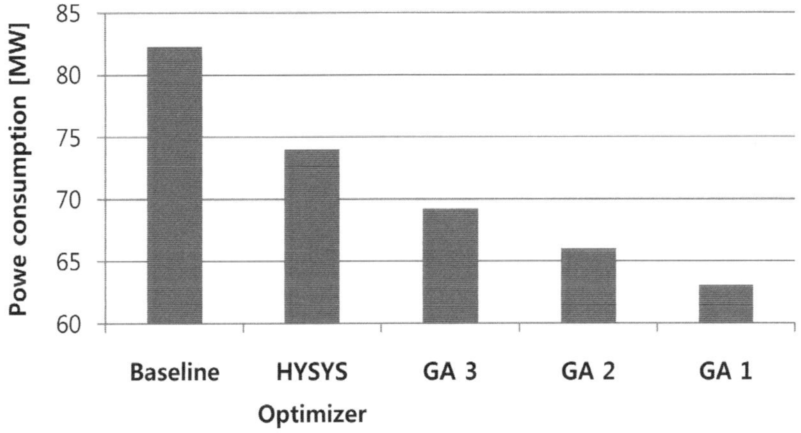

Figure 6 Comparison of power consumptions of baseline and optimization results

Summary

In this study, the SMR process installed on the LNG FPSO was optimized by connection Aspen HYSYS® with GA in MATLAB Global Optimization Toolbox. The result shows the power consumption of optimized process can be reduced by 15.9% when compared with the baseline case. It is derived that there is room for lessening the power consumption additionally by adjusting minimum approach temperature. In the future, the existing SMR process will be consistently developed and optimized by considering more realistic constraints.

References

1. Shukri, T. (2004). LNG technology selection. Hydrocarbon Engineering, 9(2).
2. Lee, G. C.; Smith, R.; Zhu, X. X. (2002), "Optimal synthesis of mixed refrigerant systems for low-temperature processes," *Industrial and Engineering Chemistry Research*, 41, pp. 5016–5028.
3. Jensen, J. B. (2008), "Optimal operation of refrigeration cycles," Ph. D. Thesis, Norweglan University of Science and Technology.
4. Chang, H. M., Chung, M. J., Kim, M. J., & Park, S. B. (2009), "Thermodynamic design of methane liquefaction system based on reversed-Brayton cycle," *Cryogenics*, 49(6), pp. 226–234.
5. Meiqian Wang, Jian Zhang, Qiang Wu, Kuyen Li. (2011), "Thermodynamic-Analysis-Based Energy Consumption Minimization for Natural Gas Liquefaction," *Industrial and Engineering Chemistry Research*, 50, pp. 12630-12640.
6. Meiqian Wang, Jian Zhang, Qiang Wu. (2012), "Optimal design and operation of a C3MR refrigeration system for natural gas liquefaction," *Computer and Chemical Engineering*, 39, pp. 84-95.
7. Sanggyu Lee, Nguyen Van Duc Long, Moonyong Lee. (2012) "Design and Optimization of Natural Gas Liquefaction and Recovery Processes for Offshore Floating Liquefied Natural Gas Plants," *Industrial and Engineering Chemistry Research*, 51, pp. 10021-10030.

8. Nogal, F. D., Kim, J. K., Perry, S., & Smith, R. (2008), "Optimal design of mixed refrigerant cycles," *Industrial and Engineering Chemistry Research*, 47, pp. 8724–8740.
9. Alabdulkarem, A., Mortazavi, A., Hwang, Y., Radermacher, R., & Rogers, P. (2011), "Optimization of propane pre-cooled mixed refrigerant LNG plant," *Applied Thermal Engineering*, 31, pp. 1091–1098.
10. Ji-Hyun Hwang, Myung-Il Roh, Kyu-Yeul Lee. (2013) "Determination of the optimal operating conditions of the dual mixed refrigerant cycle for the LNG FPSO topside liquefaction process," *Computer and Chemical Engineering*, 49, pp. 25-36.
11. Douglas, J. M. (1988). Conceptual design of chemical processes. New York, NY, USA: McGraw-Hill.
12. Nexant, Inc. (2007) "LNG: The Expanding Horizons of Liquefaction Technology and Project Execution Strategies."

Constructability Issues of FLNG for Design and Project

Moon, Herbert
Technical Director, Project Development & Technology

Kim, Ingu
Engineering Manager, Project Development & Technology

DAEWOO Shipbuilding and Marine Engineering LTD, Korea

AIChE Spring Meeting, April 2013
13th Topical Conference on Natural Gas Utilization
San Antonio, April 28 – May 2, 2013

Introduction

Constructability issues shall be reviewed and incorporated in the Project Development Plan and Design of the FLNG. This paper addresses important features on the FLNG Constructability, and recommends significant aspects from the Pre FEED to Detail Engineering Stages.

FLNG Projects Status and Brief Comparisons between FLNG and Oil FPSO.

Several FLNG projects are under the investigation and study, and some of them are in construction phases. Overall view of the Current FLNG Projects and comparison between the FLNG and Oil FPSO is presented.

Overall review of the construction schedule and man-hour will be done, and several recommendations will be also described.

DSME's Yard Capacity Overview

The brief summary of DSME's Okpo Construction Facilities is presented, and brief plans for construction and commissioning will be presented.

Also, new facility for FLNG construction, test and commission will be explained.

Trends of Hull Dimensions, Weights of FLNG, and Limitations

The hull of FLNG is bigger and heavier than that of Oil FPSO, and the limitations for Construction will be presented. Also, recommendations will be also identified.

Trends of Topside Module Dimension and Weights and Topside Module Integration Methodology

The Topside Module Dimensions and Weights are much larger and heavier. It is recommended that new facility will be mobilized for efficient topside module integration.

Summary of Major Utility Consumptions and its Impacts on the Hull Main Deck Layouts and Arrangements

Major utility consumptions are reviewed and the aspects of the piping, electrical layouts and arrangements are discussed and presented. Comparisons between FLNG and Oil FPSOs will be made, and several recommendations will be also proposed.

Identification of Long Lead Items and Design and Construction Issues

The delivery durations of major equipments in FLNG are expected to be more than one year, and may be more than two years for some equipment. It is required that most of the equipments shall be delivered minimum 6 months prior to module fabrication.

However, if the equipment delivery is later than construction plans, several measures and mitigations will be presented.

Construction Issues of Different Cargo Containment Systems for FLNG

For FLNG, membrane tanks and SPB are normally applied. For each Cargo Containment System, the features considering the construction plans will be presented

Review of Electrical Installations and Recommendations

Pre-Commissioning, Commissioning Methodology and Recommendations

Two major methodologies for commissioning works will be presented and compared. And, differences in the conceptual and basic design designs will be presented.

Also, several recommendations will be proposed to perform the commission works efficiently.

Methodology of Gas Trial and Near Shore Commissioning

Unlike the LNGCs, the methodology of Gas Trial and near shore commissioning for FLNGs are not finalized yet. Several restrictions on these issues will be presented, and several future plans for maximizing the tests and near commissioning activities will be presented.

Conclusions

Constructability is one of the key considerations for FLNG Project. From the conceptual stage, it must be reviewed and implemented in the design and development plan.

References Cited

Not Applicable

Biography of Speaker

Herbert Moon is a Technical Director and in charge of Technical Proposals for FLNG Projects in DSME. He has been working with DSME for over 26 years and has been involved in many offshore projects. He was graduated from Seoul National University and received BS and MS in Naval Architect.

(57a) Challenges in Syngas Preparation for Different Applications

Vk Arora, Kinetics Process Improvements, Houston, TX

The syngas composition & purity requirements differ significantly for different applications. Depending on the end application of the syngas, the required configurations could be from a simple scheme to a very complex and capital intensive one. As the size of the syngas derivative units are getting much larger in scale- there is all the more the need to minimize the complexity & the cost of producing syngas- which is major contributor in the overall economics.

This presentation will examine the challenges in design and relative economics of various syngas configurations for different applications including world scale Methanol, Ammonia, Oxo-syngas, CO & GTL plants. It will also include key measures in energy savings including co-production of Methanol & Ammonia & their benefits.

(57b) Topsoe Tigas Technology for Conversion of Abundant Low Cost Resources Into High Value Synthetic Fuels

Livia Sierra Llorens, Haldor Topsoe, Houston, TX

In times with increasing oil prices it becomes very attractive to consider alternative sources than crude oil for transportation fuel. The Topsoe Improved Gasoline (TIGAS^{TM}) technology is designed to produce high quality and valuable gasoline from natural gas, shale gas, coal, petcoke or biomass.

The process converts natural gas to gasoline, and the gasoline product properties are such that it can readily be used as a drop-in fuel or used as a perfect blend stock in the gasoline pool. Plant capacities can range from as small as 1,500 bbl/d to above 30,000 bbl/d and can easily be matched to the available gas resources.

A 23 bbl/d TIGAS^{TM} demonstration plant, DOE-funded project, will start-up early 2013. Data and results from the initial operation will be presented.

(57c) Value Additiion by Integrating C3 Derivatives with PDH Units

Vk Arora, Kinetics Process Improvements, Houston, TX

With rapid development of Ethane based Ethylene production in Middle East and switch to lighter feedstock in US crackers have been mainly responsible for shortfall in Propylene and C4s. To fill this gap in Propylene supply, a number of new on-purpose Propylene production units (PDH) are being built in US, China & other parts of the world. With so many PDH units planned in US, an excess of propylene is already being speculated while most of the Propylene is used up in polypropylene units. The projected scenario of excess propylene will impact the long term profitability of propylene producers and will likely promote and justify propane export – which will not be in the best interest of US producers using propane feedstock. To sustain the long term profitability of on-purpose propylene producers, downstream integration with the relevant propylene derivatives is called for. Based on the increasing global market demand of those derivatives with better returns- such integration units are already being planned.

This presentation will examine the relative economics of various propylene derivative combinations-including Acrylic Acid & its derivatives as well as Oxo-Alcohols/aldehydes & its derivatives. Other propylene derivatives (like Acrylonitrile, PO, Cumene/Phenol will be briefly included but not for the economic comparison). A brief relevance of the market will also be included in the presentation.

(57d) Opportunities for Modular Gtl in North America

Jeff McDaniel, Velocys, Inc., Plain City, OH

Modular Gas to Liquids (GTL) plants can bring stranded, associated or undervalued gas to market and reduce the flaring of associated gas. Of particular interest is the opportunity for using small scale distributed GTL plants located near the feedstock resource to monetize shale gas fields in North America. Velocys patented technology for the Fischer-Tropsch process, the central step of GTL, has been commercially demonstrated in the field. The unique combination of microchannel reactors and super-active catalysts provides superior economics, especially in the range 1,000-15,000 barrels per day, or 10-150 mmscfd of gas. The modular process produces clean-burning, premium diesel or jet fuel: valuable products which are easy to transport and serve a global market.

The presentation will include the latest results from an integrated GTL pilot plant that includes a steam methane reformer and microchannel FT reactor. The pilot plant is designed to provide operational data to support a variety of process configurations and conditions that may be found in commercial GTL facilities.

(57e) Compact Heat Exchange Reactors in Pilot Plants for Fischer-Tropsch Synthesis and Mixed Alcohol Production

Hani Gadalla, Steven J. Vallee and Zhijun Jia, Chart Energy & Chemicals, La Crosse, WI

Chart's Compact Heat Exchange Reactor (CHER) is a brazed metal reactor with integrated heat exchange capabilities. Because of its high metal surface area and short distance for the heat to travel, the reactor can control highly exothermic and endothermic reactions that are otherwise hard to control with conventional reactors. CHER can be constructed out of a variety of metals, including aluminum and stainless steel. The choice of material is based on the operational conditions that the reactor will experience. Recent data from two pilot plants will be presented to demonstrate the effectiveness of aluminum and stainless steel CHERs to control the reactions.

The first pilot plant is a 1bpd Waste-to-Energy plant where the CHER is used for the highly exothermic Fischer-Tropsch (FT) reaction. The reactor is constructed out of an aluminum core that is housed in a steel vessel (based on Chart's Core-in-Kettle patented design). The second pilot plant uses a stainless steel CHER for the production of mixed alcohols from syngas. The alcohol production reaction is also highly exothermic and is operated at higher temperatures and pressures than FT.

A comparison between aluminum and stainless steel reactors will be reviewed. Data for the reactions will be presented and the performance of the reactors will be analyzed. Scalability to commercial plants will also be discussed.

Solid State Carbon Sequestration by Utilization of Agro-wastes to Produce Value-added Products

Soh Fong Lim[1], Mohammad Omar Abdullah[1], Andrew Ragai Henry Rigit[2]

[1] Department of Chemical Engineering and Energy Sustainability, University Malaysia Sarawak, 94300 Kota Samarahan, Malaysia

[2] Department of Mechanical and Manufacturing System Engineering, University Malaysia Sarawak, 94300 Kota Samarahan, Malaysia

Abstract

Solid state carbon sequestration via utilization of agro-wastes to produce value added products such as biofertilizer and biosorbent is studied in this paper. The carbon sequestration is a sustainable development process which can be used to curb the negative impacts of human activities by reducing the emission of CO_2 via locking the carbon in solid form. Analytical analysis has found that both of the products contain relatively high percentage of organic carbon compare to other organic material such as soil. These might be due to the locking of carbon in solid form during the application of these products.

Keywords: agro-wastes, biofertilizers, biosorbents, solid state carbon sequestration.

I. INTRODUCTION

Around the globe, waste is piling up at an alarming rate and we need an effective waste disposal schemes very urgently while at the same time making use of the unwanted waste to provide clean energy or new materials via smart and efficient composting or recycling methods. The second important issue is linked to the goal towards global sustainability in terms of reducing greenhouse gas emissions, in particular CO_2 (Abdullah, 2012).

Solid state carbon sequestration is used to reduce carbon footprint and to minimize waste. The carbon footprint is formed by the existence of carbon dioxide (CO_2) gas in the atmosphere. CO_2 which is well known for driving global warming and climate change is the most important greenhouse gas emitted by industries and human activities. In this instance, agricultural activities are increasingly being required to account for its contribution to the release of CO_2. Thus, sustainability in the agricultural activities is important to conserve the natural environment and resources for our future generation. The sustainability should also extend beyond the economic well-being of the current generation and reflects the ability of future generations to meet their needs. These could be achieved by steering the agricultural activities along a more sustainable path using green technology or sustainable technologies.

Carbon sequestration is a sustainable development process that removes carbon footprint. This process can be used to curb the negative impacts of human activities by reducing the emission of CO_2 via locking the carbon in solid form such as in trees, soils, and other value-added solid products (eg. bio-fertilizer, bio-sorbent, etc.). There are numerous agro-wastes being produced in plantation and processing such as skins/peels, cores, pits, leaves, brunches, and pericarp. Agro-wastes are recognized as having hidden economic values which could be benefited by transforming the agro-wastes into useful products such as biofertilizer and biosorbent (Fortuna et al., 2003; El-Haggar, 2007; Kukal et al., 2009). Rivera-Cruz et al. (2008) reported that banana wastes are effective biofertilizer carriers for promoting plant growth and soil sustainability. Sud et al. (2008) reviewed the potential of agricultural waste materials in producing biosorbent for sequestering heavy metal ions from aqueous solutions. These agro-wastes largely compose of hemicelluloses, lignin, extractives, lipids, proteins, simple sugars, water hydrocarbon, starch, and etc. (Sud et al., 2008). Thus, these agro-wastes will be targeted for potential green technology application and sustainable development/practices.

2. MATERIALS AND METHODOLOGY

2.1 Materials

Banana peels (*Musa Acuminata*) that were used in this study were agro-wastes that were produced from local industry.

2.2 Biofertilizer

Composting method was used to produce biofertilizer from banana peels. 500 kg of banana peels were placed in a blender with 200 ml of water. Then, the blended banana peels were exposed to the sunray for 2 days (dehydration process). The dried banana peels were then put into a bin together with 50 g of sawdust, 250 g of lignite, and 25 ml of nipah palm sugar. The mixture was stirred for 5 minutes. After that, the bin was lid tightly and kept for 3 weeks for composting process purposes. Within the 3 weeks of composting process, compost pile was stirred frequently to allow the mixture to be well mixed and reach optimum temperatures as well as not overheat. After 3 weeks of the composting process, all the compost materials were removed from compost bin and the dehydration process was repeated for 4 hours. This is to ensure that the biofertilizer are in dry solid state (Matu, 2010; Muhammad, 2011).

2.3 Biosorbent

The banana peels were washed thoroughly with tap water to remove the adhering dirt and dried in a laboratory oven at 60°C for 58 hours. Then, the dried banana peels were crushed into a powder form using a laboratory grinder. The powdered banana peels were sieved to obtain a uniform size of the crushed banana peel. Finally, the powdered banana peels were thoroughly washed with deionized water to remove physically adsorbed contamination and dried in a laboratory oven at 80°C for a period of 24 hours (Sugumaran, 2010; Meikle, 2011).

3. RESULTS AND DISCUSSION

3.1 Value-added Products

Agro-wastes from banana industry mainly banana skins are used to produce two types of useful products; biosorbent and biofertilizer (Figure 1). These products are fabricated using simple and cost-effective methods. Both of the products have shown significant results in their applications. The biosorbent is able to remove crystal violet dye from aqueous solution. The maximum uptake of the crystal voilet dye is 245.74 mg/g. Removal of the dye is very fast, as it reached equilibrium in 3 hours of contact time. It is also found that the removal of the dye is highly pH dependent. Within the tested pH range (pH 2 to 12), the maximum removal of the dye is between pH 6 and pH 8. On the other hand, the biofertilizer which was fabricated using solid state fermentation method is able to increase the yield of crop which was proven by the plantation of the mustard and water spinach vegetables.

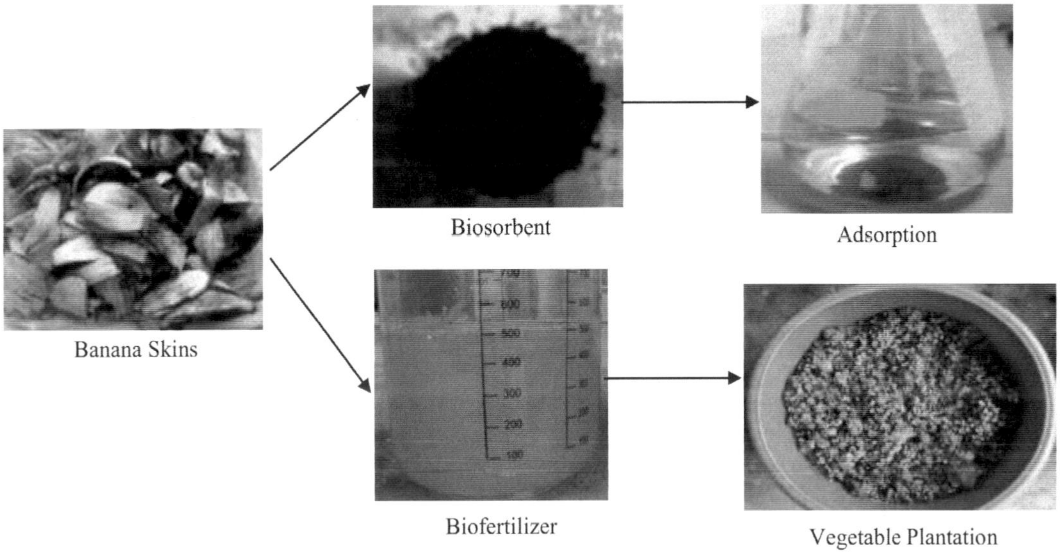

Figure 1 Flow diagram for value-added products from banana skins

3.2 Carbon analysis

Samples from the biosorbent and biofertilizer were analyzed to determine the amount of carbon content using Walkley-Black method (Saarani, 2011). Walkley-Black method is chosen based on its simplicity and rapidity (Islam and Weil, 1998; Schumacher, 2002; Krishan *et al.*, 2009). Walkley-Black method is wet combustion method where the amount of carbon is determined by the reaction of $Cr_2O_7^{2-}$ and H_2SO_4 without external heating. In this method, the amount of organic carbon content in the sample is determined from the difference of amount $FeSO_4$ used in the blank titration with the amount of $FeSO_4$ used for reaction with the remaining un-reacted part of $K_2Cr_2O_7$. The chemical reaction that involved in Walkley-Black methods is shown in equation (1).

$$2\, Cr_2O_7^{2-} + 16H^+ + 3\, C \rightarrow 4\, Cr^{3+} + 8\, H_2O + 3\, CO_2 \qquad (1)$$

Desired amount of sample (W g) was placed in the 250mL Erlenmeyer flask. 5.0mL of 1.0*N* $K_2Cr_2O_7$ solution is added into the flask that containing the sample and the flask is swirled to mix the sample with $K_2Cr_2O_7$ solution. Then, 10.0mL of concentrated H_2SO_4 is added rapidly to the mixture and followed by adding 100mL of deionized water. The suspension is filtered out and ferroin indicator solution is added into the flask containing the solution. Then the solution is titrated with 0.5N $FeSO_4$ from 50 mL burette. As the endpoint approach, the solution takes on green cast and change into dark blue green. Then additional 0.2 mL of indicator is used to sharpen the endpoint. At this point, the $FeSO_4$ solution is added drop by drop until the colour change from blue green into brownish (maroon in reflected light) and the amount of ferrous sulfate solution used is recorded (T mL). The experiment is run again with blank and the amount of FeSO4 solution used is recorded as (B mL). The amount of

$FeSO_4$ solution in the back titration which is obtained from the analytical procedure is then be used to calculate the amount of carbon in the sample in percentage using equation (2).

Where,

B	=	the amount of $FeSO_4$ solution used to titrate blank
T	=	the amount of $FeSO_4$ solution used to titrate sample
N	=	the normality of $FeSO_4$ solution
W	=	weight of the sample in mg
3	=	equivalent weight of carbon in mg.
y	=	oxidation factor

Carbon analysis is performed on both of the value-added products. Table 1 demonstrates the percentage of carbon in biofertilizer using Walkley-Black analytical method. Based on Table 1, the amount of carbon in the biofertilizer is approximately 30.09 – 35.10 wt%.

Table 1 Amount of $FeSO_4$ used in back titration and percentage of carbon in biofertilizers

SAMPLE	Total amount of $FeSO_4$ used in back titration (mL)	Percentage of carbon (wt%)	
		y = 1.14	y = 1.33
Biofertilizer 1	5.60	31.16	36.35
Biofertilizer 2	2.58	29.01	33.85
Average		30.09	35.10

Figure 2 shows an increase of carbon content in biosorbent at different crystal voilet (CV) dye concentration. For both oxidation factors, the amount of carbon in the biosorbent was significantly increased as the dye concentration increased until it reached saturated state which is at CV dye concentration of 120 mg/L and above. This saturated state shows that the biosorbent is not able to absorb the CV dye anymore. Thus, the adsorbed carbon which is in CV dye structure remains constant when the adsorptions reach saturated state (Andersen *et al.*, 2009). Manocha *et al.* (2011) also stated that the formation of sample sizes and the interconnectivity between adsorbate and adsorbents can affect the rate of adsorption capacity. The weight of the biosorbents also can affect the percentage of absorption of the dye where as the weight of biosorbents increased, the absorption rate also increased (Raymundo, 2010).

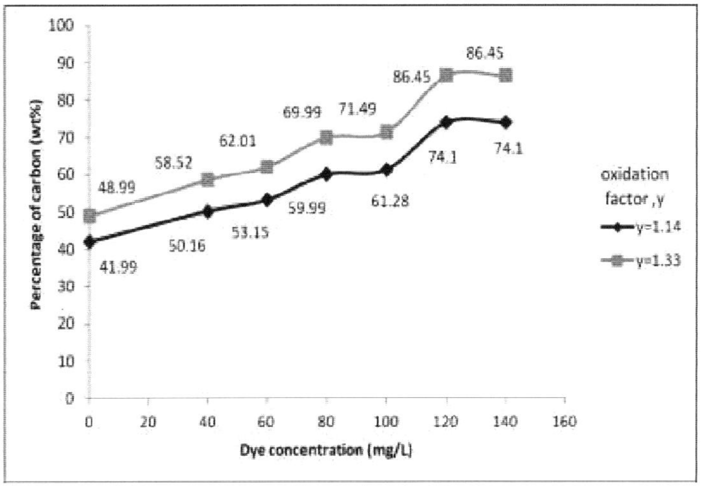

Figure 2 Percentage of carbon versus dye concentration in (mg/L)

Thus, analytical analysis has found that both of the useful products contain relatively high percentage of organic carbon compare to other organic material such as soil (Kim, 2005). Thus, the carbon is locked in solid form which could reduce the emission of CO2. This process is called carbon sequestration which is a sustainable development process to curb the negative impacts of human activities.

ACKNOWLEDGEMENT

The authors would like to thank Universiti Malaysia Sarawak for the use of laboratory facilities and equipments. This project is funded by Fundamental Research Grant Scheme (FRGS/02(17)/738/2010(24)) from the Ministry of Higher Education, Malaysia.

REFERENCES

A. Fortuna, R. Harwood, K. Kizilkaya, E. A. Paul, "Optimizing nutrient availability and potential carbon sequestration in an agroecosystem." *Soil Biology and Biochemistry*, Vol. 35, 1005-1013, 2003.

A. S. Raymundo, R. Zanarotto, M. Belisário, and M. de Godoi, "Evaluation of Sugar-Cane Bagasse as Bioadsorbent in the Textile Wastewater Treatment Contaminated with Carcinogenic Congo Red Dye," *Brazilian Archives of Biology and Technology* ,vol. 53, no.4, pp. 931-938, 2010.

B. A. Schumacher, *Methods For Determination of Total Organic Carbon (TOC) in Soils and Sediments* , pp. 1-18, 2002.

D. Sud, G. Mahajan, and M. P. Kaur, "Agricultural Waste Material as potential adsorbent for sequestering heavy metal ions from aquenous solution – A review." *Bioresource Technology*, vol. 99, no. 14, pp. 6017-6027, 2008.

F. S. Saarani, "Elucidation of Carbon Content via Laboratory Analysis," Bachelor of Mechanical and Manufacturing Engineering Thesis, Department of Mechanical and Manufacturing Engineering, Faculty of Engineering, Universiti Malaysia Sarawak, 2011.

G. Krishan, S. K. Srivastav, S. Kumar, S. K. Saha, and V. K. Dadhwa, "Quantifying the underestimation of soil organic carbon by the Walkley and Black technique – examples from Himalayan and Central Indian soils," *Current Science* , vol. 96, 2009.

H.T. Kim, *Soil Sampling, Preparation, and Analysis* (2nd Edition ed.). Boca Raton: Taylor & Francis Group, 2005.

K. R. Islam and R. R. Weil, *A Rapid Microwave Digestion Method for Colometric Measurement of Soil Organic Carbon*, 1998, pp. 2269-2284.

M. C. Rivera-Cruz, A. T. Narcía, G. C. Ballona, J. Kohler, F. Caravaca and A. Roldán, "Poultry manure and banana waste are effective biofertilizer carriers for promoting plant growth and soil sustainability in banana crops." *Soil Biology and Biochemistry*, Vol. 40, pp. 3092-3095, 2008.

M.O. Abdullah, *Applied Energy: an introduction.* New York: CRC Press (Taylor & Francis Group), 2012.

N. Muhammad, "Evaluation of Bio-Fertilizer from Agricultural Waste," Bachelor of Mechanical and Manufacturing Engineering Thesis, Department of Mechanical and Manufacturing Engineering, Faculty of Engineering, Universiti Malaysia Sarawak, 2011.

S. Manocha, J. Bhagat, M. Patel, N. Patel, and L.M. Manocha, Adsorption Behaviour of Carbons from Bio-Mass, pp. 727-731, 2011.

S. M. El-Haggar, "Sustainable Industrial Design and Waste Management." Academic Press, Chapter 7, pp223-260, 2007.

S. U. Matu, "Utilization of Agro-waste to Produce Biofertilizer," Bachelor of Mechanical and Manufacturing Engineering Thesis, Department of Mechanical and Manufacturing Engineering, Faculty of Engineering, Universiti Malaysia Sarawak, 2010.

S. S. Kukal, Rehana-Rasool, D. K. Benbi, "Soil organic carbon sequestration in relation to organic and inorganic fertilization in rice–wheat and maize–wheat systems." *Soil and Tillage Research*, Vol. 102, pp. 87-92, 2009.

T. A. Sugumaran, "Utilization of Agro Waste to Produce Biofertilizer," Bachelor of Mechanical and Manufacturing Engineering Thesis, Department of Mechanical and Manufacturing Engineering, Faculty of Engineering, Universiti Malaysia Sarawak, 2010.

V. Meikle, "Removal of Contaminants Using Biosorbent," Bachelor of Mechanical and Manufacturing Engineering Thesis, Department of Mechanical and Manufacturing Engineering, Faculty of Engineering, Universiti Malaysia Sarawak, 2011.

W. C. Andersen, S. B. Turnipseed, C. M. Karbiwnyk, R. H. Lee, S. B. Clark, W. D. Rowe, M. R. Madson, K. E. Miller, Quantitative and Confirmatory Analyses of Crystal Violet (Gentian Violet) and Brilliant Green in Fish. *U.S. Food and Drug Administration, Animal Drugs Research Center,* 2009.

(105b) Ni-CaO-ZrO2 Nanocomposite Catalyst for CO2 Reforming of CH4

Qingjun Chen[1], Jun Zhang[1], **Tiejun Zhao**[1] and Yuhan Sun[2], (1)Low Carbon Conversion Center, Shanghai Advanced Research Institute, CAS, Shanghai, China, (2)State Key Laboratory of Coal Conversion, Institute of Coal Chemistry, Chinese Academy of Sciences, Taiyuan, China

1. **1. Introduction**

Recently, utilization of natural gas to fuels and chemicals receives more attention due to its abundance over the world and less CO_2 emission. The production of syngas by CO_2 reforming of methane is one of the attractive routes to reach this goal. Compared with conventional steam reforming, CO_2 reforming reaction tends to easier coking formation and needs extra 20% heat to obtain similar conversion. Nickel-based catalysts have shown an excellent behavior in this reaction with a comparable activity to noble metal catalysts [1]. However, coke formation and sintering of nickel particles at a high temperature condition are often observed on Ni catalysts [2]. Developing a high-temperature stable catalyst is highly interesting for this reaction by controlled synthesis of catalyst. In this contribution, a stable Ni-CaO-ZrO_2 nano-composite catalyst for CO_2 reforming of CH_4 is prepared by a coprecipitation method. The effect of preparation condition is carefully investigated. The relationships between catalyst structures and activity of nanocomposite catalysts are discussed.

1. **2. Experimental**

Ni-CaO-ZrO_2 catalysts were prepared by a coprecipitation method with a reflux-digestion process. After precipitation, the obtained slurry was washed and refluxed in de-ionized water, then filtered, dried and calcined at 600 °C for 5 h. The catalysts prepared at different reflux-digestion condition were assigned as NC-X-Y, where X and Y are the reflux temperature and time, respectively. The catalyst without reflux was assigned as NC-0. The reforming reaction was conducted at 850-900 °C in a quartz fixed-bed reactor at atmospheric pressure. Obtained catalysts are tested by feeding the reaction mixture of CO_2 and CH_4 (molar ratio 1.2: 1) without dilution and a gas hourly space velocity (GHSV) of 79 000 ml/(hg) . In addition, comparison of calciantion procedure and coking quantity was performed at different operating conditions.

1. **3. Results and discussion**

3.1 **Physical and chemical properties of the Ni-CaO-ZrO_2 catalysts**

The pore structure parameters are collected in Table 1. The catalyst without reflux-digestration has the smallest specific surface area (63 m^2/g) and pore volume (0.09 cm^3/g). With the increase of the reflux time, the specific surface area and pore volume of the catalysts increased. NC-100°C-24h exhibits the largest surface area of 245 m^2/g. Reflux temperatures also have a significant effect on the pore structures of the catalysts. At the same reflux time, the specific surface areas of the catalysts increased with increasing the reflux temperature.

Table 1. Pore structure parameters of the catalysts

Samples	S_{BET} m^2/g	V_t cm^3/g	D_m nm

NC-0	63	0.09	5.58
NC-100°C-6h	152	0.18	4.73
NC-100°C-12h	176	0.21	4.85
NC-100°C-24h	245	0.30	4.95
NC-60 °C-24h	92	0.13	5.83
NC-80 °C-24h	160	0.21	5.31

The XRD patterns of the catalysts prepared at different reflux-digestion condition are shown in Figure 1. The catalyst without reflux has strong diffusion peaks of ZrO_2 and NiO. As the increase of the reflux time and temperature, diffusion peaks assigned to ZrO_2 and NiO became weaker and wider. This result indicates that reflux digestion improves the dispersion of NiO and decreases the size of ZrO_2 crystals in the catalysts.

TPR profiles of Ni-CaO-ZrO_2 catalysts are illustrated in Figure 2. The peak or shoulder at about 400 °C is attributed to the reduction of large NiO particles, while the peaks at 600 °C and 700 °C may belong to the reduction of NiO that have the weak and strong interactions with ZrO_2, respectively. For the catalyst without reflux, NiO mostly has the weak interaction with the carrier or is in the form of bulk particles. At the reflux temperature of 100 °C, the interaction between NiO and ZrO_2 became stronger with the increase of reflux times (Figure 2 a). However, when the reflux temperature is lower than 100 °C, the interaction between NiO and carrier was not significantly enhanced (Figure 2 b). So the high reflux temperature and long reflux time are key factors to prepare the catalysts with strong interaction between metal and the carrier.

Figure 1. XRD patterns of the catalysts Figure 2. TPR profiles of the catalysts

3.2 **Activity of the Ni-CaO-ZrO_2 catalysts for CO_2 reforming of methane**

Results of the catalytic tests at 850 °C with a high GHSV (CO_2/CH_4=1.2:1) of 79, 000 ml/(h·g) are shown in Figure 3(a). The catalyst without reflux has a low activity with the CH_4 conversion of about 70 %. In addition, CH_4 conversion of this catalyst decreased with time and only 55% (conversion) is achieved after reaction on stream for 150 h. A higher conversion of CH_4 and stability (very close to equilibrium) is observed on the catalyst refluxing at 100°C for 24 h. Therefore, prolonging the reflux time will increase activity of Ni-CaO-ZrO_2 catalysts for CO_2 reforming of methane. The catalyst with 24 h reflux-digestion (NC-100 °C-24h) exhibits a highly stable activity over 1000 h without obvious deactivation at a GHSV of 79, 000 ml/(hg) at850 °C (Data are not shown) .

NC100-24 h

Figure 3. Catalytic performance of catalysts for CO_2 reforming of methane

(a) Effect of reflux temperature and time; (b) Effect of calcination temperature; (c) Effect of calcination procedure

Furthermore, the calcination temperature effect on the catalyst activity is carried out at similar operating conditions. It is found that the activity decreases when the sample is calcined at static air at a higher temperature. The NC-100-24h sample calcined at 900°C presents a very low activity (Fig.3b) . XRD examination indicates that monoclinic ZrO_2 is gradually formed and particle size of ZrO_2 becomes larger. TPR measurement indicates no obvious peaks for NiO reduction for the sample calcined at 900°C. These observations indicate that the sintering occurs at high temperatures. However, if the catalyst is directly reduced in the mixtures of H_2 and N_2 at 850°C or 900°C for 3 h, the conversion of CH_4 still can maintain a high CH_4 conversion (Fig. 3c), implying that the calcination procedure has a profound impact on the activity of catalyst. Interestingly, the catalyst performing at 900°C gives a lower carbon deposition than 850 and 800°C after time on stream for 200 h. The full operating window will be given by a more systematic study afterwards.

1. 4. Conclusions

A high-temperature stable Ni-CaO-ZrO_2catalyst is developed by the coprecipitation method with a reflux-digestion process. Reflux digestion can improve the dispersion of NiO and enhance the strong interaction between metal and support. The obtained Ni-CaO-ZrO_2 catalyst can be further improved by modifying the calciantion procedure.

References

[1] J. R. Rostrup-Nielsen, J. H. B. Hansen, ***Journal of Catalysis*** 144 (1998).

[2] C. J.Liu, J. Y. He, J. J. Jiang, Y. X.Pan, ***ChemCatChem*** 3 (2011)

(105c) Gas Cleanup Using Elastic Layered Metal-Organic Framework Adsorbents

Francisco Sotomayor and **Christian M. Lastoskie**, Civil & Environmental Engineering, University of Michigan, Ann Arbor, MI

The suitability of elastic layered metal-organic frameworks (ELM) for the removal of carbon dioxide from methane-rich gas mixtures was evaluated using molecular modeling and experimental measurements for two adsorbents, ELM-11 [$Cu(BF_4)_2(bpy)_2$] and ELM-12 [$Cu(OTf)_2(bpy)_2$]. These materials are promising absorbents for CO_2 separation from biogas and natural gas mixtures, due to their reversible gated isotherm features and their high selectivity adsorption of CO_2 over CH_4 (S = 50-100 at 298 K) from gas mixtures with CH_4 and CO_2 compositions representative of biogas / natural gas extraction streams.

Grand canonical Monte Carlo (GCMC) simulation was used to calculate the framework adsorption capacity and CO_2 selectivity from binary CH_4 / CO_2 mixtures. Simulated CO_2 capacities above the gating pressure and CO_2 isosteric heats of adsorption agree well with ELM-11 and ELM-12 experimental data. Ideal adsorbed solution theory (IAST) estimates of CO_2 selectivity are generally in good agreement with the GCMC simulations, although modifications of the IAST methodology are needed in order to take into account the co-adsorption condition for CH_4 uptake by CO_2 gate-opening. Density functional theory (DFT) calculations are also presented for CO_2 and CH_4 interactions with the ELM-11 and ELM-12 framework structures. Upon the adsorption of CO_2 in the ELM framework, the metal-anion bond of the framework was extended, whereas the same bond length is virtually unchanged upon CH_4 adsorption. DFT analysis of the adsorption binding energies of other trace components in natural gas and biogas production (e.g. H_2S, H_2O, N_2) and CH_4 will also be presented and discussed.

(105d) Effect of Preparation Parameters On Mixed Amine Functionalized Silica Sorbents for CO2 Capture

Adefemi Egbebi[1], Sonia Hammache[1] and McMahan L. Gray[2], (1)URS-NETL, Pittsburgh, PA, (2)U.S. Department of Energy, National Energy Technology Laboratory, Pittsburgh, PA

Effect of preparation parameters on mixed amine functionalized silica sorbents for CO_2 capture

Adefemi Egbebi[1, 2], Sonia Hammache[1, 2], and McMahan L. Gray[1]

[1]US DOE-National Energy Technology Laboratory, Pittsburgh, PA

[2]URS, Pittsburgh, PA

Global climate change has been linked to growing atmospheric concentrations of carbon dioxide. Therefore the development of practical methods to remove, reduce or prevent the release of anthropogenic CO_2 is prudent to slow this increase. The most mature technology for the separation of CO_2 from flue gas streams from power plants is via liquid amine scrubbing which is considered expensive and energy-intensive. The potential therefore exists for the development of alternative CO_2 removal technologies, including the use of solid amine-based sorbents among others, capable of less energy-intensive CO_2separation.

Previous work in our laboratory investigated co-impregnated mixed amine (Polyethyleneimine (PEI) + 3-aminopropyltriethoxysilane (APTES) / Silica) solid sorbents with the goal of improving sorbent performance in terms of its CO_2 uptake capacity, thermal and hydrolytic stability as well as the kinetics during sorption and regeneration cycles. Results showed improved CO_2capture capacity and stable performance of the mixed amine sorbent relative to a single amine (PEI) sorbent.

Current work investigates the effects of synthesis parameters/variables such as bath temperature, solvent type and impregnation sequence on sorbent performance. Modified Pressure Swing Adsorption (PSA) tests were conducted with a thermogravimetric analyzer (TGA) on two sorbents with same composition of 15 wt. % PEI: 15% APTES: 70 % SiO_2 but with different solvents used during their syntheses - methanol for one and methanol-hexane mixture for the other. Adsorption was performed using dry 10% CO_2 /N_2 feed gas and desorption was conducted with 100% N_2 feed gas while temperature was kept the same at 100°C. Preliminary results show negligible differences between the sorbents in terms of CO_2uptake capacity and thermal stability.

The effect of various preparation parameters will be elucidated from the results of TGA, fixed bed reactor and other characterization experiments on sorbents and will be presented.

CO_2 Reduction Efforts for Canadian Oil Sands Industry

Ravishankara Subraya, Oscar Aguilar

The Quest project is a joint venture of the Athabasca Oils Sands Project partners (Shell Canada, Chevron Canada Limited, and Marathon Oil Canada Corporation) to fully integrate CCS technology to capture, transport, inject, store, and monitor CO_2 from the Scotford Upgrader in the Industrial Heartland Region near Fort Saskatchewan, Alberta. CCS is an important part of both Canada's and Alberta's CO_2 reduction plans. Project is currently under execute phase. This facility will sequester 1.08 Mtpa of CO_2 on completion in Q1 2015. The technical presentation will cover technology selection, process & utility integration challenges and key learning from the project.

(81a) Simulation and Optimization of Tubular Fixed-Bed Fischer-Tropsch Synthesis Reactor with Recycle

Bo Liao Sr.[1], Luhaibo Zhao[1], Xiaoquan Wang[2], Xiaohao Liu[1] and Yuhan Sun[3], (1)Low Carbon Conversion Center, Shanghai Advanced Research Institute, Shanghai, China, (2)Low Carbon Conversion Center, Shanghai Advanced Research Institute,Chinese Academy of Sciences, Shanghai, China, (3)State Key Laboratory of Coal Conversion, Institute of Coal Chemistry, Chinese Academy of Sciences, Taiyuan, China

Theincreasing demand of world's energy and the fast depletion of liquid fossil fuel reserves have made natural gas, coal and biomass increasingly attractive as feedstocks for synthetic liquid fuels. For the last two decades, Fischer–Tropsch synthesis (FTS) has gained renewed interest for the conversion of syngas to liquid hydrocarbons.

In industrial FT reactors, tubular fixed-bed reactor with recycle is widely used. Although many fixed-bed models for FT process have been reported in literature, research on concentration/temperature distributions in fixed-bed reactor with recycle is very limited. In this work, a 2D heterogeneous fixed-bed reactor model with recycle is developed and the simulation results are validated by using the industrial data. The effects of operating conditions on concentration/temperature distributions in FT reactor, such as inlet temperature, pressure, recycle ratio and feed syngas composition, are investigated. The results show that inert content in feed syngas and recycle gas, such as CH_4, N_2, etc., have significant effect on the conversion of CO and temperature distributions in fixed-bed FT reactor. The increase in recycle ratio results in the decrease in the CO conversion for inert gas rich syngas, and the inert gas content in syngas plays an important role to determine the hotspot location and temperature, which can be easily applied to biomass conversion process. Finally, the model is used for multiobjective optimization to maximize CO conversion and minimize the operation cost for fixed-bed FT process with recycle.

(81b) Fischer-Tropsch Cobalt Catalyst Improvements with the Presence of

Jennifer L. Klettlinger, Combustion Branch, NASA Glenn Research Center, Cleveland, OH

The objective of this study was to evaluate the effect of titanium oxide, lanthanum oxide, and zirconium oxide on alumina supported cobalt catalysts. The hypothesis was that the presence of lanthanum oxide, titanium oxide, and zirconium oxide would reduce the interaction between cobalt and the alumina support. This was of interest because an optimized weakened interaction could lead to the most advantageous cobalt dispersion, particle size, and reducibility. The presence of these oxides on the support were investigated using a wide range of characterization techniques such as SEM, nitrogen adsorption, x-ray diffraction (XRD), temperature programmed reduction (TPR), temperature programmed reduction after reduction (TPR-AR), and hydrogen chemisorptions/pulse reoxidation. Results indicated that both La2O3 and TiO2 doped supports facilitated the reduction of cobalt oxide species in reference to pure alumina supported cobalt catalysts, however further investigation is needed to determine the effect of ZrO2 on the reduction profile. Results showed an increased corrected cluster size for all three doped supported catalysts in comparison to their reference catalysts. The increase in reduction and an increase in the cluster size led to the conclusion that the support-metal interaction weakened by the addition of TiO2 and La2O3. It is also likely that the interaction decreased upon presence of ZrO2 on the alumina, but iii
further research is necessary. Preliminary results have indicated that the alumina- supported catalysts with titanium oxide and lanthanum oxide present are of interest because of the weakened cobalt support interaction. These catalysts showed an increased extent of reduction, therefore more metallic cobalt is present on the support. However, whether or not there is more cobalt available to participate in the Fischer-Tropsch synthesis reaction (cobalt surface atoms) depends also on the cluster size. On one hand, increasing cluster size alone tends to decrease the active site density; on the other hand, by increasing the size of the cobalt clusters, there is less likelihood of forming oxidized cobalt complexes (cobalt aluminate) during Fischer-Tropsch synthesis. Thus, from the standpoint of stability, improving the extent of reduction while increasing the particle size slightly may be beneficial for maintaining the sites, even if there is a slight decrease in overall initial active site density.

(81d) Enhanced Fischer-Tropsch Catalysts for Conversion of Biomass and Coal to Transportation Fuels

Gokhan Alptekin, Ambalavanan Jayaraman, Douwe Bruinsma, Ewa Muteba and Alex Wickham, TDA Research Inc., Wheat Ridge, CO

The United States needs viable alternatives to crude oil to moderate the effect of oil price hikes and provide an interim bridge until other fuel sources can commercially supplant petroleum-based fuels. The candidate resources that can be used to produce liquid fuels include biomass, oil sands, oil shale and coal. Coal is by far the largest and least expensive resource, with over 250 billion tons of known domestic reserves. Co-gasification of coal and biomass will reduce the carbon footprint of the CBTL processes while Fischer-Tropsch (FT) synthesis is the most economically viable option to produce transportation fuels from syngas. Although the FT process is run on a commercial scale in both South Africa and Malaysia, several challenges face the production of liquid fuels from coal, including the attrition resistance of the catalyst and downstream hydrocarbon separation. The FT process could also benefit from a highly selective catalyst to produce desired range of hydrocarbon products.

TDA Research, Inc. (TDA) is developing a unique catalytic material to address three technical challenges: FT activity, catalyst selectivity and better heat management. TDA's FT catalyst will produce the desired (gasoline or diesel) range hydrocarbons (rather than producing heavy wax and then cracking it into lower molecular weight hydrocarbons). A key advantage of our approach is the reduction in the number of processing steps and process severity in the conversion of FT products into the respective transportations fuels (gasoline or diesel). In a conventional FT process, without cracker processing of the product stream, only 16% of the plant output falls into the range of a gasoline hydrocarbon cut. Even with further processing, this production rate only increases to 26% of the plant output, with a loss of 9% of the available production during the downstream processing. TDA's novel size-selective FT catalyst has achieved 68% selectivity in the naphtha range in preliminary bench scale testing experiments. Naphtha is used in gasoline blending and this cut represents the highest value product that can be produced from the FT process. The results from our FT catalyst development will be presented in the meeting.

(81e) Aqueous-Phase Fischer-Tropsch Synthesis by Using Ruthenium Nanoparticle Catalyst

Venkat Ramana Rao Pendyala, Wilson D. Shafer and Burtron H. Davis, Center for Applied Energy Research, University of Kentucky, Lexington, KY

Fischer-Tropsch synthesis has long been known as an alternative process to produce long-chain hydrocarbons for their use in transportation fuels. In the Fischer–Tropsch reaction, synthesis gas produced from natural gas reforming or gasification of coal or biomass is transformed in the presence of cobalt, ruthenium, and iron catalysts. The product distribution is highly dependent on the process conditions and the type of catalyst used. Commercially FTS is performed in the gas or slurry phase. Recent studies demonstrated that FTS can be performed in aqueous phase too [1, 2]. The present work aims to investigate the effect of reaction temperature and reaction time on the performance of ruthenium nanoparticle catalyst during aqueous-phase Fischer-Tropsch synthesis by using a stirred tank batch reactor.

World Scale Boil-Off Gas Reliquefaction

Heinz C. Bauer, Linde Engineering, Pullach, Germany
Hans Mattsson, Siemens, Finspång, Sweden
Sven-Erik Brink, Siemens, Duisburg, Germany

Abstract

Boil-off gas reliquefaction has become an industry standard for many LNG carriers, which no longer use gas as fuel for steam boilers, but tap others energy sources to generate power for propulsion. Well known examples are the ships of the Q-Flex and Q-Max class. Most of these vessels use simple nitrogen expander cycles with relatively high energy consumption (typically 800 kWh/t) and a moderate capacity of about 200 t/d LNG liquefaction.

Increasing LNG demand in Asia, especially in Japan after the nuclear catastrophe triggered innovative ideas to increase the LNG output of existing liquefaction plants. Boil-off gases and ship return gases from LNG tank farms have been identified quickly as promising source for additional LNG, as these gases are perfectly suitable for cryogenic processing. Obviously, dehydration and sweetening is not required. However, low pressure, low temperature and elevated nitrogen content pose challenges to an efficient design.

Linde Engineering has developed an innovative concept for a world scale boil-off gas reliquefaction unit with 0.3 to 1.0 mtpa capacity, which includes the following features:

- Boil-off gas compression starting from cryogenic conditions
- Mixed refrigerant cycle with two heat exchanger bundles only
- Load balancing between feed gas and refrigerant cycle compressors
- Nitrogen rejection into the gas turbine fuel gas with a double flash process
- Use of identical gas turbines as mechanical drive

Introduction

Large LNG export terminals include an LNG tank farm with large LNG tanks providing storage capacity of several hundred thousand meters cubed. Low pressure gases at tank pressure (about 100 to 200 mbarg) originate from several sources like

- Boil-off gas caused by heat ingress through the tank walls and LNG pumps
- Flash gas generated in the run-down line between liquefaction and storage
- Ship return gas / displacement gas during ship loading
- Flash gas during cool-down of warm send-out systems

In total the overall LP gas (hereinafter called boil-off gas or BOG) flow rate may accumulate to about 1,000 t/d for a 10 mtpa LNG facility (this figure can be scaled almost linearly for larger plant capacities). The quality of the BOG may raise concerns as it will be enriched with nitrogen and may contain contaminants from ship return gases. The distance between the LNG liquefaction trains and the LNG storage may add up to a few kilometers depending on the site conditions. Thus, returning the BOG to the liquefaction trains and merging it with the end-flash gas may not be the optimum solution for every site.

The decision to install a dedicated BOG reliquefaction system to the entire LNG facility will depend very much on the specific site conditions including the fuel gas balance, the capacity of the existing refrigeration systems, availability of natural gas as feed stock, distance between liquefaction trains and jetty/tank area, and flare load especially during peak BOG generation as it happens for example during ship loading. Under all circumstances, however, BOG reliquefaction will generate a high value product starting from very favorable feed stock conditions, as the absence of water, sour gas components and heavy hydrocarbons in BOG simplifies cryogenic processing significantly.

Concept Selection

A BOG reliquefaction unit with a capacity between 0.3 and 1.0 mtpa LNG can be compared with mid-scale LNG plants. All land based LNG plants in this capacity range are using a single mixed refrigerant (SMR) cycle. This choice is based on its good efficiency (typically 350 kWh/t[1]), minimum count of rotating equipment and good safety records. Considerations to use nitrogen expander cycles for floating LNG concepts are not applicable for a land based BOG reliquefaction facility, mostly as compact design and readily available make-up supply are not likewise important.

Even if there is some discussion about the use of brazed aluminum plate-fin heat exchangers (PFHE) for SMR based mid-scale LNG plants, experienced operators prefer coil wound heat exchangers (CWHE) in this service, as especially load changes during ship loading will cause extra stress to the heat exchangers. All experience shows that CWHEs are more tolerant against frequent load cycles than PFHEs. The marginal cost adder to go with CWHEs is more than offset by the peace of mind, if downtime caused by heat exchanger repair is taken into account.

BOG reliquefaction requires large compression systems for the refrigeration cycle and also for BOG compression, as liquefaction under low pressure would be very inefficient. A typical SMR process with a specific energy consumption of 350 kWh/t needs about 15 MW for a feed gas flow rate of 1,000 t/d under a pressure of at least 50 bar. Feed gas compression starting from tank pressure to at least 50 bar requires almost the same power. For compressor trains with more than 10 MW each gas turbines have been selected as driver in the past, if fuel gas is available under economically attractive conditions. Using electric motors instead of gas turbines seems to be preferable only, if and when electric power can be sourced reliably at low prices.

Process Optimization

World scale BOG reliquefaction units have not been built so far. Thus, there is no proven template for a cost effective and robust process design. It is fair, however, to start with building blocks as they are known from mid-scale or world scale LNG plants.

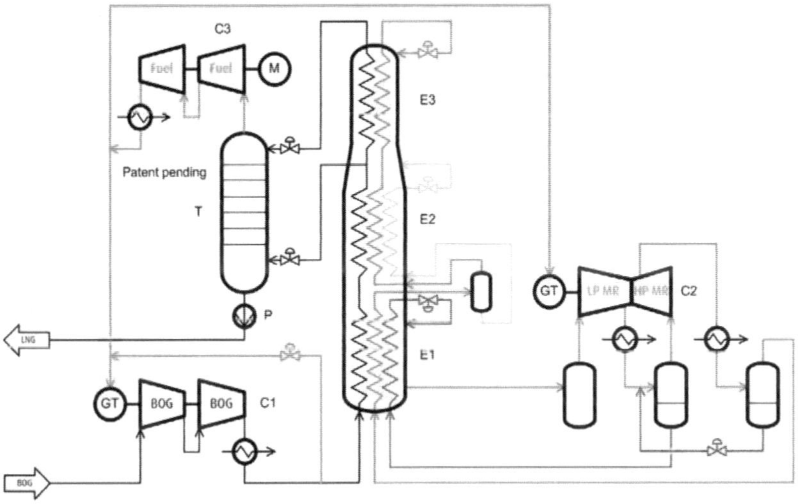

Fig. 1 Base concept with N2 stripper

Fig. 1 shows a conventional design for a BOG reliquefaction system. BOG from the LNG storage tanks has to be compressed first. The low molecular weight of the BOG and the high pressure ratio require two compressor casings for the single shaft, centrifugal compressor C1. In the absence of a need for sour gas removal and dehydration the compressed BOG will be sent directly to the cryogenic heat exchanger, which consists of a pre-cooling bundle E1, a liquefaction bundle E2, and a sub-cooling bundle E3. In most of the cases the nitrogen content of BOG will exceed the standard specification for LNG (<1 vol%) so that nitrogen rejection is required.

At elevated nitrogen levels in the BOG a nitrogen stripper T operated at low pressure will be the right solution for this purpose. An elegant manner to provide strip gas[2] for the stripper is the use of 'warm' LNG after the liquefier E2, which will generate a sufficient quantity of gas, when it is flashed into the column. On-spec LNG will be sent by pump P via run-down lines to the storage tanks. The nitrogen rich stripper overhead gas has to be recompressed in compressor C3 to the required fuel gas pressure of the selected gas turbines. The chosen single mixed refrigerant cycle is the LIMUM® process, which has been developed and patented[3] by Linde Engineering and is successfully in operation since many years in different locations. It is built around a single casing, barrel type compressor C2 with two compression stages.

This conventional design was scrutinized with value engineering methods with the following findings:

- Shaft power of C2 is about 150% of shaft power of C1, identical duties are desirable
- LP suction conditions of C3 result in an expensive multi casing design
- Stable operation of stripper T under frequent load changes is hard to achieve

The result of the subsequent process optimization is shown in Fig. 2. The MRC compressor C2 is unloaded by shifting part of its duty to the BOG compressor C1. This can be achieved by replacing the sub-cooling refrigeration duty of the MRC, which was used in E3, by an open methane cycle. Instead of sub-cooling LNG to such a low temperature that only the desired fuel gas stream is generated, a larger methane rich flash gas rate is deliberately accepted, which is recycled to the suction side of the BOG compressor C1. Thus, a very good match between the shaft power of C1 and C2 could be achieved[4].

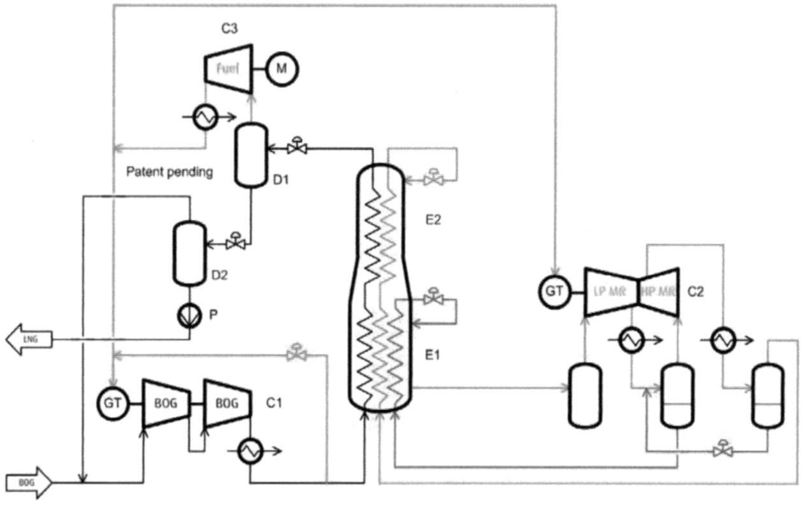

Fig. 2 Final concept with double flash system

In a second step the nitrogen stripper T was replaced by a medium pressure flash drum D1 and a low pressure flash drum D2. The operating pressure of D1 can be selected such that the fuel gas can be compressed in a single casing, single stage compressor C3, which can be operated optionally at ambient temperature suction conditions, as the elevated suction pressure allows for a warm-up of the fuel gas before it is compressed.

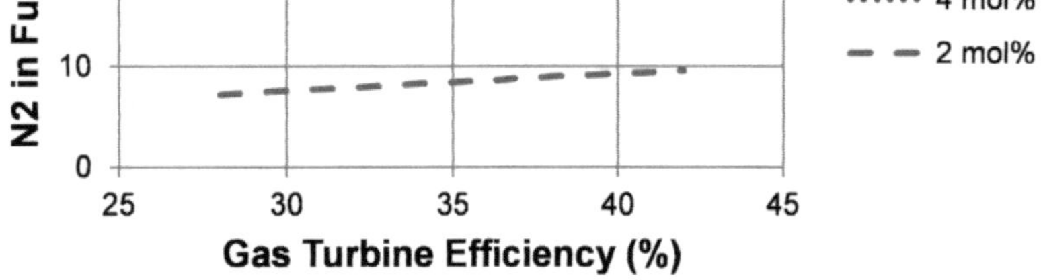

Fig. 3 Correlation between GT efficiency and fuel gas quality

In total the value engineering exercise helped to eliminate the sub-cooling bundle E3 and the low temperature LP casing of the fuel gas compressor C3. The stripper column T was replaced by two simple flash drums D1 and D2 and the main compressors C1 and C2 can be hooked up to drivers of identical rating.

After a successful streamlining of the process the suitable equipment had to be selected. One very complex decision was the selection of the best fitting gas turbine. In a situation, where the BOG is worked up into only two products, LNG and fuel gas, obviously most of the nitrogen has to go with the fuel gas, as LNG can accept only a very small amount (max. 1 vol% N2). As a result highly efficient gas turbines with a low heat rate have to be operable with higher nitrogen content in the fuel gas and vice versa. This correlation is plotted in Fig. 3 for different BOG compositions.

Low BTU gas turbines

The combination of DLE (Dry Low Emission) requirements with a low heating value of the fuel gas requires certain measures to guarantee the flame stability. Fig. 4 shows a suitable concept as it is used e.g. in the Siemens industrial gas turbine family SGT-600/700/800 being equipped with eighteen 3G fuel nozzles, also called burners, in an annular combustor. The burners use pilot gas, which is fed at the burner outlet and is burned around the main flame for flame stabilization. Typical for DLE combustion, which is used by Siemens for more than 20 years, fuel is burned with twice the amount of air in comparison to stoichiometric combustion. The large amount of air keeps the flame temperature low reducing emissions of nitrous oxides, but on the other hand increasing the risk of combustion instabilities and incomplete combustion leading to emissions of CO.

At load levels exceeding 50 % most of the gas is burned in the main flame, which has a low ignition temperature producing low concentrations of CO and NO_x. At lower power output, more gas is burned in the pilot flames, which operate at a higher temperature and thus create more NO_x. At all levels of power output the amount of pilot flame gas is limited to the amount needed to ensure stable operation of the turbine.

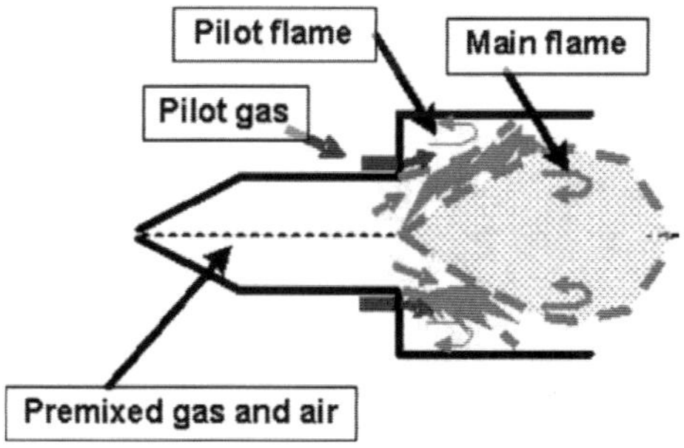

Fig. 4 3G burner schematic

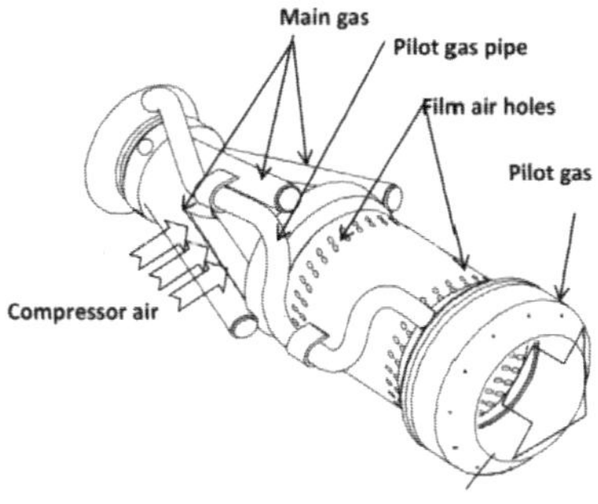

Fig. 5 3G DLE burner details

The unique fuel nozzles (see Fig. 5) of the Siemens industrial gas turbines have been proven to be very flexible in terms of the heating value of the fuel, respectively the Wobbe index. A high N_2 content in the fuel acts just as ballast since N_2 does not contribute to the combustion. Results from test beds and from actual operation in LNG plants[5] show that up to 40% N_2 in fuel gas can be accepted without changes in the standard configuration. This feature comes with the full capability to accept substantial changes in Wobbe index. However the starting gas needs to be correlated with the control system for safe start-up, but once in operation a very reliable operation can be achieved with this system.

BOG compressor

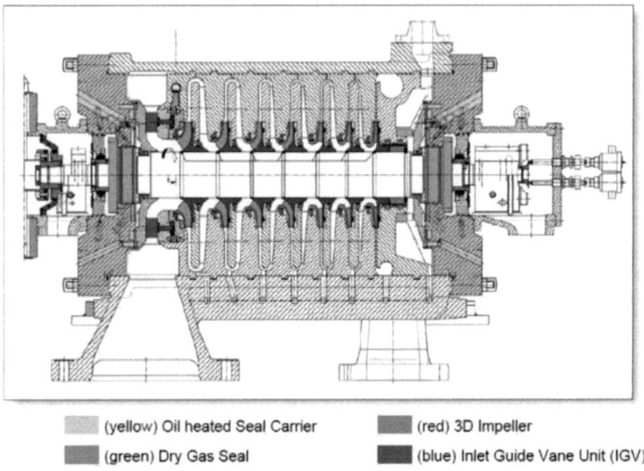

Fig. 6 Turbo compressor cross sectional drawing

Among all turbo compressor applications the cryogenic BOG compressor service is one of the most demanding applications. Due to the nature of the BOG service the compressor trains are periodically exposed to large fluctuations regarding volume flow and suction temperatures. The proprietary Siemens design features (see Fig. 6) include variable Inlet Guide Vane (IGV) control (blue) in front of the impeller (red), protected dry gas seals (green) in a heated and cooled seal carrier (yellow), extensive expertise in material selection, compressor design, manufacturing and testing.

The IGV unit (Fig. 7) is an aerodynamic device and induces a vortex onto the flow increasing the compressor head respectively adjusting the angle of attack (incidence) to the impeller blading. The IGV concept improves the turbo compressor performance control and increases the operating range (turndown). The IGVs provide a map of individual performance curves within the IGV setting range, where the volume/head adjustment is quick and smooth. Main advantages of IGV control for BOG applications are

- Direct Online (DOL) start-up possible (start-up without any prior static cool-down, therefore no gas needs to be flared)
- Compact design, less space required, less weight, minimum equipment count, less spare parts, minimized footprint
- The IGV concept for cryogenic BOG compressor applications is well backed-up by references
- Largest turndown (highest process flexibility)
- Lowest maintenance

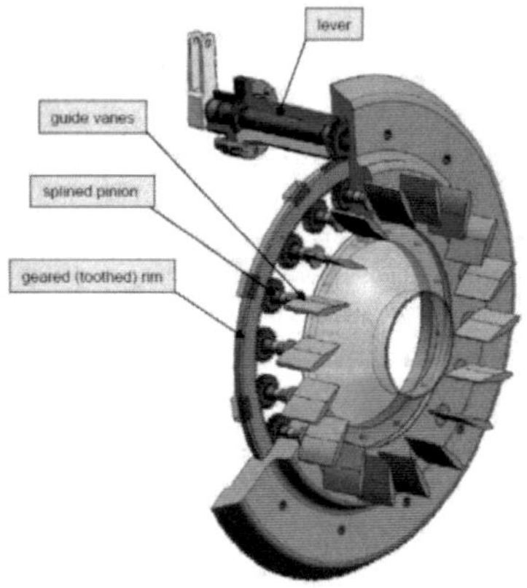

Fig. 7 variable IGV unit

This design was first executed in 1972 for the first single shaft compressor train supplied for BOG service (Brunei) and in 1992 the first time also for cryogenic (-160 degree C) temperatures (Qatar). Since then, this design has become industrial standard and today all new BOG single shaft centrifugal compressor trains have being equipped with this advanced technology.

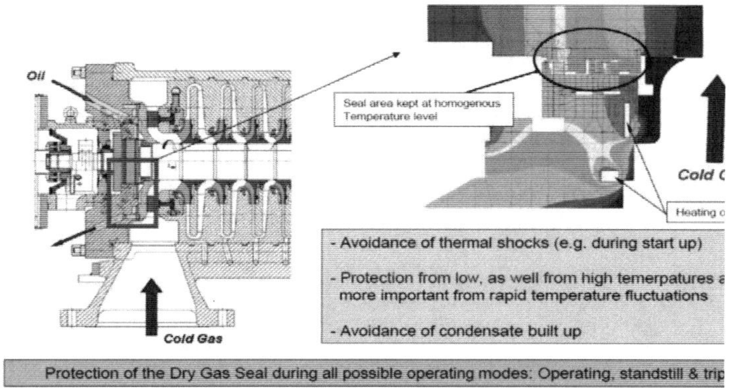

Fig. 8 heated/cooled dry gas seal carrier

Due to the nature of the BOG service the turbo compressor train is periodically exposed to very low suction temperatures (even down to -170 degree C) as well as to ambient and even high discharge temperatures during start up, compressor run in, or performance test. In order to eliminate or minimize thermal tension/stress for the dry gas seals caused by the extremely low suction temperature and periodically occurring temperature fluctuations (e.g. start/stop) resulting out of the different operation modes it is essential to keep the seal area at an almost constant temperature level during all possible operating modes. The required constant temperature level is achieved by providing oil heated/cooled seal gas carriers (see Fig. 8, the oil heated/cooled seal gas carrier is marked in yellow colour on the left hand side drawing).

The heating oil is split from the regular lube oil flow and the fluid is circulating around the dry gas seal cartridge in seal-welded channels inside the compressor seal carrier (casing cover) forming a thermal shield. This design was first executed in 1991 for the first single shaft compressor train supplied with dry gas seals for cryogenic temperature BOG service. Since that time all new single shaft BOG centrifugal compressors provided with dry gas seals are equipped with oil heated/cooled seal carriers (in 2013 total 83 compressor casings) ensuring highest reliability (safety) and start-up availability, long-life service periods and lowest operation, inspection and maintenance cost.

Conclusions

World-scale BOG reliquefaction has become reality. Linde Engineering has successfully developed a new product for the LNG industry. Siemens contributed with innovative compressor and gas turbine designs, which are proven in service and are highly respected worldwide. The first plant, which is based on this novel concept is under execution for Malaysia LNG and is scheduled for start-up in less than two years[6].

References

---

[1] Tariq Shukri, Michael Barclay, Single mixed refrigerant process has appeal for growing offshore market, LNG Journal, June 2007, pp.35-37

[2] Patent pending

[3] e.g. US patent 6,334,334

[4] Patent pending

[5] Development of Medium-sized Gas Turbines for LNG Applications, Gudmundsson B., Larfeldt J. , Troger C., Cox G. 16th International Conference & Exhibition on Liquefied Natural Gas, 18-21 April 2010, Oran, Algeria

[6] http://www.the-linde-group.com/en/news_and_media/press_releases/news_121126.html

(89b) Tuning and Optimization of Compressor Controls Using Dynamic Simulation

Sanjay Ganjam and Mohit Thakur, Process Engineering, Bechtel OG&C, Inc, Houston, TX

Compressor systems are critical and integral part of an LNG plant for safe, productive and cost effective operations. Robust control design and strategies are essential for stable operation and for protection of machines during abnormal operations. With advancements in dynamic modeling and its use in designing, testing and troubleshooting of plant operations it is important to apply it in designing and testing of control strategies of compressor systems.

In this paper we present and discuss a case study based on a high fidelity dynamic simulation model integrated with CCC emulator for centrifugal compressors. Integration of dynamic model with CCC controls eliminates emulated controllers in the model and brings the actual controller functionality. The integrated dynamic model augments the capability and accuracy of results in dynamic simulations.

Placement of SCL is done by balancing two different objectives – placed to the right of Surge Line for better controllability and as close to Surge Line as possible for production. Location of the SCL is usually determined at site during startup. Purpose of this work is to demonstrate how to achieve this at the EPC phase of the project using dynamic simulation model.

A high fidelity dynamic model gives a good representation of the plant leading to pre-determination of Surge Control Line (SCL), adequacy of SCL for compressor control, pre-tuning of the compressor controllers, check the adequacy of the limit controls.

Opportunity of compressor control tuning at site is less due to risk and complication of compressor surge. Therefore the final controls are usually sub-optimal. Main benefit of performing this activity at EPC stage of the project is to achieve optimal compressor control with minimal compressor surge tests at site during startup.

AIChE 2013 Spring Conference

Topical 6: 13th Topical Conference on Gas Utilization – LNG Simulation and Control

Considerations in Designing a Closed Loop Heating Medium System

Benjamin Gross
Process Engineering
Bechtel Oil, Gas & Chemicals, Inc.

Debby Sielegar, P.E.
Process Engineering
Bechtel Oil, Gas & Chemicals, Inc.

Abstract

A closed loop heating medium system using synthetic heat transfer fluids, such as hot oil, has become a desirable alternative to the conventional boiler feed water / steam in the oil, gas and petrochemical industry. Whether it is for onshore or offshore applications, such systems are reliable, low-maintenance and can be implemented at relatively low cost.

Closed loop heating medium systems can be designed using various configurations depending on the process heat demand, availability of waste heat recovery, desired process control schematic and system reliability, and other operational preferences. In an LNG plant, several factors can affect the required process heat input requirement, such as the feed gas sources and quality, operating conditions, and operating objectives. Waste heat recovery improves the overall plant thermal efficiency by utilizing the waste heat from the exhaust of refrigeration gas turbines or power generation units. While waste heat recovery reduces fuel consumption, increases LNG production and environmental emission benefits, and therefore the NPV of the plant, other factors such as plot space, local environmental regulation differences, and operational preferences may lead to other simple alternative closed loop heating medium system utilizing conventional fired heater.

These various system configurations along with equipment / mechanical design, process control and reliability, and relief system design challenges for a closed loop heating medium system integrated with or without waste heat recovery are discussed in this paper.

1. Introduction

At the dawn of macro-scale natural gas liquefaction, it was not uncommon to utilize direct fired heating to meet the plant's heat requirement. However, process and equipment technology has evolved, design capabilities have improved, and environmental regulations regarding emissions have become more stringent and concerned. For all of these reasons, along with our desire to maximize overall plant thermal efficiency, and minimize capital and operating costs, the methods by which heating demands are met in an LNG plant have become increasingly important.

The logical next step was to implement a robust heating system that could satisfy both the needs of a more efficient process and stricter environmental regulations. A closed-loop heating medium system integrated with a fired heater and/or a Waste Heat Recovery Unit (WHRU) installed from the exhaust of refrigeration gas turbines or power generation units is often chosen to cater to the needs of increasingly efficient processes. Such systems are stable and flexible, and thus respond favorably to process upsets and turndown scenarios.

So we turned to the conventional boiler feed water / steam system, a widely and commonly used system for all industrial applications. A typical boiler feed water / steam system integrated with a waste heat recovery unit is shown in Figure 1 below.

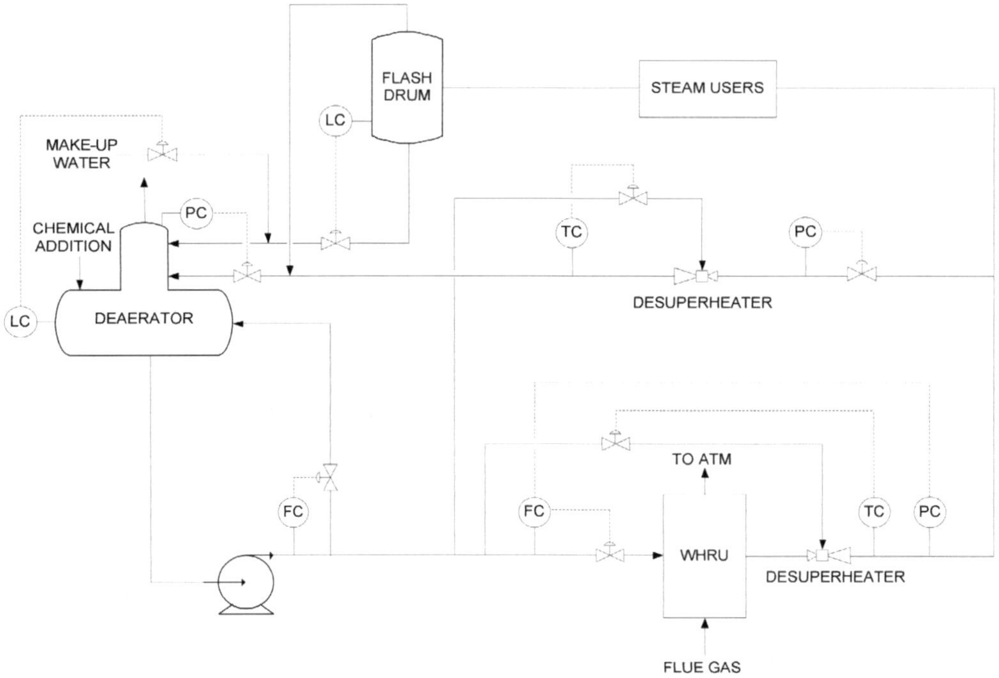

Figure 1. Closed Loop Boiler Feed Water / Steam System Integrated with Waste Heat Recovery Unit

Boiler feed water from a deaerator enters the Waste Heat Recovery Unit (WHRU) coils to generate MP steam. To maintain MP steam supply temperature and pressure, temperature control utilizing a desuperheater and pressure control regulating incoming boiler feed water flowrate are normally applied. LP steam from MP steam let-down and condensate return from MP steam users are routed back to the deaerator. The LP steam is used to strip the condensate return to prevent build-up of impurities.

As a heat transfer medium, steam is ideal. It has a relatively high heat transfer coefficient, unlimited fluid life, and is economical and readily available. However, the boiler feed water / steam system is anything but ideal. The excerpt below from "Waste Heat Recovery Considerations in LNG Plants" by Wiles, D., Kwan, S., and Rajkumar, V. [2] summarizes the disadvantages of such systems best:

> *"The preparation and production of steam requires significant monitoring and treatment of the boiler feed water to remove minerals which could cause deposition and therefore fouling in heat exchangers and pipes. In addition, steam systems with cogeneration require piping and equipment rated for high pressures. In addition, there are high maintenance costs for the removal of scale, the repair of steam traps and leaks."*

The industry has since answered the call for an alternative heating medium, such as hot oil, a synthetic hydrocarbon-based heat transfer fluid, lube-oil based heat transfer fluids, glycol solution, and many others. This paper primarily focuses on hot oil system.

While hot oil has a lower heat transfer coefficient and is more hazardous than steam, and is subject to thermal degradation, it has several very favorable properties: low vapor pressure, high boiling point, low pour point, and is non-compressible. Furthermore, a hot oil system operates at a relatively low pressure and runs at relatively lower operating and maintenance costs. And because hot oil system does not rely on latent heat transfer by condensation, such a system is efficient and reliable with proper design and temperature applications; thus, hot oil is becoming a more desirable heating medium in the oil, gas, and petrochemicals industry.

Factors to be considered in designing a heating medium system may include system configuration selection, equipment / mechanical design, process control and reliability, and relief system design. These factors with focus on LNG plant application are to be discussed in this paper. In an LNG plant, typical hot oil users may include feed gas preheater, amine regenerator reboiler, fuel gas heater, heavy removal reboiler(s), and many others.

2. System Configurations

There are various closed loop hot oil system configurations. Each system configuration is designed to fit the plant requirement and conditions, such as process heat demand, availability of waste heat recovery, desired process control schematic and system reliability, and other operational preferences. Each of these system configurations are outlined as the following:

2. 1. Closed Loop Hot Oil System Integrated with Fired Heating

Fired heating is inherently unsafe, and with regard to fuel consumption and high system design temperatures, it can be expensive given the number of services in an LNG plant (i.e. feed gas heater, molecular sieve dehydrator regeneration gas, amine regenerator reboiler, fuel gas heater, fractionation reboilers, etc.). However, a hot oil system still requires a primary source of heat, and a fired heater is a suitable source. Figure 2 below shows a closed-loop hot oil system integrated with fired heating.

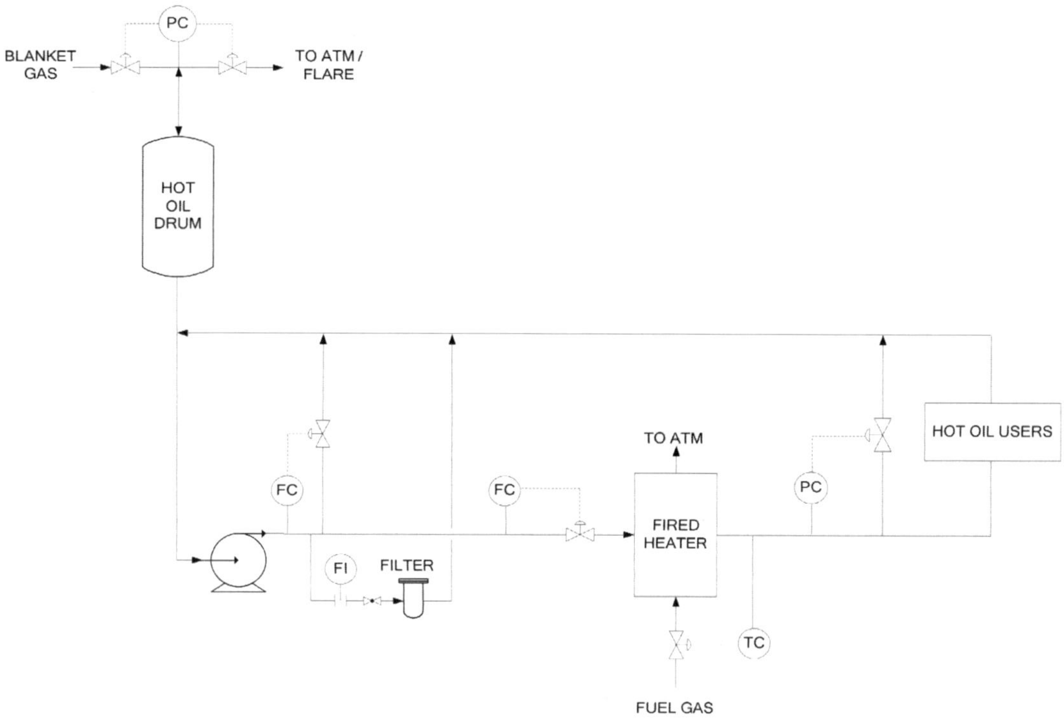

Figure 2. Closed Loop Hot Oil System Integrated with Fired Heating

If it looks like a closed-loop hot oil system is an oversimplified boiler feed water / steam system, that's exactly what it is. The advantageous physical and chemical properties of hot oil – no chemical treatment and no phase changes – allow for this simplicity.

The hot oil supply temperature is directly controlled by regulating the fuel gas flow rate to the fired heater. Note that fuel oil can also be used as a source of ignition for the fired heater. The bypass pressure control is used to provide system balance. If multiple fired heaters are installed in parallel, hydraulics must be symmetrical, and the hot oil flow rate through each unit must be individually controlled to ensure even distribution of the heating medium. The flow control shall have a minimum set point to protect heaters.

Fired heaters typically come equipped with O_2 and combustibles (i.e. CO, H_2) analyzers, which are used to monitor the completeness of combustion taking place inside the heater.

A hot oil system shall include a filter, which is implemented to remove contaminants and/or scales that result from fluid thermal degradation. Normally, these filters are designed to remove 98% of scales greater than 5-10 microns. There are two types of filter installations, which are in-line and side stream. In-line filters are used only with positive displacement pumps, while side stream filters are used with centrifugal pumps. Considering hot oil recirculation rate is high, side-stream filters are normally used. Flowrate through the side-stream filter is typically in the range of 3 – 10% of the recirculation design flowrate. [3]

Also, a hot oil drum is normally located at the highest point in the system. It serves as a main venting point of the system and provides adequate volume for thermal expansion. This vessel must be blanketed by a *dry* inert gas such as nitrogen, fuel gas, or natural gas in order to prevent the fluid from coming into contact with oxygen or moisture. Oxygen tends to accelerate heat transfer fluid degradation, while moisture ingress may lead to sudden pressure surges upon vaporization of the water. Therefore, if nitrogen is used, it is imperative that oxygen and moisture either be removed or reduced to a negligible concentration. The hot oil drum can be a horizontal or a vertical vessel. Hot oil return line can be routed back to the drum or to the suction of the hot oil pumps. Also, a tank, make-up unit, or sump drum is usually provided as part of the hot oil system for provision of filling at least one heat transfer service during pre-commissioning after maintenance shutdown.

2. 2. Closed Loop Hot Oil System Integrated with Waste Heat Recovery and Fired Heating

Whether taken from the front-end or the back-end, fuel gas consumption is an additional operating cost, and a hot oil system integrated with fired heating does not necessarily consume less fuel than an LNG facility that uses fired heating only. So how can we reduce fuel gas consumption, thereby increasing overall plant thermal efficiency? The answer is waste heat recovery.

Waste heat recovery is not a new concept. It has been used in steel mills, power plants, and refineries for decades. In an LNG plant, heat from refrigeration gas turbine and/or power generation exhaust gas can be recovered via coils in the turbine exhaust stacks that service hot oil and even regeneration gas. Figure 3 below shows a hot oil system integrated with both fired heating and waste heat recovery.

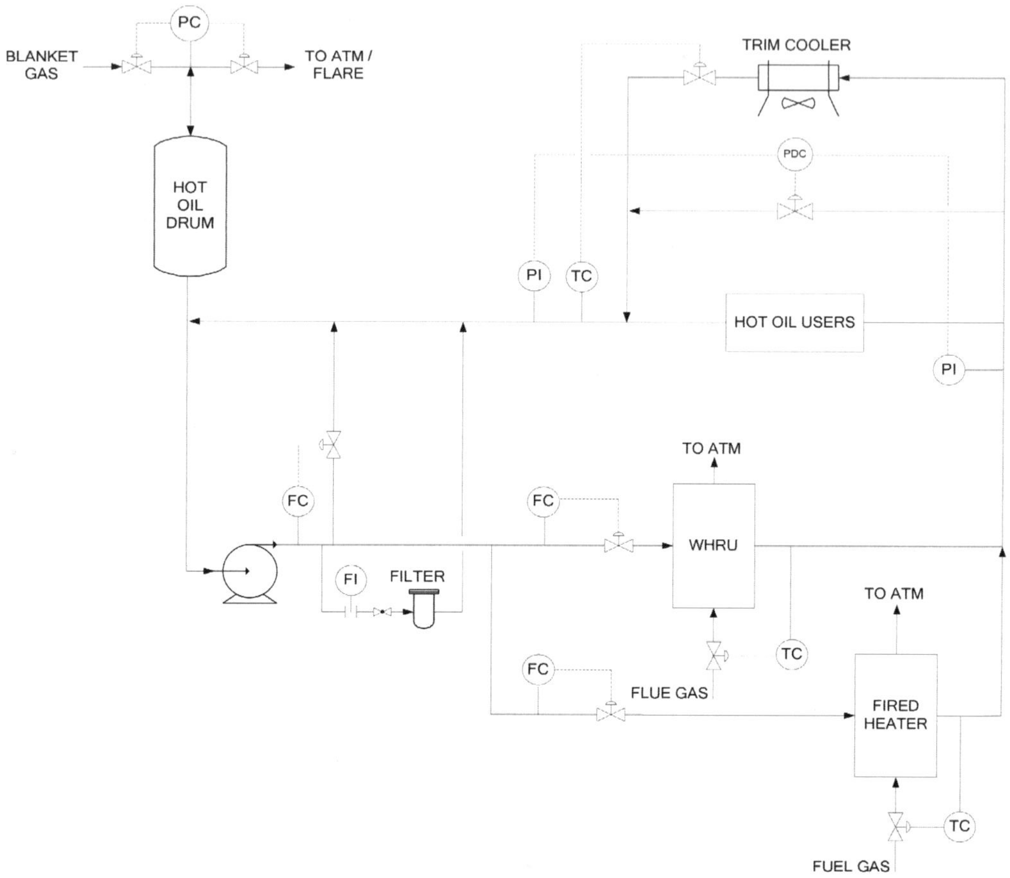

Figure 3. Closed-loop Hot Oil System Integrated with Waste Heat Recovery and Fired Heating

In this scheme, fired heating is typically only required for regeneration gas and during start-up and stand-by, when refrigeration compressors are offline. However, fired heating can be used to supplement the process heat demand if there is insufficient net recoverable heat.

The Waste Heat Recovery Unit (WHRU) process outlet temperature is normally controlled by regulating the flue gas flow rate between the main and bypass exhaust stacks via a damper valve. As damper valves are not as reliable as regular control valves, a certain hot oil bypass flow around the waste heat recovery unit may also be established under temperature control to provide flexibility and robustness in controlling the hot oil supply temperature, considering variability in the gas turbine and power generation turbine exhaust. Again, if the system requires that multiple WHRUs be installed in parallel, each unit must be hydraulically balanced.

Another important consideration in WHRU design is acid gas condensation, which may lead to corrosion in the exhaust system. The WHRU vendor shall consider the stack gas dew point when designing the coils taking into account all operating cases and apply a temperature approach well above the dew point for the worst case. Nevertheless, it is typical to monitor stack gas temperature in the waste heat recovery unit. For environmental regulation reasons, exhaust sample ports are provided on both the WHRU stack and bypass stack, from which exhaust gas can be tested.

In addition, turndown requirements and/or process upsets are a key consideration. The other major difference between this scheme and the one depicted in Figure 2 – aside from the heat source obviously – is the addition of a trim cooler, which helps to maintain a constant hot oil return temperature, especially during turndown. Furthermore, variance in hot oil supply and return pressures associated with fluctuation in process heat users and with turndown are managed by using pressure differential control. Finally, the trim cooler requires a minimum continuous hot oil flow rate in order to permit rapid ramp-up without risk of thermal stress damage. Note that air cooler fans may be turned off completely or turned down to achieve the desired minimum flow, when required.

2. 3. Closed Loop Hot Oil System Integrated with Waste Heat Recovery Unit (WHRU)

A hot oil system integrated with waste heat recovery only is the final stage in the evolution of this closed-loop heating medium system. Figure 4 below shows a closed-loop hot oil system integrated with waste heat recovery only.

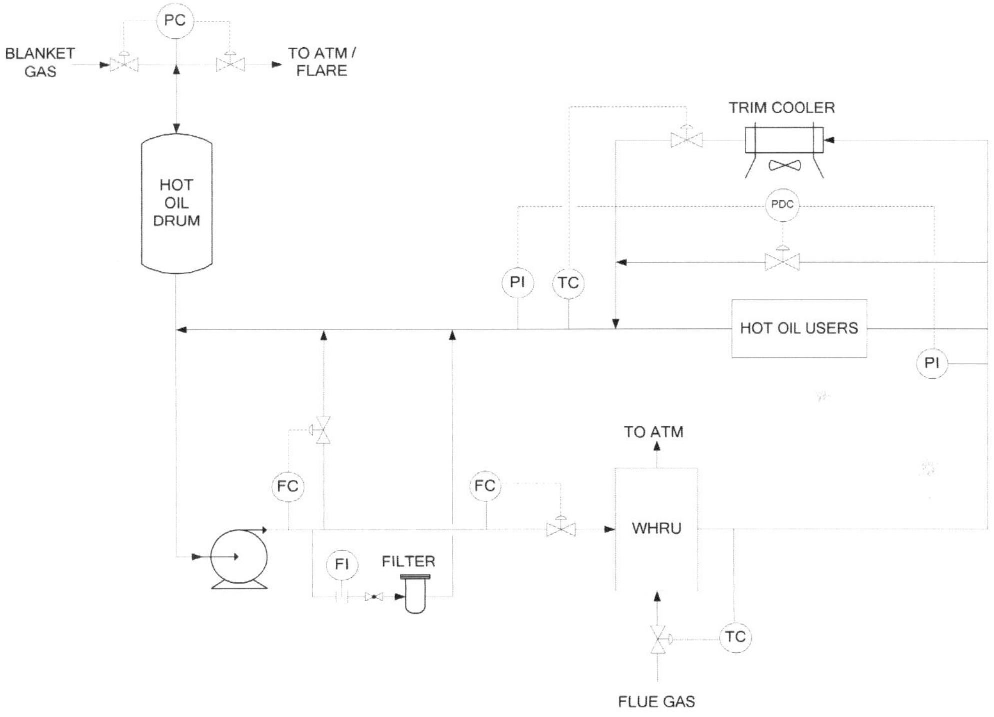

Figure 4. Closed-loop Hot Oil System Integrated with Waste Heat Recovery

The differences between this schematic and the one illustrated in Figure 3 – aside from the absence of a start-up fired heater – are not obvious. In both cases the waste heat recovery unit supplies 100% of the process heat requirement *during normal operation*, and the control strategy is generally the same. But removal of the start-up heater has introduced a new dynamic: If there are no power generation units, how do we heat the heating medium that is required for acid gas removal that must take place prior to liquefaction, which utilizes gas turbines whose exhaust heats the heating medium? It is important to consider the start-up procedure when applying this particular process scheme.

Inclusion of the regeneration gas coil in the hot oil WHRU has been considered. Regeneration gas heating is an intermittent service, and as such, it will introduce fluctuation in the WHRU duty requirement. But with proper mechanical design and coil module arrangement, this strategy further improves plant overall thermal efficiency and reduces installation costs associated with multiple waste heat ducts.

It is also important to evaluate the process heat demand for various design and rating cases, to ensure that there is sufficient recoverable heat and appropriate design approaches. Factors impacting the net recoverable heat include turbine engine degradation, fouling, erosion, radiation heat loss, and gas leakage.

2. 4. Direct Heating

When a heater is only required by a single unit operation, direct heating may be applied instead of a closed loop heating medium system due to its simplicity. For example, assuming only the amine regenerator reboiler requires heat input and there is no source of waste heat (i.e. all refrigeration compressors are electric driven), direct heating is often chosen as a preferred heating method for the reboiler. In this case, closed loop heating medium system is no longer required and a fired heater can be installed. Another example is molecular sieve dehydration regeneration gas heater application. It has been a common practice that the regeneration gas is directly heated via a separate coil in the waste heat recovery unit or via an independent fired heater, as heating medium supply temperature is typically lower than the required regeneration gas temperature.

3. Waste Heat Recovery Unit versus Fired Heating

Waste heat recovery might seem like the optimal choice when considering the reduction in environmental emissions and *potential* improvement in plant overall thermal efficiency; however, additional consideration and evaluation are needed during conceptual design phase. For instance, withdrawing fuel gas from the back-end can "...provide a purge for unwanted inert gases, such as nitrogen." [2] Nitrogen can be present in the feed gas at various concentrations. It accumulates by entering the refrigeration circuit via flash and boil-off gas and thus "...consume[s] unnecessary compression power, reducing overall plant thermal efficiency." [2] LNG product and fuel gas specifications may require that a Nitrogen Removal Unit (NRU) be installed. Also, if there are no or limited exhaust sources – all rotating equipment is electrically powered and all electricity comes from the grid – then waste heat recovery would not be an option.

Also, the extensive ductwork associated with the waste heat recovery unit along with the additional equipment (i.e. trim cooler) and piping require an higher installation cost and more plot space (this is an especially important consideration during a retrofit). Finally, longer exhaust ducts means more head loss, which means increased turbine heat generation and therefore less power available. [2]

Fired heating may be simpler from a controllability standpoint, but that controllability comes with increased environmental emissions and loss of feed and/or product. As discussed above, fired heating control scheme does not require constant return temperature to the hot oil drum, thus trim cooler is not required. Fired heater's fuel gas control valve is more robust in controlling supply temperature to the hot oil users than waste heat recovery unit's damper valve.

4. Relief System Design Challenges

The relief scenarios to consider when designing a closed-loop hot oil system are for the most part standard (i.e. pump shut-off, abnormal heat input, fire, etc.) with one exception: exchanger tube rupture. A tube rupture that causes leakage of hot oil to the process side may lead to loss of hot oil inventory. On the other hand, if an exchanger is servicing a high pressure vapor (i.e. feed gas or high pressure fuel gas) on the tube-side, and a tube ruptures, the result, which is analogous to water hammer, can be catastrophic. Subject to the piping component design, transient analysis can be used to determine which mitigation strategy is appropriate in conjunction with overpressure protection. For example, provision of fast acting valves at the inlet and outlet of the exchanger can be used to isolate the exchanger in the event of a rupture, preventing translation of a pressure wave to downstream equipment and piping.

When the hot oil recirculation is routed directly back to the drum, the drum should be sized adequately for vapor disengagement in the event of an exchanger tube rupture at one of the hot oil users. On the other hand, when the hot oil recirculation is routed to the hot oil pumps suction, provision of a high pressure interlock can be used to shut down the hot oil system immediately, thereby protecting the pumps from cavitation.

5. Conclusions

Just as with any process in the oil, gas, and petrochemicals industry, safety, cost, control, maintainability, and operability are the factors which should be considered for the design of a closed-loop heating medium system. And just as with any process, there is no compromise on safety and operability.

Fired heating requires fuel consumption that is ultimately a loss of revenue – lost feed, product, energy, and potential carbon tax benefits – but fuel consumption can help reject inert gases that have accumulated in the liquefaction system, which may require removal by other means (i.e. NRU). Waste heat recovery *can* increase overall plant thermal efficiency and the NPV of the facility, but this comes with increased installation cost, loss of turbine power, and a larger plot space requirement.

Fired heating can be dangerous but has simple, robust control. Waste heat recovery does not come with the dangers associated with an open flame, but the system configuration, control, and start-up are as a result more complex.

The owner's priorities, environmental emission regulations, economic evaluations, and ongoing turbine driver and waste heat recovery developments must be considered in the design of a closed-loop heating medium system for an LNG facility.

6. Acknowledgement

The authors would like to thank Stanley Kwan for his constructive inputs and David Messersmith for reviewing the paper.

7. References

1. Therminol, Liquid Phase Systems Design Guide, Pub. No. 7239128D, 2008
2. Wiles, D., Kwan, S., Rajkumar, V., Waste Heat Recovery Considerations in LNG Plants, AIChE Spring Conference, April 2012
3. MultiTherm Heat Transfer Fluids, TechTeam Report Volume 2 Issue 1, March 2005

8. Biography

Benjamin Gross joined Bechtel Oil, Gas, and Chemicals, Inc. as a Process Engineer in January of 2012 subsequent to completing his Bachelors of Science in Chemical Engineering from the University of Texas at Austin. As a Process Engineer at Bechtel OG&C, Inc., he has completed assignments in utility system design, lube oil, and LNG technology.

Debby Sielegar, P.E. is a licensed engineer in the state of Texas. She completed her bachelor's degree in Chemical Engineering from Purdue University in 2007 prior to joining Bechtel OG&C, Inc. While working for Bechtel as a process engineer, she pursued her Master's degree in Chemical Engineering from University of Houston and completed the degree in 2011. In Bechtel, she has satisfactorily completed assignments on various projects such as lube oil refinery, LNG liquefaction, offshore FPSO crude production, and delayed coking.

(89d) How Glycols Affect the Acid Gas Removal Process

Georg Sieder, BASF SE, Ludwigshafen, Germany, Torsten Katz, Gas Treatment Process Technology, BASF SE, Ludwigshafen,, Germany, Justin Hearn, BASF, Ludwigshafen, Germany and **Jay Habayeb**, BASF Corporation, Houston, TX

In some natural gas applications glycols, such as MEG, DEG or TEG, are added into the pipeline or the gas conditioning process, either as hydrate inhibitors or for dehydration purposes to protect downstream pipelines. These substances have high boiling points and are not supposed to be carried over into the downstream process. Proper separator design shall avoid carry-over of these substances into the Acid Gas Removal plant.

Some amine plants however report glycol build-up in the amine solution over time, which can lead to performance losses and ultimately to unexpected plant shutdowns. This paper deals with the sources and consequences of glycol build-up in the amine unit and delivers some understanding of how the glycol content affects the acid gas capture performance. It provides recommendations on how to control the full acid gas capture performance over time and how to avoid operational surprises.

13 AIChE
Spring Meeting, San Antonio, TX

CONDENSATE STABILIZATION: HOW TO GET THE MOST FOR YOUR MONEY

Mahdi Nouri, John Rizopoulos
Process Engineering & Technology
CH2M HILL Energy & Chemicals
Houston, TX

Prepared for Presentation at
American Institute of Chemical Engineers
2013 Spring Meeting
9th Global Congress on Process Safety
San Antonio, Texas
April 28 – May 1, 2013

CONDENSATE STABILIZATION: HOW TO GET THE MOST FOR YOUR MONEY

Mahdi Nouri, John Rizopoulos
Process Engineering & Technology
CH2M HILL Energy & Chemicals
Houston, TX

ABSTRACT

Condensate stabilization is one of the most common operations in the gas processing world. The principal considerations affecting the performance of the plant when debottlenecking or designing new condensate stabilization units are the configurations, product quality, energy saving, and equipment size. Sound process design can significantly reduce capital investments and operating costs. The purpose of this paper is to review and illustrate the fundamentals of this process, including flow-sheet configuration and key operating parameters, and to show how they affect the plant performance, including energy consumption and economics.

INTRODUCTION

The incoming gas from offshore or onshore pipelines carries condensates generated during transmission by heat transfer and pressure drop. The two phases are typically separated in the slug catcher or inlet separator. The gas stream is treated before feeding to the cryogenic system for extracting NGL. It is first routed to the acid gas removal unit, then dried in a Glycol unit, and polished in a molecular sieve dehydration unit, (Figure 1). The recovered hydrocarbon condensate is sent to the condensate stabilization unit for vapor-pressure reduction. Stabilized condensate can be stored in floating roof storage tanks, and safely transmitted in pipelines. The stabilized condensate, typically C5+, is sold as final product or as petrochemical feed-stock to olefins and aromatic plants. Figure 1 illustrates how the vapor from a condensate stabilization unit is compressed and recycled to the inlet feed separator gas stream before entering the gas treating system. The presence of contaminants (water, salts and suspended solids) in the slug catcher can cause problems on the condensate stabilization unit performance, including long-term maintenance issues. Water carryover from the slug catcher and hydrocarbon/water separator can lead to off-spec stabilized condensate.

This study (1) evaluates six alternative process configurations and key operating parameters for meeting the stabilized condensate specifications and saving energy, (2) develops optimized utility consumption for the alternatives, (3) develops comparative capital and operating cost appraisals, and (4) presents the optimum alternative in terms of process efficiency and economic analysis. An appropriate process simulator with the right thermodynamic package, optimum flow sheet configuration and equipment sizing is a very important tool for achieving the desired results.

Figure 2 illustrates the integrated engineering and economics work flow. In this paper, HYSYS and Process Economic Analyzer (APEA) is used for simulation and economic analysis. This Aspen Technology, Inc., software is very effective in carrying out comparative evaluation of process designs. Based on a HYSYS flow-sheet simulations, APEA determines the capital expenditures, operating costs, and finally the profitability of the proposed designs. APEA has a system which links process simulation programs to the estimating tool, though a "process engineer's window" to the "Icarus Estimating engine". It is used to (1) extend the results of process simulation, (2) generate sizing and cost estimates for process equipment, (3) perform preliminary mechanical designs, and (4) estimate materials and labor costs, indirect costs, total capital investment, and (5) the EPC schedule. [1]

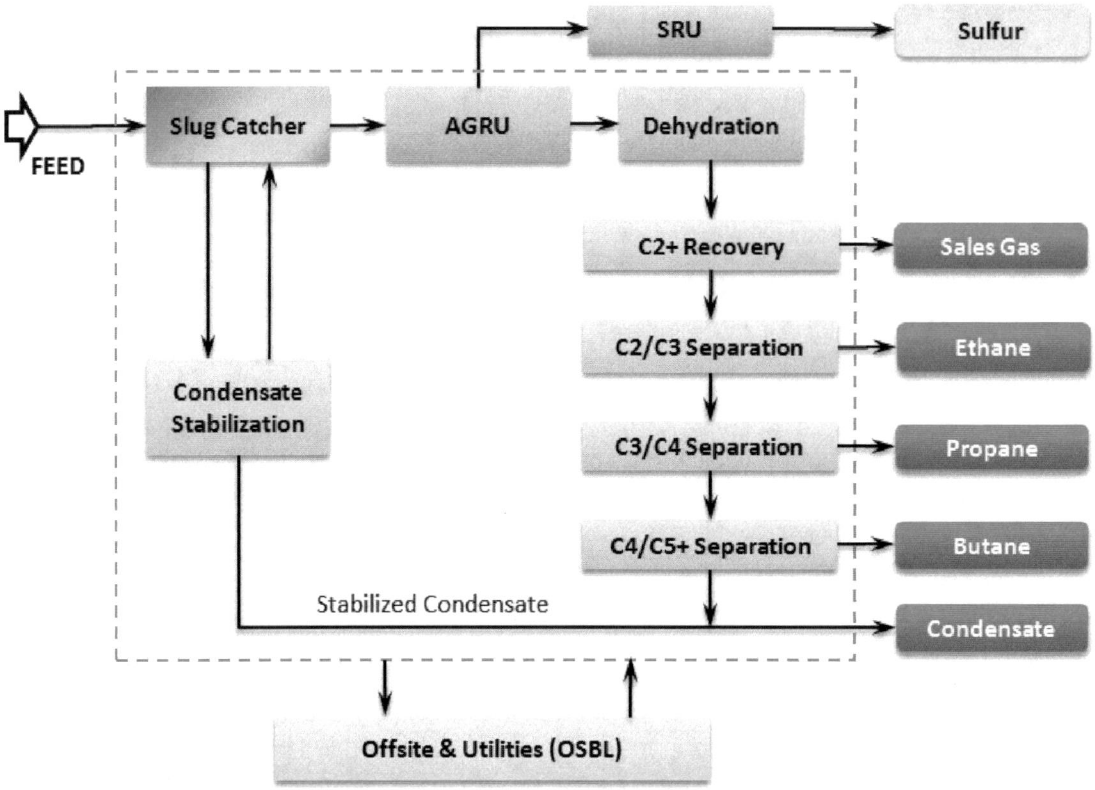

Figure 1 – Typical Block Flow Diagram for a NGL Recovery Plant

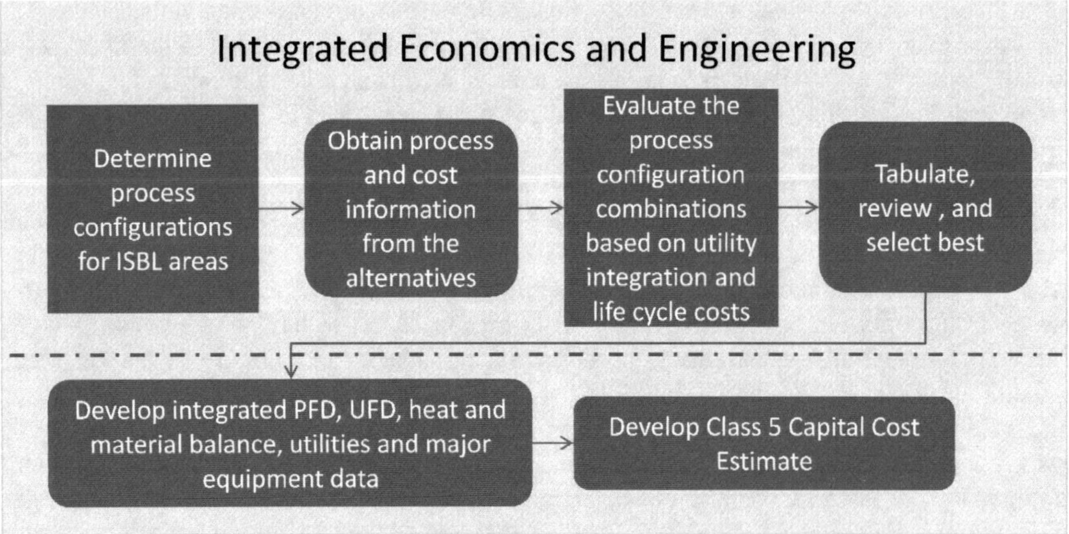

Figure 2 – Process Configuration Screening Work Flow Process

BASIS of DESIGN

Feed characteristics

The single-phase liquid feed properties used for the case studies are shown in Table 1.

Table 1 – Unstabilized Condensate Characteristics

Component	Molar Flows (lbmole/hr)
H_2O	1449.7
N_2	16.1
CO_2	63.5
H_2S	0.0
C1	1317.2
C2	271.3
C3	218.2
i-C4	99.6
n-C4	165.6
i-C5	146.8
n-C5	126.4
C6+	2330.0
Total	6204.3
Pressure, Psig	600.0
Temperature, °F	90.0
RVP, Psia	250.0

Products specifications

The products of condensate stabilization unit are:

- Stabilized condensate meeting the required specifications (Table 2),
- Off-gas compressed and recycled to the gas treating system unit.

Table 2 – Stabilized Condensate Specifications

Component	Specification	Test Method
RVP, Psia	<8.5	ASTM D-323
Color	Not Less than 25 Saybolt	ASTM D-156
Water Content	Nil	

Battery limit conditions

The operating conditions of the major streams to the condensate stabilization unit are:

Table 3 – Battery Limit Conditions

		Operating Conditions	
	Description	Pressure, Psig	Temperature, °F
Feed	Liquids from slug catcher	600	90
Products			
Stabilized condensates	To floating roof storage tank	Atm.	130
Off-gas	2nd stage aftercooler outlet	1100	130

Turndown ratio

The Condensate Stabilization unit is designed to operate 40-110% of the normal continuous flow rate.

CASE STUDIES

The objective of this case study is to evaluate different condensate stabilizer configuration alternatives and operating parameters. Table 4 describes the six process configurations compared in this paper. In all alternatives, the stabilizer is operated at pressure of 145 Psig except for Alternative 4, which is operated at 87 Psig. In all studied cases, 10 theoretical-stages of valve trays are used as additional stages appear to offer no further benefit. The following assumptions are made for the economic analysis based on U.S. Gulf Coast:

- Currency Conversion Rate = U.S. Dollar
- Stabilized condensate = $ 102 per barrel
- Gas Price = $ 3.30/1000ft^3
- Steam cost @ 400 Psia = $ 5/klb
- Electricity cost = $ 0.07/KWh
- Plant availability = 355 days per year
- Annual rate of return = 20%
- Plant life = 25 years

Table 4 – Process Configuration Alternatives for Comparison

	Refluxed Column	Non-Refluxed Column	Non-Refluxed LP Column	Feed Preheating	Feed Splitting	Side Reboiler
Alternative 1		X				
Alternative 2		X		X		
Alternative 3		X		X	X	
Alternative 4		X	X	X	X	
Alternative 5		X		X	X	X
Alternative 6	X			X		

Alternative 1: non-refluxed column, no feed preheating, no split feed & no side reboiler

The condensate from the slug catcher is flashed again at lower pressure in the hydrocarbon/water separator. All hydrocarbon condensate leaving the condensate/water separator is sent to the top of the column. The stabilized condensate leaving the column is cooled before storage, which increases the bottom condensate cooler duty. Stabilizer overhead vapor is compressed by a two-stage reciprocating motor-driven compressor. The first-stage discharge gas is cooled down to 130 °F in an air cooler and is then sent to the three-phase separator for removing any condensate and water. Overhead gas is then mixed with gas from the hydrocarbon/water separator prior to feeding the 2nd stage overhead compressor suction drum. Gas is further compressed by the second stage, cooled in the after-cooler, and sent to the acid gas removal unit for further processing. Figure 9 shows the relationship between the inter-stage compressor pressure and the compressor power.

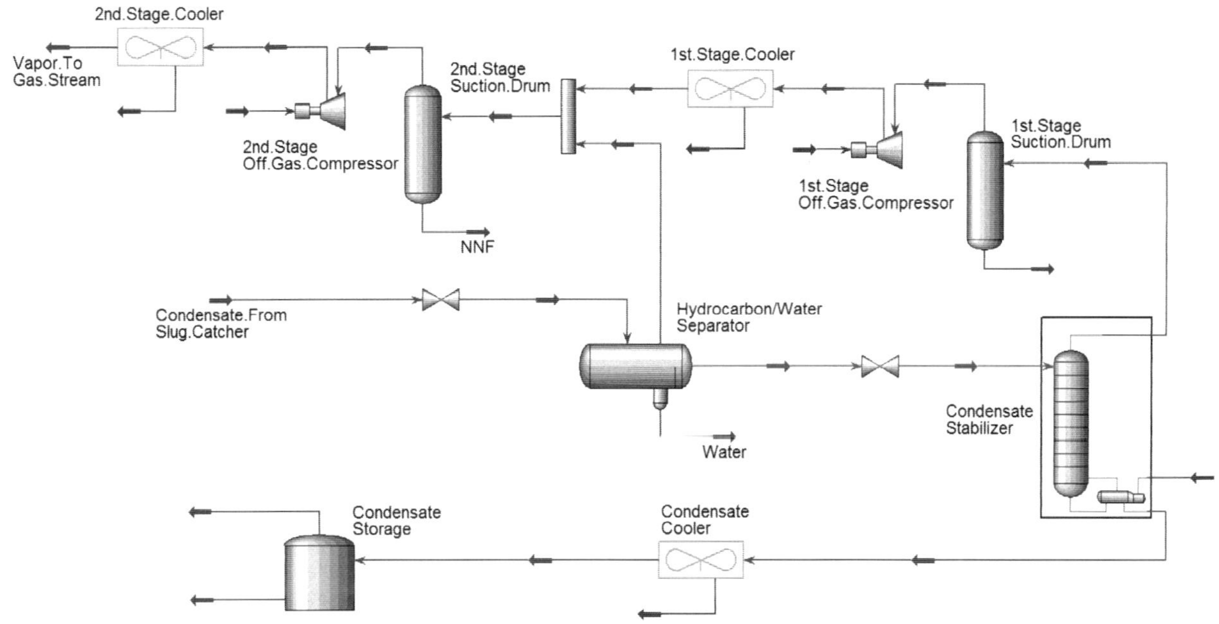

Figure 3 – Alternative 1 Process Flowsheet

Alternative 2: non-refluxed column with feed preheating, no split feed, and no side reboiler

This scheme is similar to Alternative 1. The difference is the presence of a feed/bottom exchanger as a preheating system. The entire hydrocarbon condensate stream from the hydrocarbon/water separator flows through the feed/bottom heat exchanger and enters to the top of the column. The vapor recovery compressors have relatively high power requirements, and reboiler has higher thermal duties. In addition, the bottoms from the 2nd stage suction drum are routed directly to the hydrocarbon/water separator to remove water.

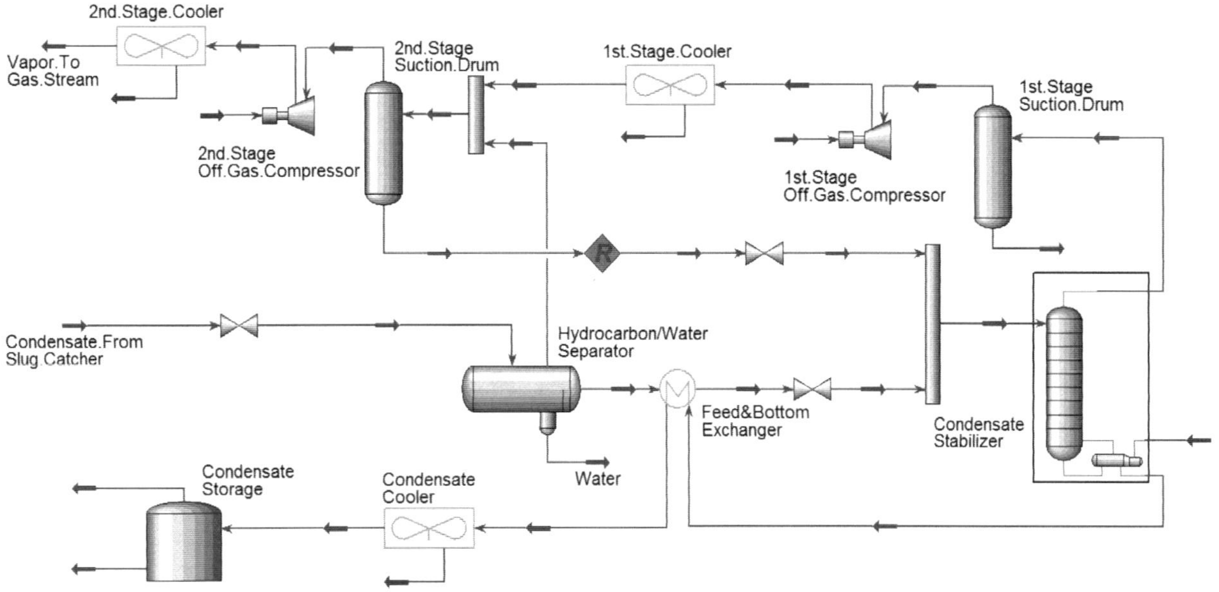

Figure 4 – Alternative 2 Process Flowsheet

Alternative 3: non-refluxed column with feed preheating, split feed, and no side reboiler

The process scheme detailed in Alternative 2 is modified so that the feed is split, with 17% sent cold to the top of the column. The residual feed passes to the feed/bottom exchanger before entering the column at the 4th theoretical stage. Figure 10 illustrates the impact of splitting feed and feed stage location on reboiler duty.

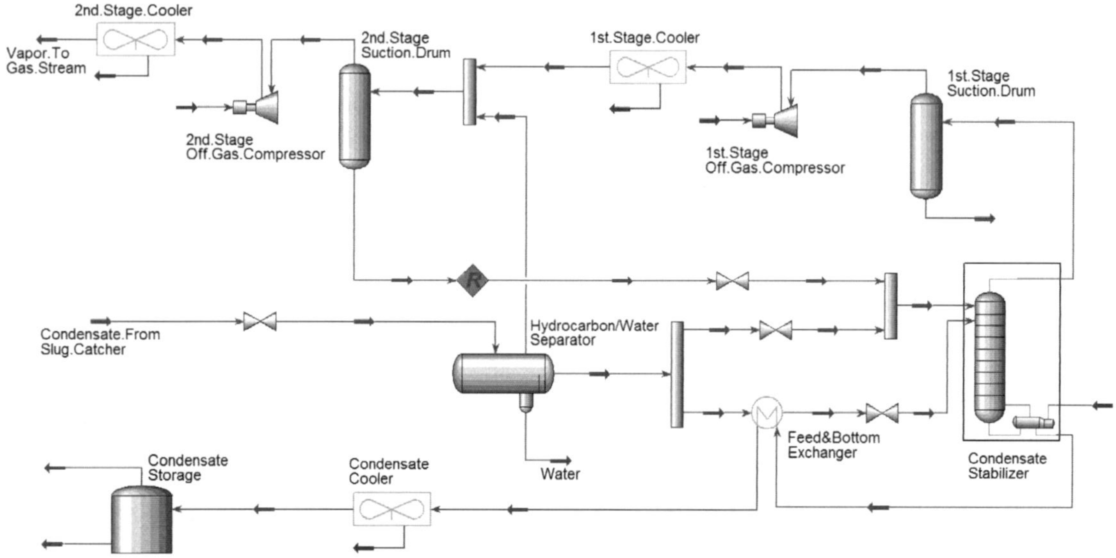

Figure 5 – Alternative 3 Process Flowsheet

Alternative 4: non-refluxed low pressure column with feed preheating & split feed & no side reboiler

This configuration is similar to Alternative 3, except that the stabilizer is operated at lower pressure (87 Psig), which requires an additional recompression stage. In this scheme, the feed is split, with 11% sent cold to the top of the column.

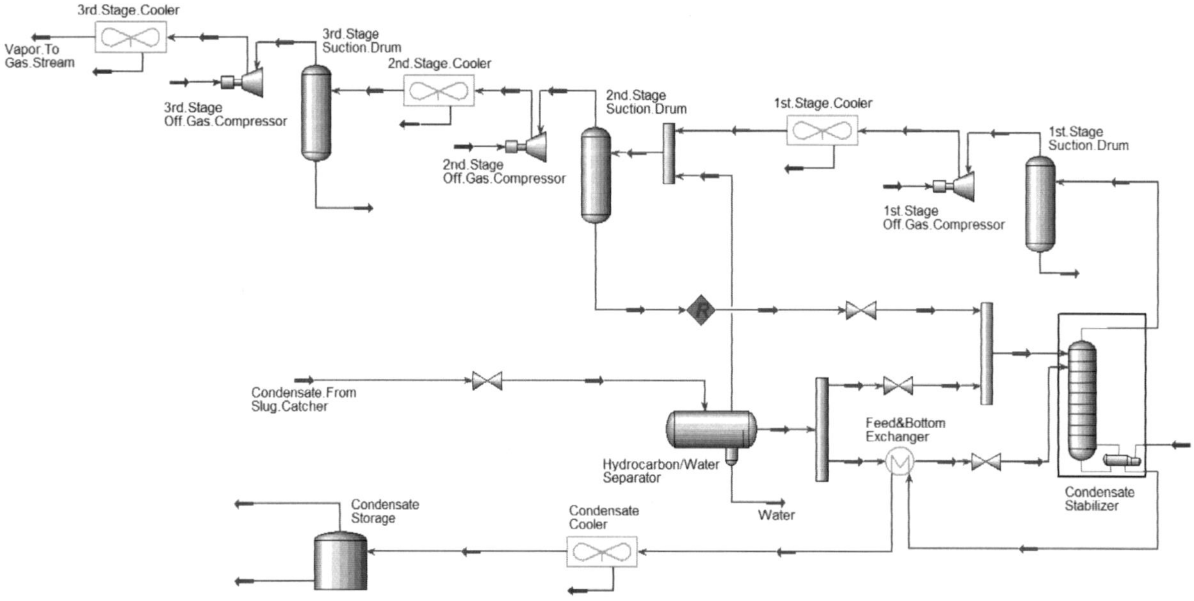

Figure 6 – Alternative 4 Process Flowsheet

Alternative 5: non-refluxed column with feed preheating, split feed, and side reboiler

This scheme of Alternative 4 is modified by the addition of a side reboiler at the 7th theoretical stage. The feed is split, with 10% sent cold to the top of the column. The residual feed is routed through the feed/bottom heat exchanger before entering the column at the 4th theoretical stage.

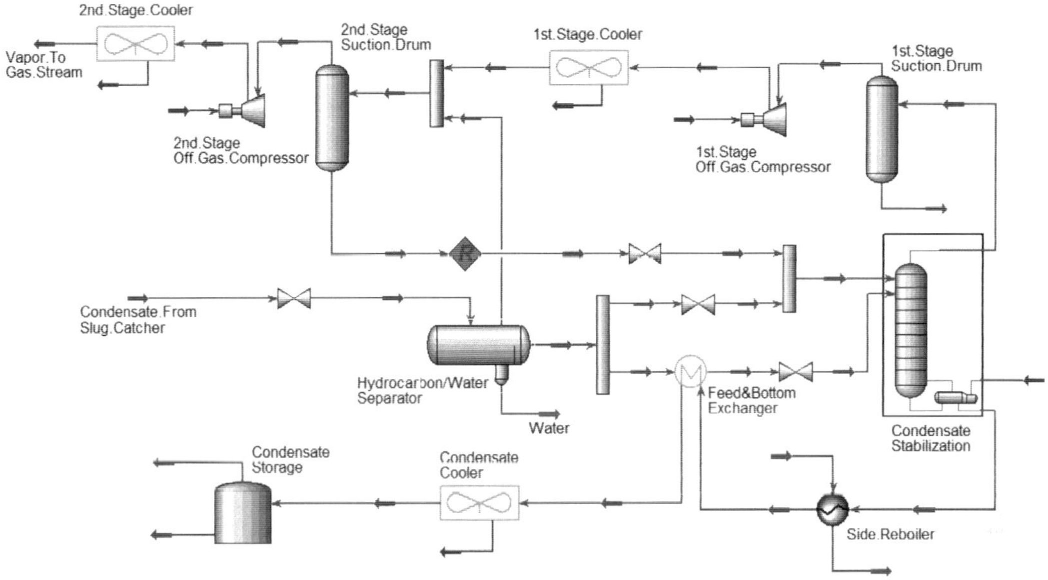

Figure 7 – Alternative 5 Process Flowsheet

Alternative 6: refluxed column with feed preheating, no side reboiler

The condensate leaving the hydrocarbon/water separator is heated using the feed/bottom heat exchanger, and enters the 4th theoretical stage of a conventional distillation column.

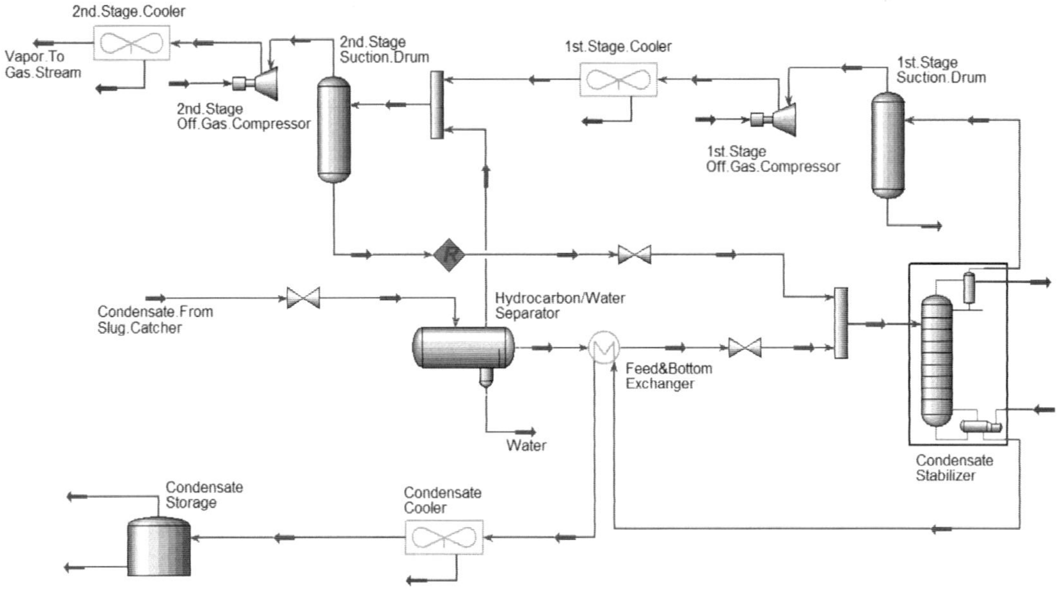

Figure 8 – Alternative 6 Process Flowsheet

RESULTS

Plant designers consider major design parameters for increasing process efficiency and saving energy. The following discussion is limited to the findings of the sensitivity analysis. Sensitivity analyses are performed by investigating and focusing upon various operating parameters, process configurations, and robustness of columns, by observing differences from simulation outputs, and economic analysis outputs. Table 5 presents the optimum alternative in terms of CAPEX and OPEX. Figure 11 illustrates that Alternatives 2 to 6 produce essentially the same stabilized condensate. The provided capital costs are a rough order of magnitude on a U.S. Gulf Coast basis. These estimated costs have been calculated by APEA, and cover engineering, procurement, and installation of process and utility equipment; storage, metering, and loading facilities; plus, construction indirect, spares, engineering, and commissioning. Items such as OSBL costs, owner's costs, land costs, escalation, and contingency are excluded.

Alternative 1 does not have the heat recovery system, and the hydrocarbon condensate with feed operating temperature is sent to the stabilizer. Operational experiences show that the product specifications are achieved with difficulty in terms of water content, because water separation from hydrocarbons is not easy at low temperatures. Figure 11 shows stabilized condensate flow rate, however, the highest reboiler, and condensate cooler duties in comparison to the other cases as shown Figure 13. Figure 9 shows the minimum off-gas compressors power at the lower operating temperature. This process configuration has a lowest capital cost because of the simplest design, fewest pieces of equipment, and less severe operating conditions, however, the operating and utilities costs are the highest, as reported in Table 5.

Alternative 2 can meet the product specifications with an efficient column temperature profile. The reboiler duty is significantly decreased by heat recovery (Figure 13).

Alternative 3 features a traditional design approach in terms of meeting product specifications, and operating cost optimization. Figure 13 shows that the reboiler duty is reduced by feed splitting in comparison to Alternative 2. Figure 10 shows the relationship between feed splitting, location of feed tray, and operating costs, including steam consumption and electricity usage.

Alternative 4 requires an additional compression stage because of the lower column pressure condition. Table 5 reports the highest capital investment; however, the reboiler duty is reduced in comparison to Alternative 3.

Alternative 5 shows an optimum design approach in terms of meeting product specifications, and operating cost optimization. The operating and utilities costs are minimized by the presence of the side reboiler exchanger for further heat recovery; however, the capital cost is slightly increased in comparison to Alternative 3 (Table 5).

Alternative 6 is a conventional distillation column. This case is the most complex system. CAPEX, OPEX and utilities cost are more expensive than Alternative 3 and 5 because of the additional reflux system, reported in Table 5.

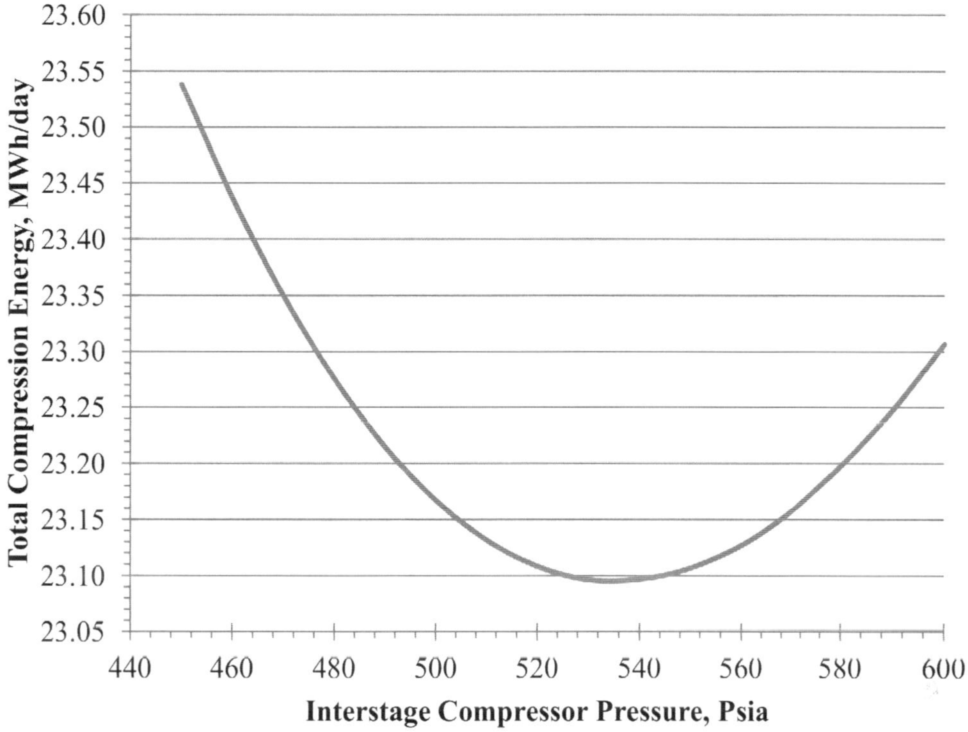

Figure 9 – Off Gas Compressor Power Optimization for Alternative 1

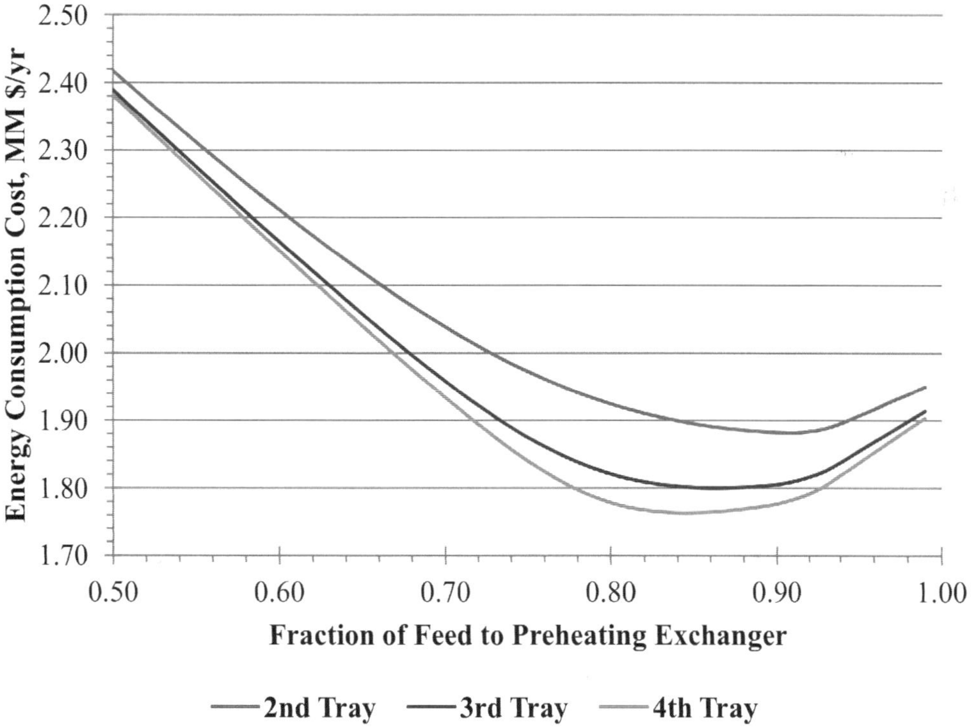

Figure 10 – Reboiler Duty Optimization by Splitting Feed & Feed Stage Sensitivity for Alternative 3

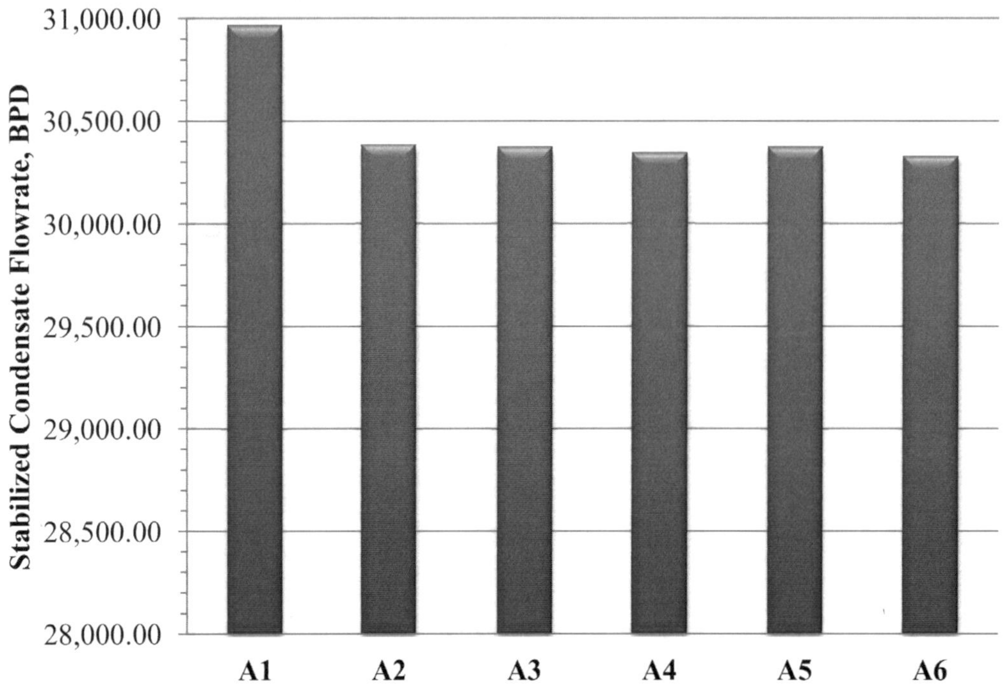

Figure 11 – Produced Stabilized Condensate for Each Process Configuration

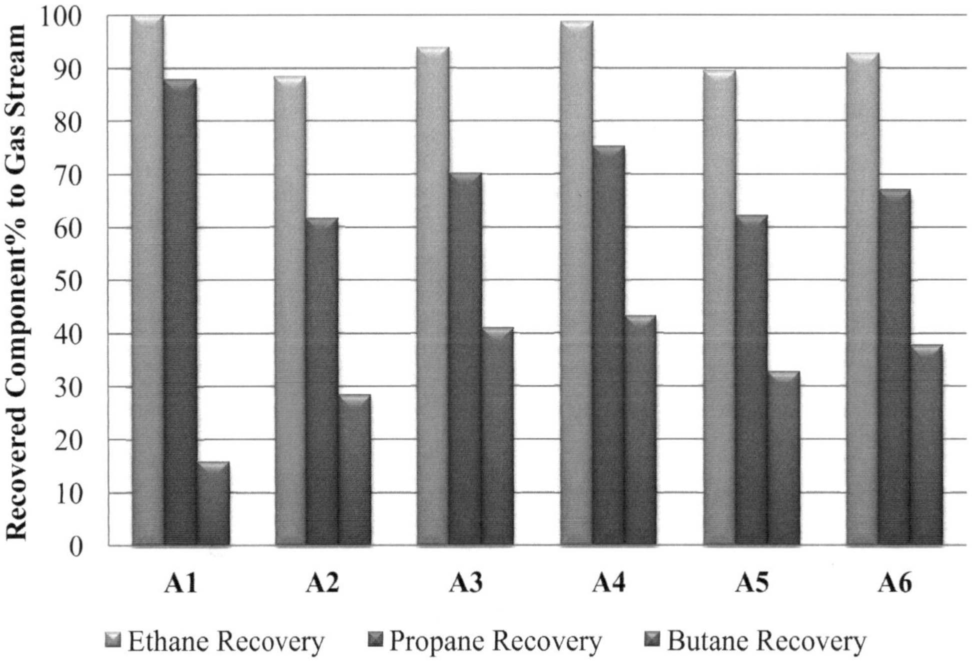

Figure 12 – Products Recovery for Each Process Configuration

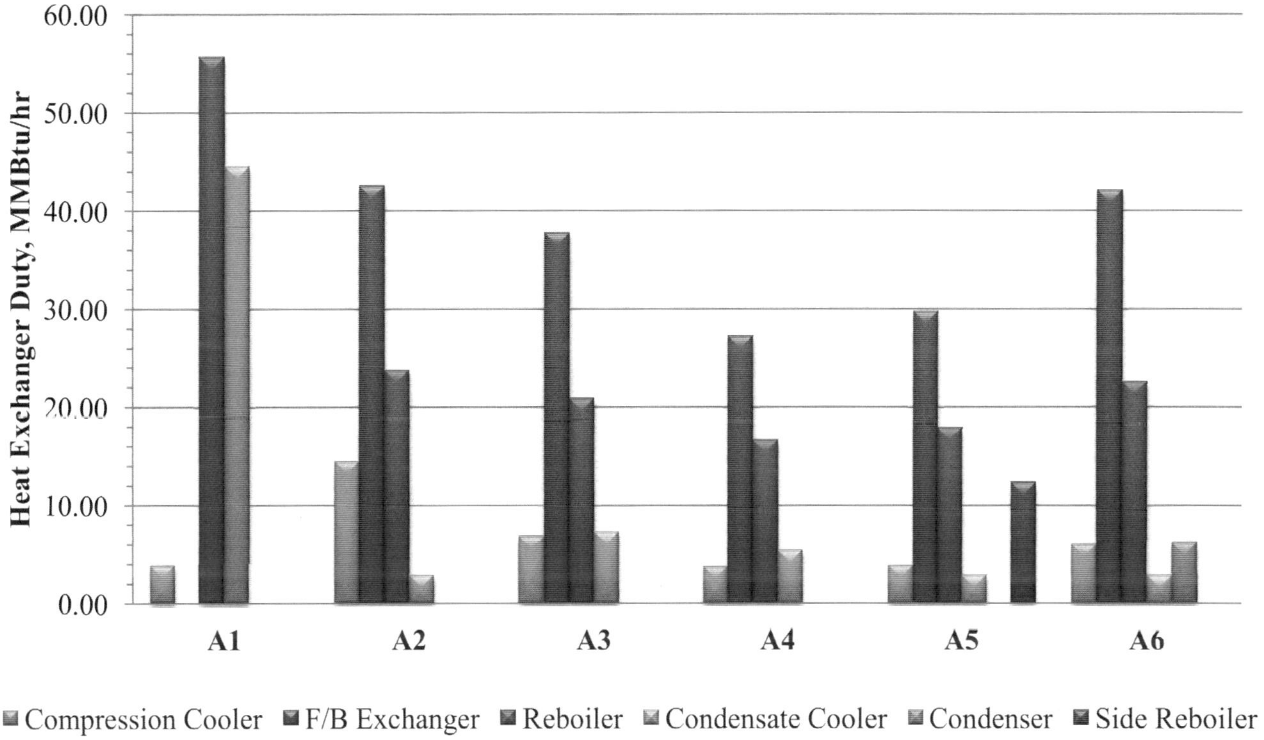

Figure 13 – Heat Exchangers Duty for Each Process Configuration

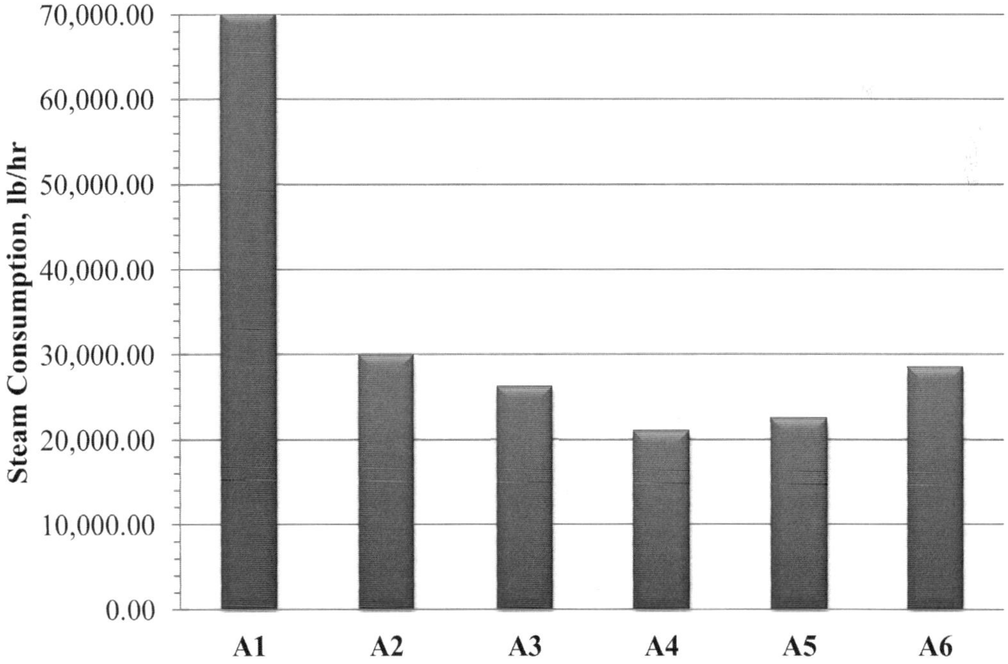

Figure 14 – Steam Consumption for Each Process Configuration

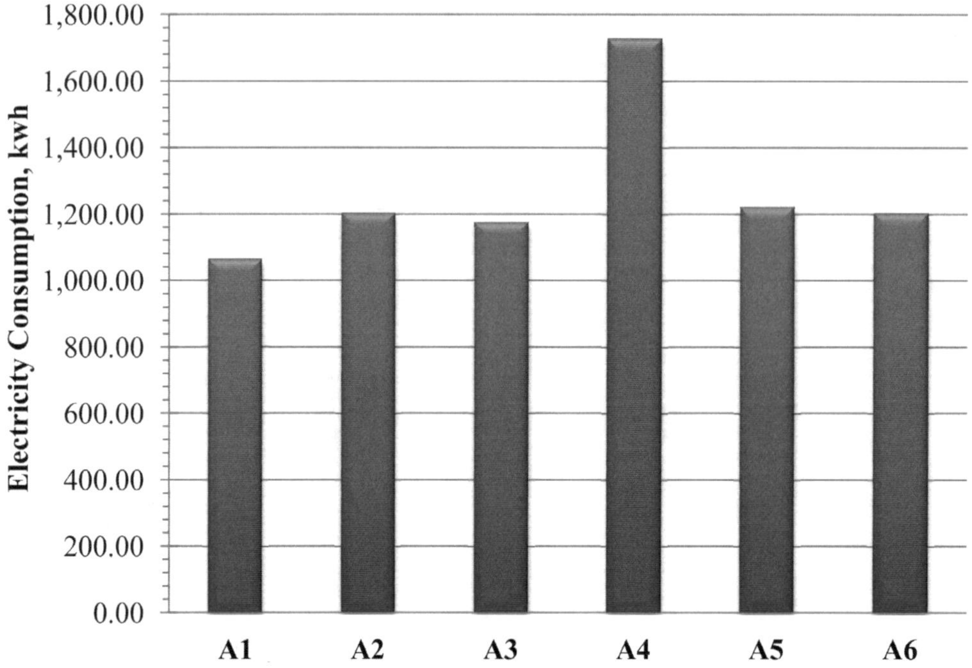

Figure 15 – Electricity Consumption for Each Process Configuration

Table 5 – Cost Comparison for Alternatives

	Capital Cost	Operating Cost (1)	Utilities Cost (2)	Product Sales
Alternative 1 (3)	0.89	1.68	2.16	1.02
Alternative 2	0.89	1.07	1.09	1.00
Alternative 3	0.97	1.06	1.08	1.00
Alternative 4	1.24	1.18	1.07	1.00
Alternative 5	1.00	1.00	1.00	1.00
Alternative 6	1.07	1.08	1.14	1.00

(1) The operating costs cover the raw materials, operating labor costs, maintenance, plant overhead, utility costs, electricity, fuel gas, potable water, and instrument air.

(2) The utilities costs cover the heating and cooling.

(3) Operational experiences show that the stabilized condensate specifications are achieved with difficulty with higher water content.

SUMMARY & CONCLUSIONS

Condensate stabilization units can be optimized by selecting the best process configuration and operating conditions depending on feed characteristics and required product specifications. A number of optimization runs are required using a steady-state simulator in order to determine the appropriate column pressure, column temperature, feed split ratio, and the feed tray location in terms of energy optimization. It is recommended that appropriate equipment design margins and sizing be used in order to ensure acceptable operational flexibility (particularly for water separation from hydrocarbon condensate). Table 5 summarizes the results of this study. Alternative 5 was chosen over the other process configurations through screening following on the following criteria:

- Acceptable capital cost, within the limits of study accuracy
- Lowest operating cost in terms of steam and electricity consumption
- Relatively low process complexity, which should lead to higher availability and lower maintenance costs in the long term, yielding the highest potential return to plant stakeholders.

Figures 9 – 14 illustrate sensitivities of various operating parameters. Based on this analysis and alternatives comparison, the following conclusions can drawn through the selection of the optimum process configuration:

- Optimizing of feed splitting and feed tray location:
  a. Reduces reboiler duty
  b. Reduces tower size
- Off-gas compressor power is optimized by adjusting the interstage pressure
- Off-gas after cooler duty is optimized by adjusting the interstage pressure
- Adding a bypass to the preheat exchanger and changing the feed tray location are obvious zero cost options.

REFERENCES

[1] Warren D. Seider, "Product and Process Design Principles" Second Edition, John Wiley, 2002

(91b) Simulation and Optimization of Internal-Loop Airlift Slurry Reactor Design Using Computational Fluid Dynamics

Luhaibo Zhao and Bo Liao Sr., Low Carbon Conversion Center, Shanghai Advanced Research Institute, Shanghai, China

Internal-loop airlift reactors (ILAR) are widely used in biotechnology and chemical industry, because they are suitable for processes with good mixing and close contact between the phases. However, the fundamental hydrodynamic behaviors in ILAR, which are essential for reactor scale-up and design, are still not fully understood. In the paper, the three-dimensional computational fluid dynamics (CFD) modeling is used to study the performance of three-phase ILAR with multi-layer ring gas distributor. The global flow characteristics, such as gas holdup, liquid velocity and solid particles distribution, are simulated, and the results agree well with the experimental data. Then, the effect of the sparger structure of gas distributor, including the number of layers, orifice size and angle, porosity, on hydrodynamic behaviors in ILAR is investigated based on CFD model. The optimal configuration of the multi-layer ring distributor is obtained and verified by cold model experiment, which results in even gas distribution and promotes the gas-liquid-solid mixing in ILASR, thus improves its yield and productivity.

(58a) Tunable Polyimide and Resulting Carbon Molecular Sieve Membranes for Gas Separation

Wulin Qiu, School of Chemical and Biomolecular Engineering, Georgia Institute of Technology, Atlanta, GA and William J. Koros, School of Chemical & Biomolecular Engineering, Georgia Institute of Technology, Atlanta, GA

Polymer membranes have been used for natural gas and olefin-paraffin gases separations successfully. However, the separation performance is limited by a productivity and efficiency trade-off upper bound. Therefore, advanced carbon molecular sieves (CMS) materials were developed. These materials show separation performance that exceeds the polymer "upper bound" for gas separations, while potentially maintaining economically attractive manufacturing. CMS membranes can be formed by thermal decomposition of polymer precursors, and contain "micropores" of 6–20 Å connected by smaller so-called "ultramicropores". The combination of ultramicropores and micropores provides the molecular sieving function and high permeability characteristics of these unusual CMS membranes.

We synthesized diverse but systematic sets of co-polyimides, containing symmetric and asymmetric polymer backbone structure, to explore effects of matrix rigidity and packing of precursor polyimides on final CMS properties. We also explored the interaction of pyrolysis conditions on final CMS properties. Permeation was tested using both pure gases (He, O_2, N_2, CO_2, CH_4,C_2H_4, C_2H_6, C_3H_6, C_3H_8) and mixed gases for both the polyimide precursor membranes as well as the resulting CMS membranes. We correlated separation performance of precursor polyimides and formation conditions with the CMS membranes' gas separation performance. This research provides an improved fundamental framework to guide the systematic understanding and fabrication of CMS membranes for important energy intensive separations.

(58b) Post-Spinning Infusion of Poly(ethyleneimine) Into Polymer/Silica Hollow Fiber Sorbents for CO2/N2 Gas Separation

Ying Labreche[1], Ryan P. Lively[2], Fateme Rezaei[3], Grace Chen[4], David S. Sholl[5], Christopher W. Jones[5] and William Koros[2], (1)School of Chemical and Biomolecular Engineering, Georgia Tech, Atlanta, GA, (2)Chemical & Biomolecular Engineering, Georgia Institute of Technology, Atlanta, GA, (3)School of Chemical & Biomolecular Engineering, Georgia Institude of Technology, Atlanta, GA, (4)Georgia Tech, Atlanta, GA, (5)School of Chemical & Biomolecular Engineering, Georgia Institute of Technology, Atlanta, GA

Post-spinning infusion of poly(ethyleneimine) into polymer/silica hollow fiber sorbents for CO_2/N_2 gas separation

Ying Labreche[1], Ryan P. Lively[2], Fateme Rezaei[1], Grace Chen[1], David S. Sholl[1], Christopher W. Jones[1,*], William J. Koros[1,*]

[1] Georgia Institute of Technology, 311 Ferst Drive NW, Atlanta, GA 30332-0100, USA

[2] Algenol Biofuels, 28100 Bonita Grande Drive, Bonita Springs, Florida 34315, United States

Abstract

Amine-loaded hollow fiber sorbents for post-combustion CO_2recovery were created by utilizing a novel post-spinning infusion technique. This technique infused poly(ethyleneimine) (PEI) into cellulose acetate/mesoporous silica hollow fiber sorbents during the solvent exchange after spinning. A suitable post-spinning infusing solution was found to be 10% PEI in methanol with an infusion time of 4 h.

After amine infusion, the 51 wt% silica hollow fiber sorbents were found to have a nitrogen loading of 0.52 mmol/g-fiber and a CO_2 uptake of 1.2 mmol/g-fiber at equilibrium. The amine-loaded fibers were formed into a shell-and-tube module and exposed on the shell side at 1 atm and 35 °C to simulated flue gas with an inert tracer (10 mol% CO_2, 80 mol% N_2 and 10 mol% He at 100% R.H.). The fibers were found to have a breakthrough CO_2 capacity of 0.58 mmol/g-fiber, with a mass transfer propagation front velocity of 1.12 cm/s. The CO_2 "equilibrium" uptake at 1 atm and 35 °C is 0.92 mmol/g-fiber. The sorption enthalpy release was monitored and the thermal front propagation velocity was 0.75 cm/s. Finally, the water uptake was found to be 3.2 mmol/g-fiber. The results indicate the post-spinning infusion method provides a new platform for ultimate scale-up of practical fiber sorbents for flue gas application.

(58c) Strategies for Acid Gas Enrichment and Tail Gas Treatment:

Tofik K. Khanmamedov, TKK COMPANY, Sugar Land, TX

Strategies for Acid Gas Enrichment and Tail Gas Treatment:

Special Solvents or Better to Stick With Generic Solvents

Tofik K. Khanmamedov

TKK COMPANY, Sugar Land / Houston, Texas

Abstract

The quality of the acid gas feed stream to sulfur recovery units is crucial for proper operation. In this regard, acid gas streams with low H_2S concentration should be enriched with H_2S. This is also true for the amine section of tail gas treating units. There are different approaches for acid gas enrichment. One approach is based on using generic N-Methyl-Diethanol Amine (MDEA) with the family of patented HIGHSULF™ processes. Other approaches use more selective and very expensive solvents.

Also there are claims that some specialty solvents allow designing units using relatively smaller equipment than in case of MDEA, there are limited advantages in long term to say the least.

Due to depreciation of the unit in most cases the capital cost is not the primary concern of the clients. Even operating cost based on the use of local sources of power is not the major issue. However, they do seriously care about the costs of the solvents, that they have to purchase annually from the single foreign vendors and this factor limits their freedom of the choice. While this approach might reduce initial capital cost, it could significantly increase the operating costs over time due to the necessity to continue to purchase the expensive specialty solvents from the single vendor.

In most cases, switching from special solvents to generic, less expensive, solvents in the future would require a reduction of the overall unit capacity. Units designed for MDEA based HIGHSULF™ technologies can be easily switched to other specialty solvents within certain capacity constraints.

This paper focuses on advantages of the MDEA based family of HIGHSULF™ processes that require no additional equipment.

*** Formatting Issue - Best Available Paper/Graphic ***

(58d) Synthesis of Ion-Exchanged SAPO-34 Zeolite Membrane for CO2/CH4 Separation

Yanfeng Zhang, Meng Li and Yuhan Sun, Low Carbon Conversion Center, Shanghai Advanced Research Institute, Chinese Academy of Sciences, Shanghai, China

Synthesis of ion-exchanged SAPO-34 zeolite membrane for CO_2/CH_4 separation

***Yanfeng Zhang*[1], Meng Li[1], Yuhan Sun[1]**

[1]Shanghai Advanced Research Institute, Chinese Academy of Sciences, 100 Haike Rd, Pudong New District, Shanghai 201203, P.R China

Introduction

Natural gas usually is contaminated with CO_2, which must be removed to meet pipeline requirement. The presence of CO_2 in natural gas decreases the energy content of the gas, and causes pipeline corrosion in the presence of water. Amine adsorption is a mature technology for CO_2 removal from natural gas, but it suffers from high energy cost for solvent regeneration, especially for natural gas with high CO_2 content. Membrane separation has many advantages, such as low energy cost, low capital investment, flexible size, etc, and polymer membranes have been used for natural gas sweetening at commercial scale. However, polymer membranes cannot treat natural gas with high CO_2 content (>20%), since plasticization effect decreases membrane selectivity significantly. Zeolite membranes have great chemical and thermal stability, and their uniform pore size. Zeolite membranes[1¨C4] have potential for CO_2/CH_4 separation owing to their great thermal, mechanical, and chemical stability, and stability at high CO_2 pressures. SAPO-34, a silicoaluminophosphate with chabazite (CHA) type framework having pore diameter 0.38 nm, is perfect for CO_2/CH_4 separation because CO_2 (0.33 nm kinetic diameter) and CH_4 (0.38 nm) can be separated by molecular sieving mechanism. The selective adsorption of CO_2 in SAPO-34 pores also favours the separation of CO_2. Many factors affect the performance of SAPO-34 membrane, including feed pressure [1,2], Si/Al ratios [3,4], seed size [5], template types [5], membrane thickness [4], cation forms [6], CO_2/CH_4 feed ratio, support properties [7], concentration polarization [8] and template residue [9]. Mei et al prepared ion-exchanged SAPO-34 zeolite membrane and CO_2/CH_4 selectivity was increased by 30%. Here we report a simple one step method of preparing ion-exchanged SAPO-34 membrane by combining template removal and ion-exchange in to one step. Obtained Na, K and Li-SAPO-34 membranes have improved separation performance.

Experimental

Detailed information about synthesis, calcination and gas separation test of SAPO-34 membrane can be found in ref 9. Uncalcined SAPO-34 membrane was soaked in 1wt% $NaNO_3$, KNO_3 or $LiNO_3$ solution for 5 min and dried at 383K for 2 h. Then the SAPO-34 membrane was calcined in vacuum at 673 K for 4 h with 1K/min heating and cooling rates. For comparison, blank SAPO-34 membranes were also calcined in vacuum at the same condition.

Results and discussion

Fig 1 shows the SEM images of SAPO-34 membrane. Typical cubic crystals with perfect intergrowth were found in Fig 1a and membrane thickness is ~ 5 μm (Fig 1b). SAPO-34 crystals with thin plate morphology, as shown in Fig 1c, were used as seed for membrane synthesis. XRD patterns indicate that the obtained crystals and membranes have CHA structure (not shown here).

*** Formatting Issue - Best Available Paper/Graphic ***

(a) **(b)**

(c)

Fig 1 SEM images of SAPO-34 membrane and SAPO-34 seed, (a)top view and (b)cross section view and (c)SAPO-34 seed.

Table 1. Separation performance of SAPO-34 membranes (50/50 molar CO_2/CH_4 mixtures, room temperature, 4.0 MPa feed pressure).

Membrane treatment	CO_2 Permeance mol/(Pa m^2 s)	CO_2/CH_4 Selectivity
1wt% $NaNO_3$ in H_2O	2.1°Á10^{-7}	75
1wt% $LiNO_3$ in Acetone	3.9°Á10^{-7}	70
1 wt% $NaNO_3$ in H_2O	2.6°Á10^{-7}	81
1 wt% KNO_3 in H_2O	2.8°Á10^{-7}	73
Eight membranes calcined directly	3.0°Á10^{-7}	52

*** Formatting Issue - Best Available Paper/Graphic ***

Eight SAPO-34 membranes were prepared and calcined directly in vacuum to remove the template. The average CO_2 permeance and CO_2/CH_4 selectivity are 2.8°Á10^{-7} mol/(Pa m^2 s) and 52 respectively (feed pressure 4.0MPa and 50/50 CO_2-CH_4 mixture). Four membranes were soaked in 1wt% salt solution for 5 min and dried at 383K. Then these membranes were calcined at 673K in vacuum to remove template. The melting points of $NaNO_3$, KNO_3 and $LiNO_3$ are 581, 607 and 528K respectively, all lower than calcination temperature of 673K. This indicates that the deposited salt on the membrane surface will be melted during calcination. After template removal, the as prepared H-SAPO-34 membrane will be exchanged to Na, K and Li-SAPO-34 membranes. All four membranes exhibits 40 to 60% increase in CO_2/CH_4 selectivity, compared with control experiment. Ion-exchange of H-SAPO-34 increases basicity of SAPO-34 crystals, which favours the adsorption of CO_2. Ion-exchange may also reduce pore size of SAPO-34 crystals, which might limit the diffusion of CH_4. This result is consistent with Mei's result [6].

Conclusions

A simple method was developed to prepare ion-exchanged SAPO-34 membranes in one step. By combining template removal and ion-exchange into one step, Na, K and Li-SAPO-34 membranes were prepared. The obtained SAPO-34 membranes have much higher CO_2/CH_4 selectivity, which is the result of increased basicity from ion-exchange. The combination of template removal and ion-exchange eliminate one step and greatly reduce synthesis cost. This method can also be applied to other zeolite membranes.

REFERENCES

[1] S. Li, J.G. Martinek, J.L. Falconer, R.D. Noble and T.Q. Gardner, Ind. Eng. Chem. Res., 44 (2005) 3220.

[2] M. A. Carreon, S. Li, J.L. Falconer and R.D. Noble, J. Am. Chem. Soc., 130 (2008) 5412.

[3] S. Li, J.L. Falconer and R.D. Noble, Microporous Mesoporous Mater., 110 (2008) 310.

[4] S. Li, J. L. Falconer and R. D. Noble, Adv. Mater., 18 (2006) 2601.

[5] M. A. Carreon, S. Li, J.L. Falconer and R.D. Noble, Adv. Mater., 20 (2006) 729.

[6] M. Hong, S. Li, H. F. Funke, J.L. Falconer and R.D. Noble, Microporous Mesoporous Mater., 106 (2007) 140.

[7] S. Li, J.L. Falconer and R.D. Noble, J. Membr.Sci., 241 (2004) 121.

[8] A. M. Avila, H. H. Funke, Y. Zhang, J.L. Falconer and R.D. Noble, J. Membr. Sci., 335 (2009) 32.

[9] Y. Zhang, A. M. Avila, H. H. Funke, J.L. Falconer and R.D. Noble, J. Membr. Sci., 363 (2007) 29.

(58e) An Experimental Study of Effect of Cyclone Separator On Transmission of Pressure Fluctuation in a CFB

Chen Yong, Zhang Qiong, Yan Chaoyu and **Wei Yaodong**, China University of Petroleum-Beijing, Beijing, China

Abstract: In large scale circulating fluidized bed device, in order to analyze the mechanism and regulation which cyclone separator effect on transmission of pressure fluctuation,dynamic pressure of before and after cyclone separator and different locations in cyclone separator was measured under many kinds of conditions according to change the solid circulating rate at the entrance of the cyclone separator. The particles used were FCC catalyst. The experimental results show that in the axial direction of the cyclone separator, with the same solid circulating flow rate, pressure fluctuations of the cyclone side walls are same as pressure fluctuation of the top of the dipleg but different with the inlet's. Pressure fluctuations of the cyclone center have differences with both the top of the dipleg's and the inlet's; In radial direction, from the side walls to the center, the differences of pressure fluctuations between the cyclone and the inlet and the top of dipleg are gradually increasing, and the tendency increased with increasing particle circulating flow rate; In circumferential direction, There are a slight differences of pressure fluctuations in different angles near the cyclone center. Near the side walls, when the solid circulating rate is low, pressure fluctuations in certain areas are different with other areas'. While when the solid circulating rate is high, pressure fluctuations have no difference substantially.

Small and Mid-Scale LNG Pumps for an Evolving Market

Steve Rush
Director of Sales, Ebara International Corporation – Cryodynamics Division

Natural gas has become more and more prominent around the globe as a key fuel source. The need for a more clean burning fuel and the abundance of natural gas has increased the need for smaller systems to deliver LNG to customers for their use. As more countries develop an infrastructure to utilize natural gas as a fuel for trucks, buses, small power stations and local buffer storage for small communities, the need for small and mid-scale equipment is increasing.

In its infancy, the LNG industry began as a means of delivering large volumes of gas from an export country with excess supplies to another location where the gas was needed. These systems required large equipment, large ships and large plant sites to liquefy and transport the LNG. Equipment was developed for this purpose and one of the key components in these systems is the cryogenic pumps used to move the liquefied gases at very low temperatures.

ADVANTAGES OF SUBMERGED MOTOR PUMPS

Due to the hazardous nature of handling natural gas, the type of pump developed for the LNG industry and now widely used for this purpose is the submerged electric motor pump (Figure 1). Since LNG is basically non-conductive, this pump type utilizes a submerged electrical motor which is installed inside a vessel or storage tank. With no oxygen present during operation, this is the safest design available. The entire pump and motor is submerged in the LNG and the liquid is also used for cooling the motor and lubricating the bearings. The alternative to this design is an external motor type pump using shaft seals between the motor and pump sections. The motor is outside in the hazardous area and while an explosion proof housing can be used to maintain a safe environment, there is still risk in having the motor in this environment with the potential for flammable gases present. The external motor type pump also uses a mechanical seal on the shaft where the shaft penetrates the vessel into the liquid and this design adds risk as there is the potential for leakage of the LNG into the atmosphere.

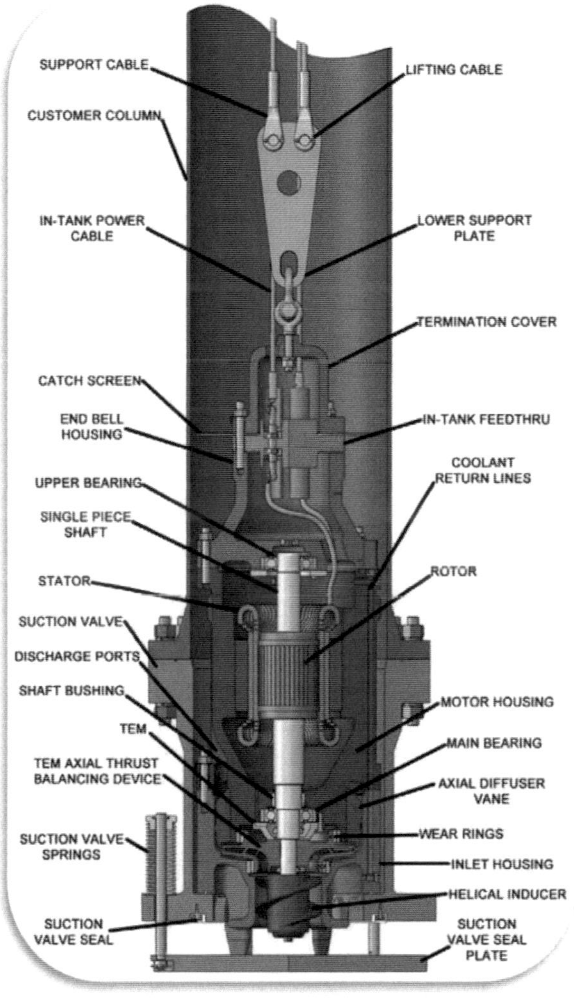

Figure 1: Typical In-Tank Retractable Type LNG Pump

Due to the safety advantages of the submerged motor pump design this pump type has become the most common type of pump used for pumping LNG as well as a variety of other liquefied gases which are non-conductive such as LPG, ethylene, ethane, propylene, LN2, etc.

THE NEED FOR MORE STANDARD SMALLER SCALE DESIGNS

The larger pumps used in many of the traditional LNG export, shipping and receiving terminal projects are typically designed to the project specifications and are manufactured when each project is built. This means that although many existing components in the manufacturer's inventory can be used, each pump is built for the project and is not a "standard" design. The delivery of these pumps is traditionally about one year from the time of order and can be very complex in their design. Many of these pumps are built in small quantities and do not lend themselves to standard manufacturing techniques. They are not built to stock and due to all of these issues, the costs are relatively high. However, the submerged electric motor

LNG pump designs have matured over more than 40 years of development and have become extremely reliable, often operating for as many as five years between overhauls.

Smaller scale LNG systems are being built with shorter delivery times and more standardized designs that require lower cost and more standardized equipment. These systems therefore require equipment that is less expensive and pre-engineered with parts that can be built to stock and assembled fairly easily and quickly.

Ebara International Corporation's Cryodynamics Division has recognized this need and has developed a line of submerged electric motor pumps which meet the increasing demand for the small to mid-scale market by adapting their experience from the larger scale pumps. This new pump model has no shaft seals like its larger cousin and uses Ebara's very reliable thrust balancing system (Thrust Equalizing Mechanism or TEM™).

EMPHASIS ON SAFETY

In the large scale LNG plants, the pumps and other equipment are monitored very closely by well-trained plant maintenance personnel. In most cases, the equipment has continuous monitoring systems which alert plant operators if any problems occur in the electrical systems or pump vibration. In smaller scale systems, which can sometimes be placed in more remote locations and may not be monitored continuously, the submerged motor design is an ideal choice as it lends itself to a very safe design with very low potential for leakage. You can imagine this type of system being similar to a gasoline station where the user is filling a vehicle with no personnel monitoring the system on a continuous basis. Although the liquid volumes are much smaller, using a pump with shaft seals and an external motor in small to mid-scale applications adds risk particularly when the equipment is not monitored in the same way as a large scale facility. This is particularly true in truck loading or other applications where a vehicle operator may be the one actually operating the system to fill a truck, bus, or other small tank.

The submerged LNG pumps developed by Ebara utilize pre-engineered and standardized designs to reduce cost and provide much shorter delivery times to customers building smaller scale systems while still providing the inherent safety and reliability of the larger scale pumps. These pumps can be assembled and shipped relatively quickly and while they can be shipped directly after assembly, if desired, they can also be performance tested in LNG in Ebara's test facility in Sparks, Nevada, in the western USA.

VARIABLE SPEED OPERATION

Another feature of the smaller scale pre-engineered pumps developed by Ebara is their ability to be connected to a variable frequency drive with no changes in the pump design or construction. In many smaller scale systems, wide ranges of pressure and flow are required due to variations in tank levels and temperatures particularly in applications such as truck trailers and truck and bus fuel tanks. When the tanks are close to empty, the temperature can be significantly warmer and a higher pressure is required during initial filling to collapse the pressure. Therefore, the pump needs to operate at a higher speed at first and after the temperature begins to drop in the tank, the pump speed can be reduced to continue filling the tank.

By introducing variable speed, fewer pump designs can be used to cover a wide range of flows and pressures using a relatively low number of common components which also helps reduce costs and parts inventory.

SMALLER SCALE PUMP APPLICATIONS AND TYPES

The LNG markets considered for this newly developed pump are as follows:

- Truck trailer loading
- Filling of truck and larger vehicle fuel tanks
- Peak-shaving
- Local municipal storage
- Small scale LNG ship or barge loading
- Small or mid-scale power plants

For many small scale systems, LNG storage tanks are built in the 5,000 to 20,000 m^3 size range. For this application, a removable type pump can be supplied in the same fashion as larger LNG storage tanks (Figure 2). This pump is installed in a pump column or tube with a suction valve mounted at the bottom. When the pump is lowered into the column, the bottom of the pump pushes the valve plate open allowing LNG to enter the pump inlet. The pump can then pump the liquid up to the top of the tank and into the discharge piping to its destination.

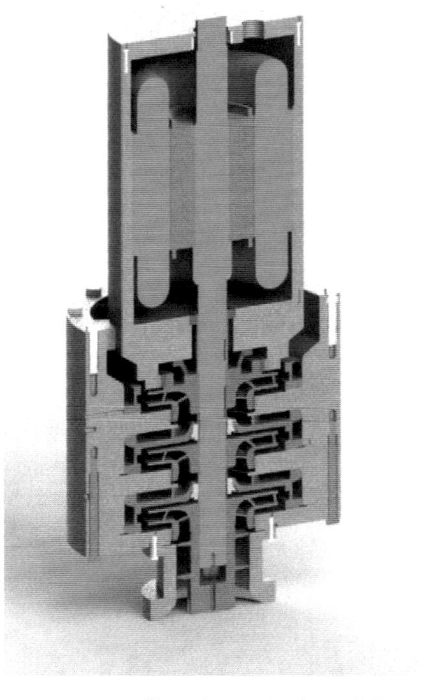

Figure 2: Small Scale In-Tank LNG Pump

If a pump is required for a pipeline type system the pump can be installed in its own stand-alone suction vessel which can be insulated using traditional insulation methods or can be supplied as a vacuum jacketed design. This type can be mounted anywhere in the piping system for LNG transfer.

In addition to the traditional land-based type pump systems, a variety of marine applications are also very suitable for the smaller scale pump designs. Ebara was the first manufacturer to introduce a design for the fuel gas pumping systems used on the Dual Fuel Diesel Electric (DFDE) LNG carriers being built today (Figure 3). This fuel gas pump can be easily adapted to any marine application where lower flows are required for off-loading LNG cargoes from barges or other smaller LNG marine vessels.

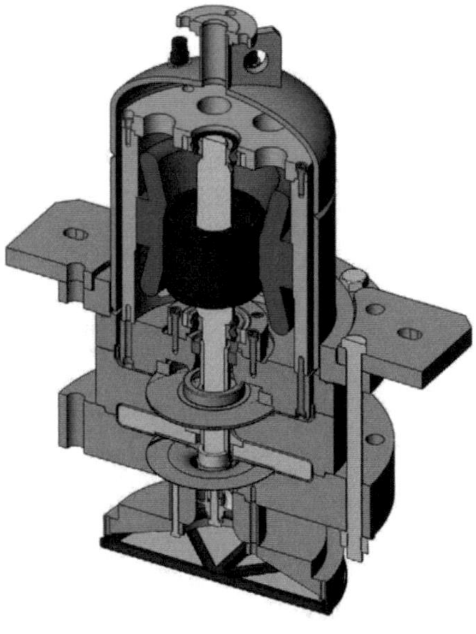

Figure 3: Small Scale Marine Pump

CONCLUSION

In conclusion, the newly developed submerged electric motor LNG pumps by Ebara International Corporation have been designed to provide safety, faster delivery and lower cost for the many customers building smaller scale systems. By providing the proven Thrust Equalizing Mechanism (TEM™) even in these lower cost pumps, customers can enjoy a significant level of reliability in addition to the inherent safety features of the submerged motor design with no rotating seals.

(116b) Relative Benefits Between Various Drivers for the Main Refrigeration Compressors for LNG Application

Patrice Bardon, Dresser-Rand, Houston, TX

The main refrigeration compressors in an LNG plant are consuming a lot of energy. The selection of the driver is a key decision and influences the future production pattern. Depending upon the driver selection the nominal plant size capacity can be affected. It is proposed to review the relative benefits of the various available drivers for the main refrigeration units: Electrical motors, Aero derivative gas turbines, industrial gas turbines and steam turbines. Two cases study will be reviewed in details: Steam turbine versus aero gas turbine for a 4 MTPA plant capacity and industrial gas turbine versus aero gas turbine for a 5 MTPA plant capacity.

(116c) Boil-off-Free LNG Transfer Between Floating and On-Shore LNG Storage Tanks

Hans E. Kimmel, Ebara International Corporation, Sparks, NV

The transfer of LNG from one tank to another tank is conducted by pumping at a discharge pressure of around 5 barg. The LNG in the tank is contained at a temperature around the boiling point stabilized by a continuous LNG boil-off to maintain vapor liquid equilibrium. Pumping the LNG at the discharge pressure increases the enthalpy of the LNG at the pump exit and during the isenthalpic expansion process inside the loading arm the LNG becomes a boiling two-phase fluid due to partial vaporization. The LNG vapor inside the pipe of the loading arm reduces the cross section of the pipe for liquid flow and substantially increases the flow velocity and the friction due to two-phase fluid flow. As a result and following the Bernoulli Equation, for a given differential pressure the mass flow will be reduced substantially.The reduction of the mass flow reduces the LNG transfer rate and increases the overall transfer time for a given quantity of LNG transfer. Also the vapor return line size will be reduced to a great extent. The paper presents a methodology and equipment to minimize or to eliminate this undesired LNG boil-off within the loading arm. It allows tight LNG flow inside inside loading arm which significantly shortens the LNG transfer time compared to current conventional methods.

() LNG Technology Development at Air Products

Yu-Nan Liu, Air Products & Chemicals, Allentown, PA

This presentation will trace the beginning of MCR (mixed components refrigerant) and use of Coil Wound Heat Exchanger, from helium recovery to the hybrid C3-MR process to AP-X® process and variation of it for natural gas liquefaction in baseload LNG plants and other applications, from Air Products' point of view.

Key Findings of Experimental and Theoretical Studies on Forced Mitigation System for an LNG Spill Emergency

*Byung Kyu Kim, Ray A. Mentzer, M. Sam Mannan**

Mary Kay O'Connor Process Safety Center, Artie McFerrin Department of Chemical Engineering, Texas A&M University System, College Station, Texas 77843-3122, USA
**Corresponding author. Tel.: +1 979 862 3985. E-mail address: mannan@tamu.edu (M.S. Mannan)*

Abstract

Mitigation measures have been categorized as one of the essential elements that require further investigation in determining the potential impact on public safety from an LNG spill. Several outdoor LNG spill experiments were conducted at the Brayton Fire Training Field to verify the effectiveness of the direct application of a water spray on LNG vapors as part of LNG Safety and Spill Response research at the Mary Kay O'Connor Process Safety Center (MKOPSC) at Texas A&M University (TAMU). Theoretical modeling was conducted of LNG forced mitigation using an upward-oriented conical water spray application. The computational fluid dynamics (CFD) model had been applied to investigate the dominant physical mechanisms involved in LNG forced mitigation and essential key design parameters were evaluated. Findings from the experimental and theoretical analysis on the LNG forced mitigation are summarized. Finally, recommendations on designing an effective forced mitigation system and potential hazards of applying the water curtain system are discussed. This work aims to provide a comprehensive understanding of the forced mitigation system with direct application to LNG facilities.

1. Introduction

The liquefaction capacity of natural gas is expected to double as the demand for natural gas expands substantially [1, 2]. The new emerging LNG supply from North America will also play a significant role in reorganizing the world gas supply chain in the near future [3]. With the transformation from an importer to exporter of LNG in North America, the new challenges rise from introduction of additional hazards from the liquefaction process and regulatory compliance to ensure safe operation [4]. NFPA 59A revised in 2013 requires all the new LNG facilities and any facilities that go through major modifications to conduct quantitative risk assessments to ensure that the facilities do not impose beyond tolerable risk to the nearby communities [5]. It is recommended to apply additional safety measures to meet the risk criteria. The effects of various mitigation techniques applicable to the LNG facilities still require further investigation to minimize the uncertainty in determining the impact of an LNG spill on the public safety and security [6]. This paper summarizes the research of water spray applications conducted at the Mary Kay O'Connor Process Safety Center (MKOPSC) at Texas A&M University (TAMU) as part of the LNG Safety Research Program.

2. Previous water spray research on LNG vapor clouds

The US Coast Guard conducted a small-scale LNG spill experiment to simulate a spill on an LNG transport ship in 1976 [7, 8]. It was concluded from the test results that the water spray curtain was effective in reducing the methane concentration and that the mechanical turbulence induced from the water droplets improved the mixing of the LNG vapors, enhancing the vapor dilution.

The Gas Research Institute conducted three-phase research, which consisted of theoretical analysis, small-scale spill tests, and wind tunnel experiments [9]. The downward spray application showed a reduction in concentration to 2–5% by volume in the downwind region. The upward sprays had diluted vapor concentrations approximately to 1 to 2% by volume; however, the effectiveness of vapor concentration reduction was not consistent with the theoretical estimation. It was concluded from the wind tunnel experiments that the upward sprays and downward-inclined sprays were the only two applications, which showed an effective dilution.

Meroney and Neff (1985) provided forced mitigation effects for both box and slab models assuming that the local entrainments increase with the application of the water curtain [10]. The enhanced air entrainment effects were incorporated into the model by applying a multiplicative or additive factor to the regular entrainment rates. The integral-type modeling provided a quick prediction compared to dense gas dispersion, and has shown reasonable predictions when compared to small-scale experiments. However, the integral-models present the fluid system in a way that limits the understanding of the complex interactions of water droplets with the gas flow. Also, there had been some concerns over the under-prediction from the integral-models in certain scenarios, mainly because of the simplified assumption [11].

3. Experimental work from MKOPSC

To investigate the complex physical mechanisms of LNG forced dispersion, the Mary Kay O'Connor Process Safety Center (MKOPSC) conducted outdoor LNG spill experiments at the Brayton Fire Training Field to verify the dominant mechanisms and evaluate the effectiveness in dilution the LNG vapor concentration [12, 13].

Fig. 1. LNG dispersion test: (a) without water curtain, (b) with water curtain (full cone) [14]*

Rana, Guo & Mannan (2010) discuss one of the experimental setups that had been conducted in 2007 [14]. Fig. 1 shows the experimental work of the water spray application on LNG vapor clouds. Two different types of water spray nozzles were mainly compared: 60° full-cone spray nozzle and 180° flat-fan spray nozzle [15]. The full-cone type produces finer droplets, while the flat-fan type creates a thin barrier in the vicinity of the nozzle and very course droplets as the bulk water travel and break into smaller droplets. The full-cone spray type provides more effective mixing with the air through high air entrainments (Fig. 2(a)). The full-cone type showed an effective dilution at all elevations. The flat-fan type creates a physical barrier in the pathway of LNG vapors and provides high momentum from the droplets created from the water pressure. The flat-fan type pushed the LNG vapors to higher elevations, resulting in an increase in concentration behind the water spray region (Fig. 2(b)).

* Reprinted figures with permission from Rana, M., Guo, Y., & Mannan, M. S. (2010). Use of water spray curtain to disperse LNG vapor clouds. Journal of Loss Prevention in the Process Industries, 23(1), 77-78. Copyright 2010 Elsevier.

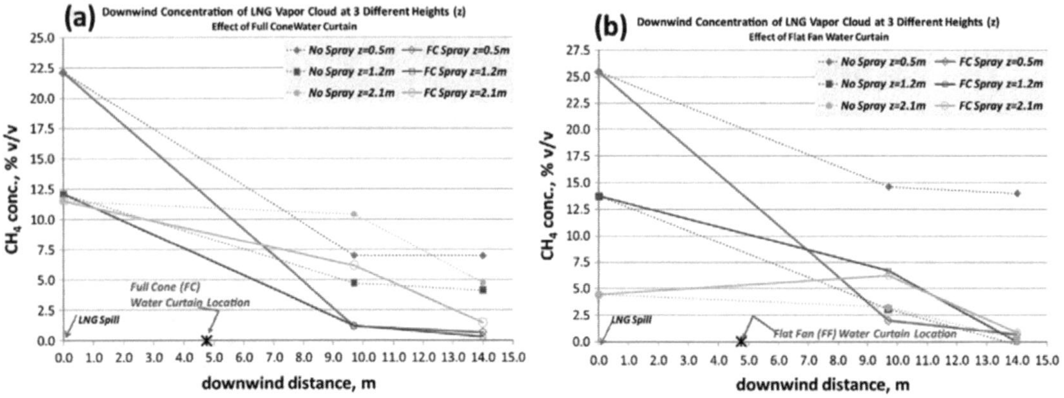

Fig. 2. Downwind concentration at three different heights with and without (a) full-cone water curtain and (b) flat fan type [14]*

The temperature changes collected from the experiments suggested that the water droplets provided heat transfer to the LNG vapors. The overall heat transfer evaluated from the experimental results showed that the droplets produced from the full-cone type nozzle provided more heat transfer to the LNG vapors than the flat-fan application (Fig. 3).

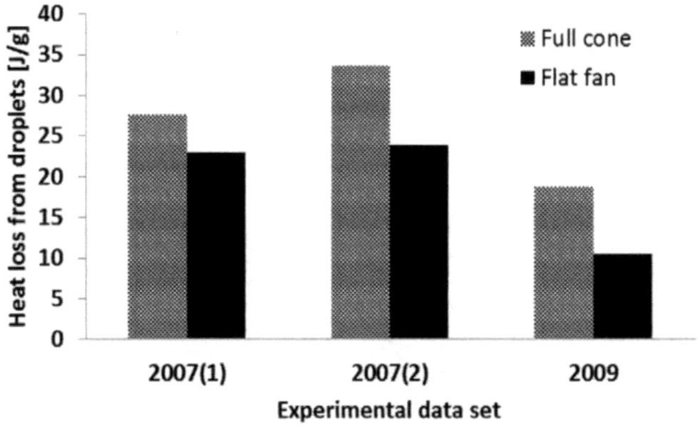

Fig. 3. Calculated heat loss by water curtain (Modified from [14])

The full-cone type nozzle produces smaller droplets, which increases the surface area. It was concluded that the mixing effects through the entrained air promotes the LNG vapors most effectively. The heat transfer from the droplets ensures that the vapors are sufficiently warmed to become positively buoyant.

4. Modeling approach from MKOPSC

The CFD tool was adopted to conduct a comprehensive assessment of the effects of the water curtain applications on LNG vapor clouds. In the present work, ANSYS Fluent was used to simulate the Eulerian–Lagrangian spray modeling coupled with the LNG vapor flows. The essential parameters for simulating the detailed description of LNG flow and interactions of the droplet-vapor system were calibrated against the experimental results obtained from the LNG spill work conducted at the Brayton

Fire Training Field. The physical parameters that are essential inputs in setting up the CFD simulation of the water-spray LNG system are discussed elsewhere [16].

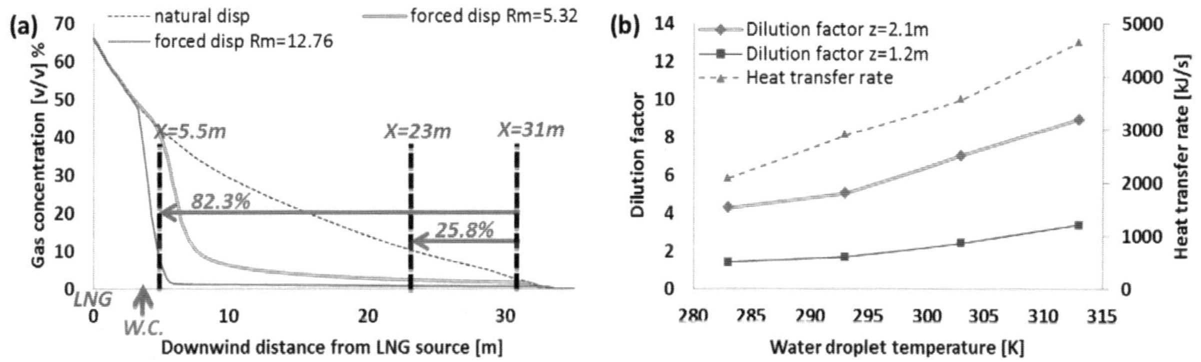

Fig. 4. (a) Volumetric concentration for natural/forced(RM=5.32)/forced(RM=12.76) dispersion and (b) heat transfer rate and dilution factor at different water droplet temperatures [16][**]

Fig. 4 shows the dilution effects of two different spray settings applied (R_M= 5.32, and 12.76) and heat transfer involved for various droplet temperatures simulated using the LNG forced dispersion model. R_M is the momentum ratio, which is the ratio of the momentum imparted from the droplets to the momentum of the gas clouds [17]. Different levels of safety distance reduction were observed for various momentum ratios applied from the water spray as shown in Fig. 4 (a). As the water nozzle discharged more momentum to the LNG vapors, the dilution effects increased. Also, the heat transfer rate increased as the droplet temperature increased as shown in Fig. 4 (b). The effectiveness of dilution improved as well, as more heat transfer was involved at different elevations (z = 1.2 and 2.1 m). These results indicate that the LNG forced dispersion modeling using the Eulerian-Lagrangian spray model is capable of simulating the dilution effects and influences on LNG vapor behavior by taking into account the detailed droplet characteristics.

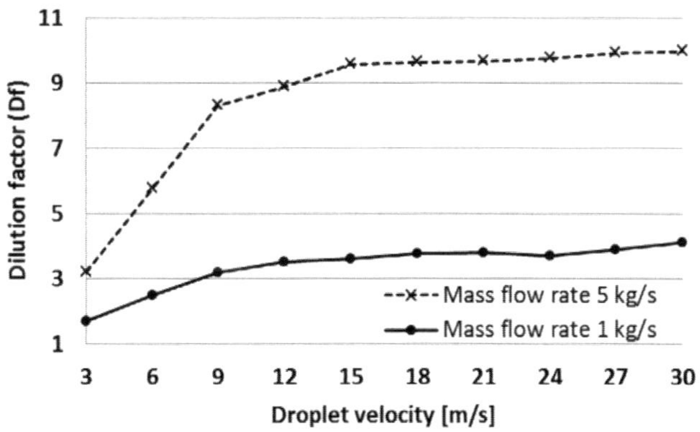

Fig. 5. Dilution effects from different mass flow rates and droplet velocities

Using LNG forced dispersion modeling, the physical mechanisms of the LNG forced dispersion were investigated. The effects of momentum imparting from the droplets to the air-vapor mixture,

---

[**] Reprinted figures with permission from Kim, B. K., Ng, D., Mentzer, R. A., & Mannan, M. S. (2012). Modeling of water spray application in the forced dispersion of LNG vapor cloud using a combined Eulerian-Lagrangian approach. Industrial & Engineering Chemistry Research, 51, 13803-13814. Copyright 2012 American Chemical Society.

thermal effects between two phases (vapor/droplet), and different level of air entrainments on the LNG vapor behavior were evaluated.

The dilution effects of LNG forced dispersion were evaluated with different ranges of droplet velocity (3–30 m/s) at two different mass flow rates (1 and 5 kg/s) as shown in Fig. 5. The dilution effects increased as the droplet velocity increased. However, the spray with mass flow rate of 1 kg/s showed only limited dilution effects even with a high droplet velocity applied. This is because even with the high droplet velocity to provide effective mixing of the air-vapor mixture, the insufficient water introduced from the nozzle limited the effective dilution of LNG vapors. This result indicates that both of the design elements, the droplet velocity and mass flow rate, must be set above a certain range to ensure an effective mitigation effect. From the water spray setting applied in this research, it was above 3 kg/s of mass flow rate with 12–15 m/s droplet velocity.

The thermal effects from the water droplets to the vapor clouds, particularly for cold gas releases, can enhance dissipation of the vapor cloud by warming to gain positive buoyancy [18]. The temperature of the water droplet was varied from 283 to 313 K to evaluate the turbulence effects induced from different levels of heat transfer effects of the droplets, as shown in Fig. 6.

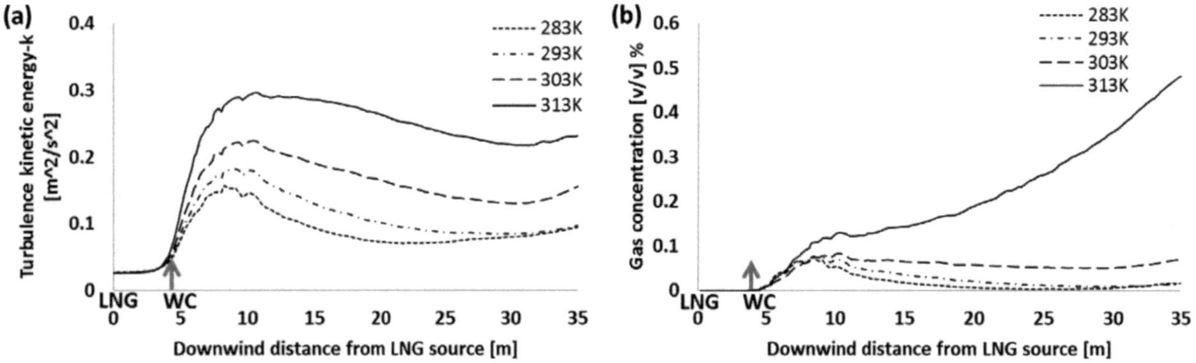

Fig. 6. (a) Turbulence kinetic energy and (b) vapor concentration from various droplet temperature applications at 8 m elevation

The turbulence kinetic energy showed the highest level being induced from the water droplet with highest temperature (313 K). The 8 m elevation was chosen to show the vapor behavior at high elevation. The concentration also increased significantly with the highest water droplet temperature, indicating that the vapors are sufficiently warmed to travel upward to the atmosphere. This result shows that the thermal effects from the droplet enhance the turbulent flow within the vapor clouds, resulting in an effective dilution of LNG vapors.

The air entrainment rate is determined by the spray characteristics: droplet sizes, droplet velocity, spray location, and configuration. Rana (2009) have verified that the entrained air rate into the spray increases as the nozzle angle size increases, while the operating pressure does not affect the rate of the air entrainment significantly [13]. A total of four different angle sizes (30°, 45°, 60°, and 75°) were simulated, as shown in Fig. 7.

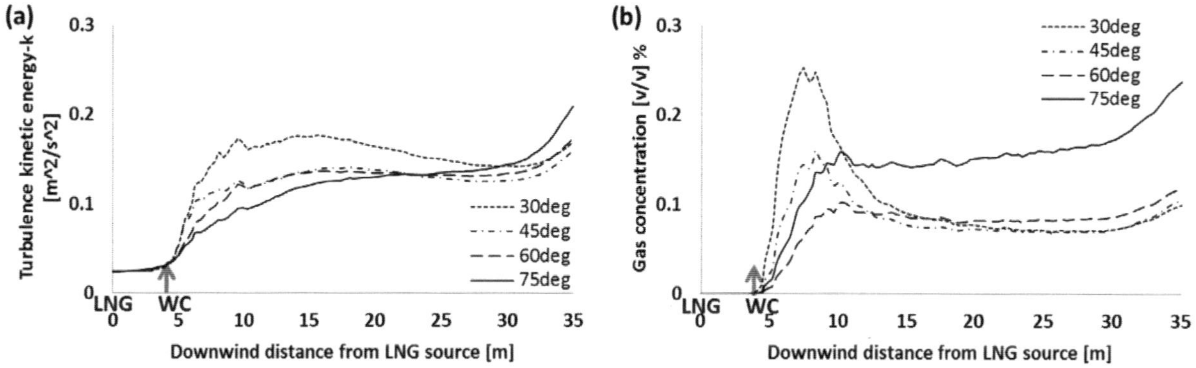

Fig. 7. (a) Turbulence kinetic energy and (b) vapor concentration from various nozzle angle size applications at 8 m elevation

The turbulence induced at 8 m elevation showed that the turbulence effects from various angle sizes were similar. On the other hand, the vapor concentration showed an increasing trend for the largest nozzle angle (75°) at 8 m elevation, while the concentration for the rest of other applications showed similar results once passed through the water spray. This indicates that the air entrainment setting that could induce a large amount of air is more effective in dispersing the LNG vapors through effective mixing, with turbulence effects playing a less significant role.

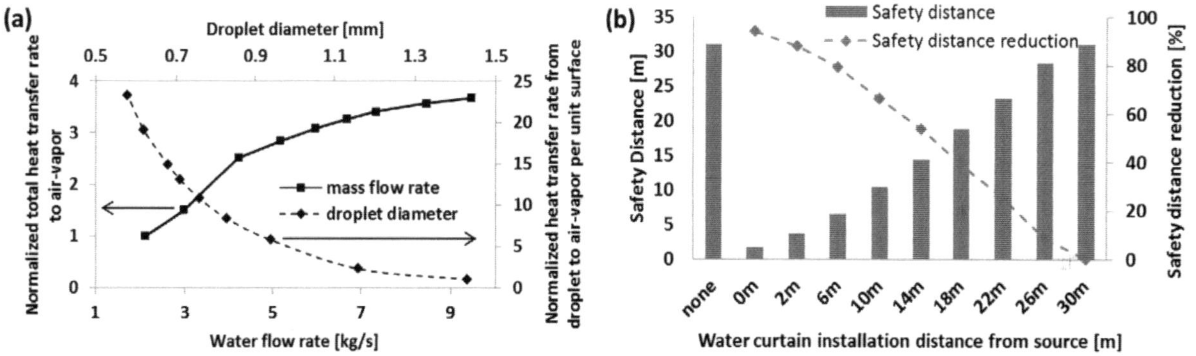

Fig. 8. (a) Heat transfer rate from droplets to air-vapor mixture and (b) safety distance evaluated from various installation distances [19]***

The size of droplets is determined by the water pressure and nozzle design, which is an intrinsic characteristic for a specific nozzle design. Different droplet sizes were simulated to evaluate the heat transfer rates (Fig. 8 (a)). The total heat transfer clearly indicated that the total amount of heat transferred from the droplet to LNG increased as the flow rate increased. However, the heat transfer from the surface decreased significantly as the droplet size increased, because of the larger droplets providing smaller surface area. The installation distance from the LNG source was investigated (Fig. 8(b)). The safe distance, defined as where the LFL is reduced by 50%, was reduced significantly as the water curtain was installed closer to the LNG source. These results can be used to evaluate the impact of the distance factor in calculating the dilution effects.

*** Reprinted figures with permission from Kim, Byung Kyu, Ng, Dedy, Mentzer, Ray A., & Mannan, M. Sam. (2013). Key parametric analysis on designing an effective forced mitigation system for LNG spill emergency. Journal of Loss Prevention in the Process Industries. doi: http://dx.doi.org/10.1016/j.jlp.2013.01.007. Copyright 2013 Elsevier.

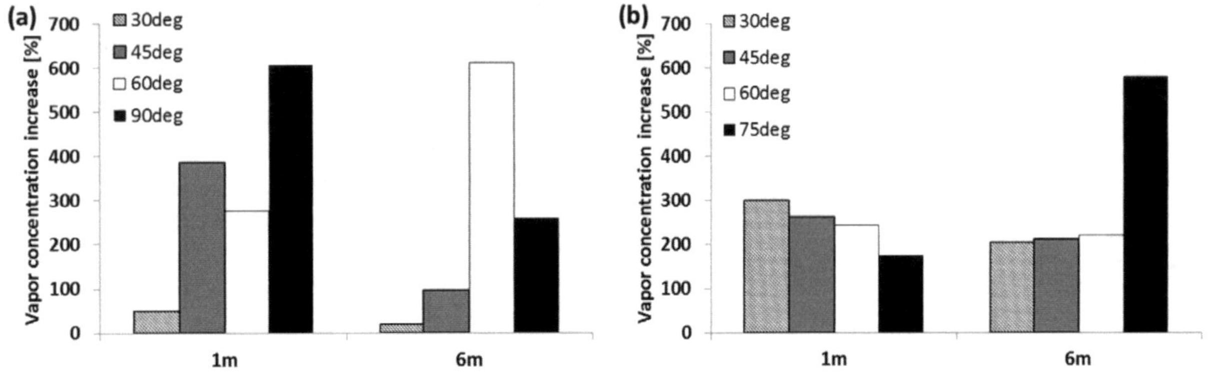

Fig. 9. Vapor concentration increase [%] at 1 m and 6 m with (a) modified nozzle installations (tilted at 30°, 45°, 60°, and 90°) and (b) different nozzle sizes (30°, 45°, 60°, and 75°) [19] [***]

The tilted installation designs at various nozzle angles were evaluated (30°, 45°, 60°, and 90°). Per Fig. 9 (a), the vertical installation showed the most effective dilution at lower heights (1 m), while at higher elevation (6 m), the nozzles tilted at 60° showed the largest increase of vapor concentration. This indicates that the nozzles tilted at 60° discharge the droplets that could effectively envelop the LNG vapors and enhance dilution of the air-vapor mixture. Various levels of air entrainments were applied to evaluate the concentration increase in the atmosphere are plotted in Fig. 9 (b). The angle sizes of the nozzle were modified in 30°, 45°, 60°, and 75° to simulate different levels of entrainments. The smaller angle nozzles were more effective in pushing the vapors to lower heights. On the other hand, the nozzle with largest angle size (75°) showed the highest increase of vapor concentration at higher elevation (6 m). This indicates that the nozzle with the largest entrainment air setting provides enhanced mixing that could effectively dilute the LNG vapors to travel upward to the atmosphere.

5. Conclusion

The engineering guidelines in designing an effective water spray system for an LNG facility are currently lacking mainly due to the existing gaps in the experimental results and limited theoretical modeling developed using integral-type models. The CFD code had been applied in evaluating the forced dispersion of LNG vapors using a water curtain application. The proposed model uses a Eulerian–Lagrangian approach, which takes into account the detailed characteristics of the water droplets in evaluating the effects on the LNG vapor behavior. The model shows a promising solution in investigating the complex interaction of the droplet-vapor system. The CFD code also can serve as a rigorous tool in evaluating the design for a site-specified mitigation system, to predict any unforeseen hazards that might rise from under-sized mitigation. The modeling approach applied in this work can evaluate the risk reduction effects from applying the water spray application, which can directly be implemented for quantitative risk assessments, as indicated by NFPA 59A.

References

1. International Energy Outlook 2011 (US Energy Information Administration), 2011, US Energy Information Administration (EIA): Washington, DC.

[***] Reprinted figures with permission from Kim, Byung Kyu, Ng, Dedy, Mentzer, Ray A., & Mannan, M. Sam. (2013). Key parametric analysis on designing an effective forced mitigation system for LNG spill emergency. Journal of Loss Prevention in the Process Industries. doi: http://dx.doi.org/10.1016/j.jlp.2013.01.007. Copyright 2013 Elsevier.

2. World Energy Outlook 2011: Are we entering a golden age of gas? 2011, International Energy Agency: Paris, France.
3. Henderson, J., The potential impact of North American LNG exports, 2012, The Oxford Institute for Energy Studies: Oxford, UK.
4. Chosnek, J. and V. Edwards. From LNG import to exports: Process safety and regulatory challenges. Presented in 15th Annual Symposium Mary Kay O'Connor Process Safety Center. 2012. College Station, TX.
5. NFPA 59A: Standard for the Production, Storage, and Handling of Liquefied Natural Gas (LNG), 2013, National Fire Protection Agency: Quincy, MA.
6. Public Safety Consequences of a Terrorist Attack on a Tanker Carrying Liquefied Natural Gas Need Clarification (Maritime Security GAO-07-316), United States Government Accountability Office, 2007: Washington, D.C.
7. Brown, L., et al., Small scale tests on control methods for some liquefied natural gas hazards, 1976, US Coast Guard: Washington D.C.
8. Martinsen, W.E., S.P. Muhlenkamp, and J. Olson, Disperse LNG vapors with water. Hydrocarbon Processing, 1977: p. 260-267.
9. Heskestad, G., et al. Effectiveness of water spray curtains in dispersing LNG vapor clouds. in In proceedings of the American Gas Association Transmission Conference Paper, Paper No 83-T-69, 169-183. 1983. Seattle, Washington.
10. Meroney, R. and D. Neff, Numerical modelling of water spray barriers for dispersing dense gases. Boundary - Layer Meteorology, 1985. 31(3): p. 233-247.
11. Gavelli, F., E. Bullister, and H. Kytomaa, Application of CFD (Fluent) to LNG spills into geometrically complex environments. Journal of Hazardous Materials, 2008. 159(1): p. 158-168.
12. MKOPSC, Data Report of MKOPSC LNG Spill Tests: 2005-2009, M.S. Mannan, Editor 2010, Mary Kay O'Connor Process Safety Center, Texas A&M Univeristy: College Station, TX.
13. Rana, M., Forced dispersion of liquefied natural gas vapor clouds with water spray curtain application, Chemical Engineering, Texas A&M University, PhD Dissertation, 2009. p. 232.
14. Rana, M., Y. Guo, and M.S. Mannan, Use of water spray curtain to disperse LNG vapor clouds. Journal of Loss Prevention in the Process Industries, 2010. 23(1): p. 77-78.
15. Rana, M. and M.S. Mannan, Forced dispersion of LNG vapor with water curtain. Journal of Loss Prevention in the Process Industries, 2010. 23(6): p. 768-772.
16. Kim, B.K., et al., Modeling of water spray application in the forced dispersion of LNG vapor cloud using a combined Eulerian-Lagrangian approach. Industrial & Engineering Chemistry Research, 2012. 51: p. 13803-13814.
17. Hald, K., et al., Heavy gas dispersion by water spray curtains: A research methodology. Journal of Loss Prevention in the Process Industries, 2005. 18(4–6): p. 506-511.
18. CCPS, Guideline for postrelease mitigation technology in the chemical process industry. 1997, New York: Center for Chemical Process Safety, American Institute of Chemical Engineers.
19. Kim, B.K., et al., Key parametric analysis on designing an effective forced mitigation system for LNG spill emergency. Journal of Loss Prevention in the Process Industries, 2013(0).

A View of the Evolving LNG Regulations and Associated Exclusion Zones from an Industry Perspective

Alfonso F. Ibarreta
Exponent, Inc.
9 Strathmore Road
Natick, MA 01760
dust@exponent.com

Ryan Hart
Delmar "Trey" Morrison
Harri Kytomaa
Exponent, Inc.

Prepared for Presentation at
American Institute of Chemical Engineers
2013 Spring Meeting
9th Global Congress on Process Safety
San Antonio, Texas
April 28 – May 1, 2013

UNPUBLISHED

A View of the Evolving LNG Regulations and Associated Exclusion Zones from an Industry Perspective

Alfonso F. Ibarreta
Exponent, Inc.
9 Strathmore Road
Natick, MA 01760
dust@exponent.com

Ryan Hart
Delmar "Trey" Morrison
Harri Kytomaa
Exponent, Inc.

Keywords: LNG, FERC, vapor dispersion, exclusion zone

Abstract

Recent growth in availability of domestic natural gas has led to a shift in the U.S. gas industry from being a net LNG importer to the prospect of becoming a net LNG exporter. As a consequence, many industry players are actively pursuing natural gas liquefaction projects at existing LNG import terminals. In concert with this growth in proposed construction projects, the LNG facility permitting process driven by Federal Energy Regulatory Commission (FERC) has changed dramatically over the past few years. Although the requirements for LNG facility siting and design are codified in 49 CFR 193 and NFPA 59A, many of the exclusion zone analysis requirements are stated broadly and involve some level of expert interpretation. FERC has in the past clarified their interpretation of the federal requirements by means of formal letters and less formal precedent-setting memoranda. These interpretations continue to evolve over time and drive new tools and considerations.

This paper examines the ramifications of consequence-based accidental release scenarios that have been historically required, describes the latest requirements imposed by FERC and explains how these came into existence. The changing requirements have ostensibly been driven toward better definition of release sources; however, FERC's requirements have challenged the available and approved analytical tools. Also with this evolution, new types of scenarios coupled with new models have heavily influenced piping and equipment design, unit layout, unit siting, and mitigation strategies. This paper will discuss the evolution and the now dominant considerations for facility siting and unit design that drive the safety aspects of LNG facility permitting process from the industry's perspective.

Specific examples of scenarios will be provided as well as their consequences in terms of the resulting ½-LFL vapor clouds as well as the distance to the 1 psi boundary for the corresponding vapor cloud explosions. This paper will also provide a summary of the effectiveness of mitigation techniques, including the use of vapor fences based on numerous case studies.

1. Introduction

Recent growth in availability of domestic natural gas has led to a shift in the U.S. gas industry from being a net LNG importer to the prospect of becoming a net LNG exporter. As a consequence, many industry players are actively pursuing natural gas liquefaction projects at existing LNG import terminals. In concert with this growth in proposed construction projects, the LNG facility permitting process driven by Federal Energy Regulatory Commission (FERC) has changed dramatically over the past few years.

The codes and standards for which LNG facilities must be designed, constructed, and operated are found in 49 CFR Part 193 [1], and the authority having jurisdiction (AHJ) for these facilities is the Federal Energy Regulatory Commission (FERC). As part of the application process, the applicants (facility owners) must perform an environmental analysis in a series of thirteen (13) Resource Reports. Included in the assessment is a reliability and safety assessment (Resource Report 11) [2]. FERC then reviews the applications and must give approval prior to the commencement of any construction activities.

Although the requirements for LNG facility siting and design are codified in 49 CFR 193 and NFPA 59A [3], many of the exclusion zone analysis requirements are stated broadly and involve some level of expert interpretation. FERC has in the past clarified their interpretation of the federal requirements by means of formal letters and less formal precedent-setting memoranda. These interpretations continue to evolve over time, and drive new tools and considerations.

2. LNG Facility Permitting Exclusion Zone Requirements

The LNG facility permitting process requires that the designs meet the siting requirements set forth in:

- 40 CFR 68 - *Chemical Accident Prevention Provisions* [4].
- 49 CFR 193 - *Liquefied Natural Gas Facilities: Federal Safety Standards* [1].
- NFPA 59A (2001) *Standard for the Production, Storage, and Handling of Liquefied Natural Gas (LNG)* [3].

These standards require the presence of certain exclusion zones separating LNG and refrigerant storage, processing and transfer areas from potential ignition sources and locations not under the control of the LNG facility (off-site properties). The siting requirements set forth in Subpart B of 49 CFR 193 must be provided with hazard exclusion zones, ensuring that the following hazard areas do not extend beyond a property line that can be built upon:

- **Flammable vapor-gas dispersion hazard:** Regulations require that a ½-LFL flammable vapor cloud must not extend "beyond the property line that can be built upon" for worst-case 10-minute release scenarios [49 CFR § 193.2059 and NFPA 59A (2001) § 2.2.3.3]. Cases to be considered include:

  - A worst-case scenario LNG or refrigerant spill.
  - A worst-case scenario LNG or refrigerant jetting and flashing release.

- **Vapor explosion overpressure hazard:** Regulations require an offsite consequence analysis of the 1 psi overpressure boundary resulting from ignition of worst-case 10-minute releases [40 CFR 68]. Cases to be considered include:
  - A worst-case scenario refrigerant jetting and flashing release.

- **Thermal radiation (pool fire) hazard**: Regulations require that a 1,600 Btu/hr/ft^2 (5,000 W/m^2) level must not be exceeded "at a property line that can be built upon" for pool fire scenarios [49 CFR § 193.2057 and NFPA 59A (2001) § 2.2.3.2]. Cases to be considered include:
  - A worst-case scenario LNG or refrigerant liquid spill.

The hazard scenarios described above are based on a 10-minute LNG or refrigerant release at the maximum leak flow rate, based on pipe break scenarios persuaded by FERC.

Based on Exponent's previous experience, the flammable vapor dispersion hazard exclusion zones resulting from the worst-case jetting and flashing scenarios in process areas normally result in the largest stand-off distance requirements. Therefore, the following sections focus on the requirements imposed by FERC for the analysis of the flammable vapor-gas dispersion hazard described above.

3. Determination of Flammable Vapor-Gas Hazard Exclusion Zone and Effective Property Boundary

The federal regulation 49 CFR 193 requires that each LNG container and transfer system existing within an LNG facility must have an exclusion zone based upon vapor dispersion analyses conducted in accordance with the prescription of sections 2.2.3.3 and 2.2.3.4 of NFPA 59A (2001). The exclusion zone, according to 49 CFR 193, must be "an area surrounding an LNG facility in which an operator or government agency legally controls all activities in accordance with §193.2057 and §193.2059 for as long as the facility is in operation." Considering the regulatory definition for exclusion zone, the property line for which the vapor dispersion exclusion zone would apply is considered to be the boundary at which the LNG owner/operator ceases to control all activities at all times.

Additionally, according to the CFR requirements, the hazard exclusion zones cannot extend beyond "a property line that can be built upon." This is normally demarcated by the facility property boundary. This area can also be extended by either (a) proving the area of interest cannot be built upon (such as a waterway), or (b) contractually ensuring that the area is under the control of the LNG facility. In the case of a railroad or roadway traversing the property, FERC has previously set a precedent to allow classification as transient travel through an exclusion zone.

4. Determination of Design LNG and Refrigerant Spill Scenarios in Process and Transfer Areas

For process areas, NFPA 59A Table 2.2.3.5 requires the design spill be selected based on single accidental leakage source scenario. Unfortunately, NFPA 59A (2001) lists a few specific scenarios but does not generally define what a single accidental leakage source entails. FERC has therefore provided guidance for selecting single accidental liquid leakage sources to analyze in order to demonstrate compliance with 49 CFR Part 193.

4.1 Types of Leaks to Consider

Based on guidance provided by FERC, the releases of LNG and refrigerants are divided into two categories that are both to be considered in an exclusion zone analysis:

(1) Liquid spill scenarios consider full pipe ruptures where liquid spills onto the ground. The liquid conveyance in trenches and troughs, and collection in impoundments, resulting in subsequent vapor cloud formation and dispersion is then considered for evaluation of the exclusion zone.

(2) Conversely, pressurized jetting and flashing release scenarios are bounded by the constraint that little or no liquid from the fluid jet reaches the ground (i.e. no rainout). Depending upon the process conditions within the pipe and the release scenario, the resulting single accidental leakage source may produce some rainout. Only jetting and flashing release scenarios that produce little or no rainout are therefore analyzed, since they tend to create the largest vapor clouds.

4.2 Prescriptive Leak Size Determination

In order to officially establish what types of leak scenarios needed to be considered, FERC staff initially sent a letter to the DOT on April 19, 2005, requesting concurrence on the procedures for determining a single accidental leakage source [5]. FERC staff originally required an evaluation of all small diameter attachments to the transfer piping for instrumentation, pressure relief, recirculation, etc., and any flanges that may be used at valves or other equipment in order to determine the largest spill rate. The DOT affirmed this approach in a May 6, 2005, letter to FERC [5]. Based on these requirements, a wide variety of single accidental leakage sources, ranging from packing and flange leaks, to full guillotine ruptures of ship unloading lines, were submitted to FERC by applicants for several years. [5]

Subsequently, in the 2011 to 2012 timeframe, FERC identified more specific criteria for selecting leak sources for piping on the order of 100 m, and later for piping on the order of 1000 m in length. FERC initially provided the following criteria in 2011 for selecting the leak scenarios in "piping lengths on the order of 100 m:" [6]

- For piping less than or equal to 6 inches in diameter, a full pipe rupture at any point along the line; and
- For piping greater than 6 inches in diameter, the greater of:

  - a full rupture of a connection less than or equal to 6 inches in diameter; or
  - a hole of 2 inches in diameter.

FERC subsequently provided the following criteria in 2012 for selecting the leak scenarios in "piping lengths on the order of 1000 m:" [7]

- For piping less than or equal to 12 inches in diameter, a full pipe rupture at any point along the line; or
- For piping greater than 12 inches in diameter, a hole equivalent to 1/3 of the diameter.

4.3 Quantitative Leak Size Determination

The initial prescriptive approach for describing a single accidental leakage source did not provide any quantitative justification for the selection of the design spill [5]. In order to establish a quantitative criterion, FERC chose to set an acceptable failure rate limit for components based on a published risk analyses for the failure of an LNG storage tank outlet line. FERC staff determined that the rupture of a storage tank outlet line is on the order of one failure every 20,000 to 30,000 years ($5x10^{-5}$ to $3x10^{-5}$ failures per year) [5, 8]. Based on this, FERC required in May of 2012 that single accidental leakage sources be selected for analysis if the likelihood of failure is greater than $3x10^{-5}$ failures per year [9].

FERC provided a failure rate table to determine the likelihood of failure of all components in the LNG and refrigerant systems [9]. The table is meant to be used to determine whether a component failure rate is higher than the $3x10^{-5}$ failures per year criterion, meaning that it would have to be considered in the permitting analysis. Failure rates must currently be considered for:

- Transfer and process piping (failure per meter of piping)
  - Catastrophic rupture of the full pipe diameter
  - 1/3 diameter hole
  - 10% diameter hole, up to 2-inches
  - 1-inch hole
- Process vessels
- Valves and joints
- Transfer arms and hoses

Figure 1 illustrates how this process works. For a 10-inch diameter transfer line that is 45 meters in length, 3 separate scenarios must be first considered: a catastrophic rupture, a 3.33-inch diameter hole and a 1-inch diameter hole. The failure scenarios are first screened to determine if they meet the failure criterion set by FERC. The largest leak size that meets the FERC criterion needs to be considered in the vapor dispersion analysis.

Piping: 150mm (6-inch) <d< 299mm (12-inch)	Failures per year of operation
Catastrophic rupture	2E-7 per meter of piping
Release from hole with effective diameter of 1/3 diameter	4E-7 per meter of piping
Release from hole with effective diameter of 25mm (1-inch)	7E-7 per meter of piping

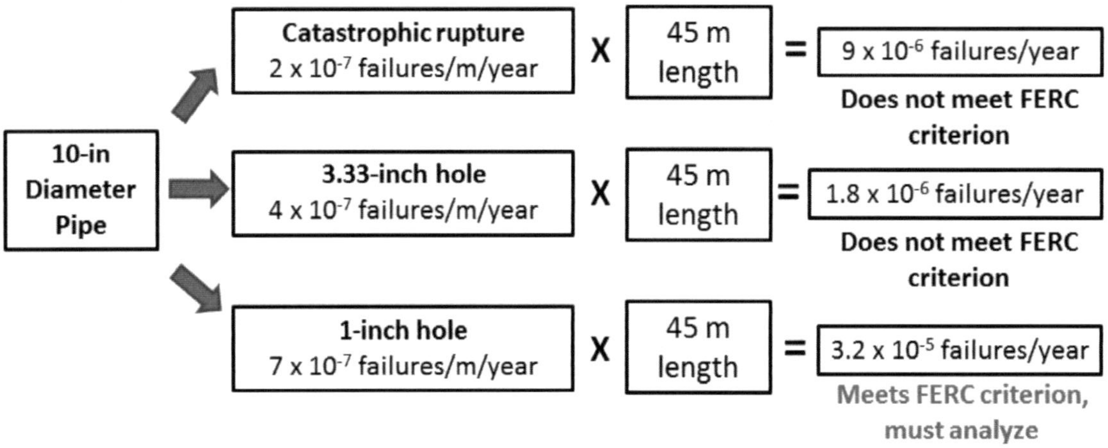

Figure 1 Example of quantitative leak size calculation logic tree.

5. Models Approved by FERC

Several computational models have been approved in the past for modeling the vapor dispersion resulting from LNG and refrigerant releases. In 2007, the Fire Protection Research Foundation issued a report which established a Model Evaluation Protocol (MEP) for LNG vapor dispersion models [10]. The computational models approved by FERC since then have been required to be first tested against the MEP. Three approved vapor dispersion models are discussed.

5.1 DEGADIS Integral Model

NFPA 59A (2001) *Standard for the Production, Storage, and Handling of Liquefied Natural Gas* §2.2.3.4 indicated that "flammable mixture dispersion distances shall be calculated in accordance with the model described in Gas Research Institute report GRI 0242, LNG Vapor Dispersion Prediction with the DEGADIS Dense Gas Dispersion Model." In the 2009 edition of NFPA 59A the prescription of DEGADIS was removed and it was required that the model used be acceptable to the AHJ based on an evaluation using the MEP [11]. FERC officially approved the model in 2011 [11]. It should be noted that 49 CFR 193 does not reference the 2009 edition of NFPA 59A.

The integral model DEGADIS was developed for the Gas Research Institute and the U.S. Coast Guard specifically to account for effects such as gravity spreading, negative or positive buoyancy effects on air entrainment, surface to cloud heat transfer, and phase change energy effects associated with air humidity in modeling dispersion of dense gases [11].

The DEGADIS model had previously been used in conjunction with the SOURCE5 model, developed originally by the Gas Research Institute. The SOURCE5 model allowed the determination of LNG evaporation rates for instantaneous or continuous releases on either land or water [12]. This model contained serious deficiencies, and DOT declared that it could no longer be used for LNG facility permitting in July 2010 [13].

5.2 PHAST Integral Model

The PHAST model was approved by PHMSA in 2011 for analyzing LNG vapor dispersion exclusion zones [14]. PHAST is commercial software developed and distributed by Det Norske Veritas (DNV). PHAST is widely used for the calculation of hazard distances from the release of several hazardous substances, including LNG. The PHAST code uses the Unified Dispersion Model (UDM) as an integral model to calculate the dispersion following a two-phase pressurized release or an unpressurised release. It consists of the following linked modules:

- Near-field jet dispersion;
- Non-equilibrium droplet evaporation and rainout, touchdown;
- Pool spread and vaporization;
- Heavy gas dispersion; and
- Far field passive dispersion.

The UDM allows for continuous, instantaneous, constant finite-duration and general time-varying releases. The UDM allows for possible plume lift-off when a grounded plume becomes buoyant. The UDM has been validated extensively with experimental data and is the subject of several peer-reviewed scientific papers.

5.3 FLACS CFD Model

The FLACS v9.1 model was validated against the experimental dataset in accordance with the LNG vapor dispersion Model Evaluation Protocol (MEP). A petition to have the FLACS v9.1 vapor gas dispersion model approved by the US DOT as an alternate model for LNG vapor dispersion under 49 CFR § 193.2059 was filed on October 27, 2010 [15]. The US DOT issued its Final Decision approving the use of FLACS v9.1 for vapor dispersion exclusion zone calculations on October 7, 2011 [15].

The FLACS CFD software is commercially-available and can model vapor dispersion scenarios and vapor cloud explosions. The CFD model provides a time-dependent solution to the Navier-Stokes equations. In the model, conservation equations for mass, momentum, enthalpy and gas concentrations are solved on a Cartesian grid using a finite volume method.

FLACS contains a liquid spill model which uses the shallow water equation to calculate the spread of LNG or refrigerants over the ground. FLACS also contains a pool evaporation model which calculates the heat transfer to the liquid from conductive heat transfer from the substrate, wind-induced convection and radiation. Currently FLACS is the only model approved by FERC that can be used to model the vapor clouds resulting from liquid spills onto trenches. FLACS is also the only approved model that can be used to determine the effect of structures and vapor fences on the flammable vapor cloud dispersion.

6. Significance of Changing Regulations

The changing LNG requirements have altered both, (a) the type and severity of scenarios considered, and (b) the tools used to analyze these scenarios. These changes have created a moving target for LNG facility permit applicants, where the latest FERC interpretations need to be reviewed and considered throughout the consequence modeling phase of a project.

Changes in the choice of leak scenarios have often resulted in having to consider larger and more severe LNG and refrigerant release scenarios. The jetting and flashing release scenarios, in particular, have transitioned to larger and larger leak sizes. This has resulted in the expansion of the size of the areas under facilities' control and the use of various types of mitigation strategies.

7. Mitigation Strategies

The growth in the severity of the scenarios to consider often results in a ½-LFL flammable vapor cloud that extends into an area that can be built upon. When this happens, some type of mitigation strategy must be employed to limit the extent of the analysis-based siting exclusion zones. Potential mitigation strategies include:

- Relocation of LNG and refrigerant storage and piping elements, in order to increase the distance to the property boundary.
- The use of insulated concrete to limit the amount of LNG and refrigerant evaporation during a spill.
- The use of vapor fences and other obstacles to contain the LNG and refrigerant vapor cloud during a release.
- Changes to the LNG and refrigerant flow design, by changing the size of piping, capacity and number of pumps, and process flow conditions in piping. This can reduce the worst case release flow-rate.
- Changes to the refrigerant storage capacity, which can limit the inventory and therefore the amount of refrigerant released.

6.1 Vapor Fences

Vapor fences can be used to significantly reduce the extent of a flammable vapor cloud. Typically, vapor fences are placed along a large portion of the facility boundary, to contain the vapor cloud within to the facility. In order to be effective, these vapor fences require large heights, up to 30 feet.

Figure 2 shows the effect of vapor fences at a sample LNG facility. The figure shows how the fences significantly shorten the downwind extent of the vapor cloud, but they increase the amount of fuel vapor present inside the facility.

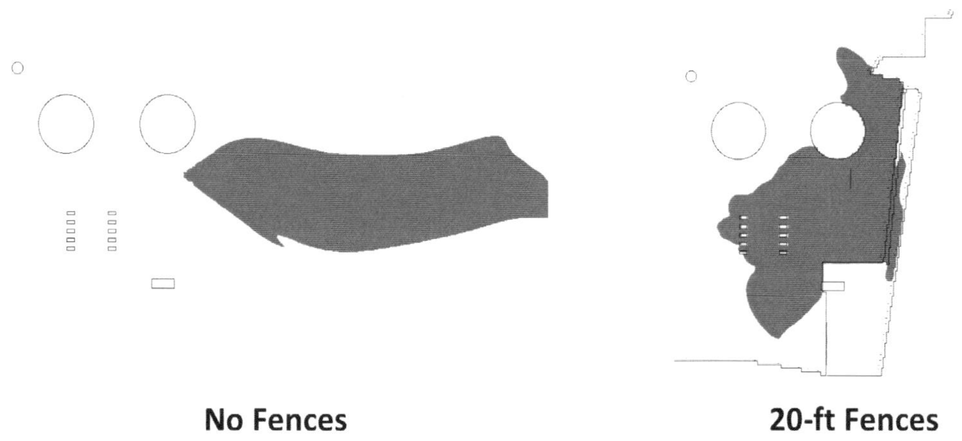

Figure 2 Effect of fences on the ½-LFL vapor cloud extent (in red) for a sample LNG jetting and flashing release scenario. Images were obtained using FLACS.

8. Conclusions

Although the requirements for LNG facility siting and design are codified in 49 CFR 193 and NFPA 59A, many of the exclusion zone analysis requirements are stated broadly and involve some level of expert interpretation. FERC has in the past clarified their interpretation of the federal requirements by means of formal letters and less formal precedent-setting memoranda. These interpretations continue to evolve over time, and drive new tools and considerations.

The changing LNG requirements have altered both, (a) the type and severity of scenarios considered, and (b) the tools used to analyze these scenarios. These changes have created a moving target for LNG facility permit applicants, where the latest FERC interpretations need to be reviewed and considered.

Changes in the choice of leak scenarios have often resulted in having to consider larger and more severe LNG and refrigerant release scenarios. The growth in the severity of the scenarios to consider has necessitated the use of mitigation strategies, including vapor fences, to limit the extent of the analysis-based siting exclusion zones.

9. References

[1] "Liquefied Natural Gas Facilities: Federal Safety Standards." Code of Federal Regulations Title 49, Pt. 193, 2012 edition.

[2] Federal Energy Regulatory Commission, "Guidance Manual for Environmental Report Preparation." FERC – Office of Energy Projects (August 2002).

[3] NFPA 59A, Standard for the Production, Storage, and Handling of Liquefied Natural Gas (LNG), 2001 Edition.

[4] "Chemical Accident Prevention Provisions." Code of Federal Regulations Title 40, Pt. 68, 2011 edition.

[5] Kohout A. U.S. "Regulatory Framework and Guidance for Siting Liquefied Natural Gas Facilities – A Lifecycle Approach". Proceedings of Mary Kay O'Connor Process Safety Center, 15th International Symposium, College Station, Texas, October 23-25, 2012.

[6] EcoEléctrica, L.P. Information Request on Exclusion Zone Modeling, Docket No. CP95-35-001, November 15, 2011.

[7] Freeport LNG Development, L.P. Environmental and Engineering Data Request, Docket No. CP12-29-000, March 20, 2012.

[8] McInerny EH, Hart RJ, Morrison DR, Kytomaa HK. "New Quantitative Risk Criteria for U.S. LNG Facilities." American Institute of Chemical Engineers, 2013 Spring Meeting, 9th Global Congress on Process Safety, San Antonio, Texas, April 28 – May 1, 2013.

[9] Corpus Christi Project LNG Engineering Conference Call, Docket No. PF12-3-000, May 7, 2012.

[10] Fire Protection Research Foundation "Evaluating Vapor Dispersion Models for Safety Analysis of LNG Facilities Research Project.",April 2007.

[11] Federal Energy Comission (FERC) "Evaluation of DEGADIS 2.1 Using Advisory Bulletin ADB-10-07." July, 2010.

[12] Fire Protection Research Foundation "LNG Source Term Models for Hazard Analysis: a Review of the State-of-the-Art and an Approach to Model Assessment." March 2009.

[13] Pipeline and Hazardous Materials Safety Administration (PHMSA) Interpretation Letter in Reference to the Proposed Waterfront Liquefied Natural Gas Plant in the City of Fall River, Massachusetts. Dated July 7, 2010.

[14] Federal Register Vol. 76, No. 137, July 18, 2011.

[15] Pipeline and Hazardous Materials Safety Administration (PHMSA) Final Decision, Docket No. 2011-0101. Dated October 7, 2011.

From LNG Imports to Exports: Process Safety and Regulatory Challenges

Jack Chosnek, Ph. D., P. E.
Knowledge*One*
President and Principal

Victor H. Edwards, Ph. D., P. E.
Director of Process Safety
IHI Engineering & Construction International Corporation

AIChE Spring Meeting
13th Topical Conference on Natural Gas Utilization
San Antonio, April 28-May 2, 2013

Introduction

As natural gas becomes more abundant in the U.S. the need for imports of gas, in the form of LNG, is disappearing and the need to find markets for the product is steadily increasing. LNG terminal operators are thus switching from LNG imports, which require re-gasification, to LNG exports, which require liquefaction resulting in dramatically changed processing of LNG. In LNG liquefaction for exporting, new flammable materials have been introduced in large quantities for cryogenic cooling: propane, ethane, ethylene, and perhaps propylene and butane. There are now present materials that can form vapor clouds that because of their low temperature and higher molecular weight will disperse more slowly reaching much further than LNG. The quantities of refrigerants that can be liberated from loss of containment can be as large as those of LNG.

In addition, as opposed to LNG vaporization when importing the material, there is significant processing which involves compression and distillation at very high pressures and low temperatures. Like LNG regasification, the natural gas and the LNG itself will be at very high pressures, on the order of 600-1,000 psig, and very high quantities will be flowing in large pipes or heat exchangers at those conditions. There is also processing of the received high-pressure pipeline gas that needs to be conditioned (e.g. removal of components such as mercury, H_2S and C2+ hydrocarbons) for the liquefaction process.

All of these represent additional hazards that typically are not known and addressed in LNG importing and re-gasification plants. There is a need to further develop data on dispersions of LNG and refrigerants at high pressures; to identify other hazards not typically found in LNG terminals; and to do facility siting studies to determine how to re-use the same facilities for two different purposes.

LNG is also heavily regulated posing challenges to producers. The Federal Energy Regulatory Commission (FERC) regulates LNG through the Department of Transportation's (DOT) Pipeline Hazardous Materials Safety Administration (PHMSA) (1-5). The main regulation is 49CFR 193 (3) which relies heavily on NFPA 59A (2) (and at that the 2001 edition and part of the 2006 edition are the only ones accepted). These regulations and standards are mainly consequence based instead of risk based. As long as there was straightforward vaporization of LNG and the only potential spills were with low pressure LNG (excluding the high pressure pumps feeding the vaporizers) there were no real problems that couldn't be solved with impoundment to comply with the regulations. This has changed significantly and guidance to and from government agencies is needed.

LNG Imports – Characteristics

LNG import facilities are comparatively simple as summarized below:

- Receive LNG from a ship
- Store LNG
- Pump to high pressure
- Heat to vaporize
- Put gas into pipeline

LNG is received from a ship and is pumped through unloading arms into pipes and then to large storage tanks that operate at low pressures. LNG is then pumped to high pressures (approximately 1,000 psi), regasified by passage through heat exchangers, and then put into pipelines.

LNG Exports – Characteristics

In contrast to LNG Imports, LNG exports are more complex. Here are typical steps in the liquefaction of natural gas:

- Receive natural gas (primarily methane) from pipeline at high pressure.

- Clean gas: Remove mercury, hydrogen sulfide, carbon dioxide, carbonyl sulfide
- Dry gas
- Remove heavies and fuel gas
- Liquefy (potentially one or more refrigeration cycles)
- Where natural gas supply contains significant nitrogen, strip nitrogen from LNG
- Send to storage tank
- Pump to ship

Natural gas, typically more than 90 percent methane is received from pipeline at high pressure (typically 1,000 psi). When processing dry pipeline natural gas, first mercury is removed by adsorption on a solid adsorbent. Then hydrogen sulfide, carbon dioxide and carbonyl sulfide are removed from the gas by absorption, typically using an aqueous amine solvent.

The wet gas is dried, usually by adsorption of water on molecular sieves. The dry gas is then cooled by heat exchange and expansion. Heavies are removed and sent to fractionation, where by-product condensate is sent out for sale and lights are used as fuel gas.

The lean gas is liquefied by refrigeration. When the lean gas contains significant amounts of nitrogen, the resulting LNG is stripped to remove nitrogen and the off-gas either used as fuel gas or flared.

LNG is then pumped to storage and then pumped to an LNG tanker for export.

Figure 1 is a block flow diagram of LNG production from natural gas.

Figure 1 - LNG Liquefaction - Linde's LIPROM Process

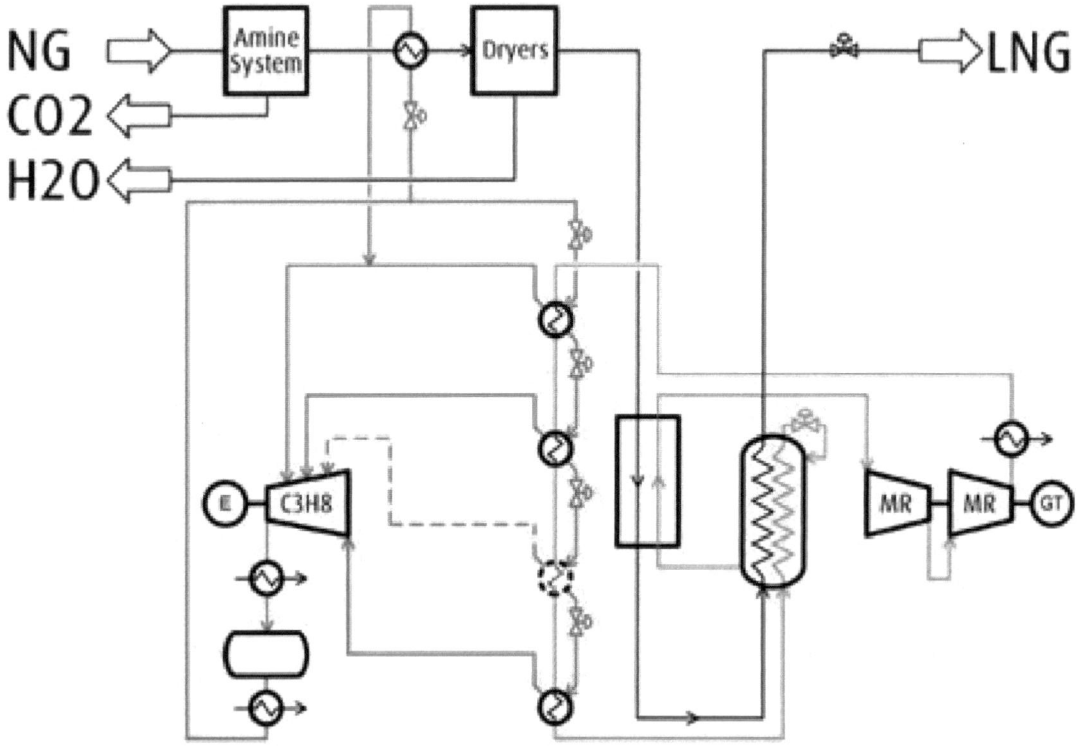

(Courtesy of Linde)

Process Chemicals – Contrasts

Table 1 contrasts the process chemicals in LNG regasification and natural gas liquefaction to LNG.

Table 1 – Contrasts in Process Chemicals in LNG Regasification and Natural Gas Liquefaction

Regasification	Liquefaction
Methane	Methane
Nitrogen	Amine
	Carbon dioxide
	Hydrogen sulfide
	Water
	Condensate (higher hydrocarbons)
	Propane, propylene
	Ethane, ethylene
	Butanes
	Mixed Refrigerant
	Nitrogen

Although LNG often contains small amounts of C2+ hydrocarbons, these are not found as a separate phase in regasification. In contrast, most natural gas liquefaction processes use and create a number of C2+ hydrocarbons streams.

Process Safety Challenges in LNG Importation

LNG import and regasification is not without its challenges. Listed below are the primary sources of process hazards:

- LNG handling, transfers and releases
- Cryogenic temperatures

- Fire
- Explosion (Low probability due to low congestion and low reactivity)
- Asphyxiation

Accidental releases of LNG pose all of the above hazards, but LNG facilities are designed to keep the risk of accidental releases very low. Explosion hazards are comparatively low because the simplicity of the process leads to low congestion and the high concentration of methane keeps reactivity low for most gas releases. (In one exception, a pool of LNG exhibits rapid differential evaporation of methane, leading to increasing concentrations of C2+ hydrocarbons in the residual pool. Explosion hazards are thus somewhat greater in weathered pools of LNG, but emergency response measures should be firmly established by the time significant weathering has occurred.)

Process Safety Challenges in LNG Exportation

Natural gas liquefaction has all of the challenges of LNG importation, plus a number of others:

- All of the process safety challenges of LNG Importation plus
- Handling of high pressure and low temperature refrigerants
- High processing temperatures (natural gas pre-treatment)
- Higher intensity fires
- Explosion (Higher probability; higher congestion and reactivity)
- Toxic exposure (H_2S) in those cases where there is significant amounts of hydrogen sulfide present in the incoming natural gas
- Training
- Facility Siting

Because natural gas liquefaction processes typically contain a number of different process streams rich in C2+ hydrocarbons, there is a higher risk of explosion caused by these more reactive compounds. In addition, the greater process complexity increases congestion due to addition of new and more complex facilities within an existing site, as well as more potential leak sites.

Where produced gas is being processed, hydrogen sulfide removed from the natural gas and concentrated during purification poses a toxic exposure hazard in the event of a release. Pipeline gas contains much less hydrogen sulfide, but in both cases common practice is to oxidize the hydrogen sulfide and the resulting sulfur oxides become an environmental emission.

The use of refrigerants or refrigerant mixtures adds the hazards of handling and storage of these materials. Propane or propylene may be used for initial refrigeration and used in mixtures with other materials such as ethane or ethylene in order to attain the low temperatures required for LNG liquefaction. These materials are typically used in closed loops where large quantities go through evaporation during heat exchange and subsequent recompression to high pressures. These materials represent also a much higher potential than methane both for the occurrence of fire and explosion and for a higher intensity it they were to occur.

The added complexity makes training of personnel more complex, with more opportunities for incidents.

The increased explosion hazards and the challenges of working within an existing site may create facility siting hazards with existing or new occupied buildings.

Regulatory Challenges – Regulatory Agencies

LNG facilities built within the United States must meet the requirements of a number of regulatory agencies (Table 2). The most specific requirements are those of the Federal Energy Regulatory Commission (FERC) and PHMSA (5). These agencies require that LNG facilities be designed to comply with NFPA (National Fire Protection Association) 59A – Standard for the Production, Storage, and Handling of Liquefied Natural Gas (LNG), as well as other applicable industry codes and standards.

Table 2 – Agencies Regulating LNG in the USA

U. S. FERC (Federal Energy Regulatory Commission)

U. S. DOE (export/import permits)

U. S. DOT (PHMSA-Pipeline & Hazardous Materials Safety Administration)

U. S. Department of Homeland Security (Coast Guard)

U. S. EPA

U. S. Fish and Wildlife Service

NOAA (in some instances)

State and Local HSE and Utility Agencies

Advocate NGOs

Although NGOs (Non-Governmental Organizations) are not governmental regulatory agencies, they often actively promote the strict enforcement of existing regulations, the aggressive interpretation of existing law, and the creation of new law.

Regulatory Challenges - LNG Imports

Current US regulations focus on LNG import and regasification facilities (Table 3). FERC is the lead federal agency and with PHMSA regulates domestic LNG facilities through 49 CFR 193 (3). This regulation draws heavily on NFPA 59A (2).

Table 3 – Principal Regulations Addressing LNG in the USA

- **FERC (49 CFR 193 relies on NFPA 59A - 2001)**

  Federal - *Veto power*
- **U.S. DOT (PHMSA)**
- **U.S. Department of Homeland Security – U. S. Coast Guard Waterway Suitability Assessment (33 CFR 127 for safety and 33 CFR 105 for security)**
- **U.S. EPA**
- **State - *Veto power***
  - **Coastal Zone Management Act**
  - **Clean Water Act**
  - **Clean Air Act**

FERC requires a detailed life cycle approach to monitor and approve the siting, engineering design, construction, and operation of LNG facilities (5). Permits to operate LNG facilities are required by FERC, and FERC has veto power over these domestic facilities. FERC normally prepares the Environmental Impact Assessment for new LNG facilities.

In addition, the US Coast Guard has regulatory authority over the waterfront LNG import and export facilities (4).

States also have veto power over LNG facilities through delegated federal regulations such as the Coastal Zone Management Act, the Clean Water Act, and the Clean Air Act.

Regulatory Challenges – LNG Exports

The existing federal regulations (2,4-6)) rely heavily on the 2001 edition of NFPA 59A (3). They mention natural gas liquefaction only briefly, and were developed when liquefaction was used primarily at peak shaving facilities. The primary focus is for LNG imports.

FERC and NFPA 59A are consequence-based (Table 4).

NFPA 59A requires that a "design spill" doesn't surpass the limits shown in Table 4 at the property line with only passive mitigations (e.g. dikes or fixed barriers, gravity-flow impoundment, etc.). The LNG tanks in the facility are also included in the exclusion zone for radiation and overpressure. The "design spill" is created from a pipe break containing LNG that will last 10 minutes, and more if there is inventory left in the pipe. This approach made sense for regasification plants where the major hazard is an LNG spill which can be controlled by impoundment.

Table 4 – Consequence-based FERC Regulatory Criteria

Exclusion Zones (at the property line) from a release caused by a "design spill"

- ½ Lower Flammable Limit
- 1,600 Btu/hr-ft^2 thermal radiation
- 1 psi overpressure

Mainly consequence based

Only passive mitigations allowed (impoundment to basins, vapor fences, fans)

FERC recently clarified (6) the conditions for the piping ruptures, providing initiating frequencies for breaks and ruptures based on valve count, pipe lengths and diameters rather than full-bore breaks for all pipe sizes. FERC based this on published data of failures of piping to/from LNG tanks and deemed this failure rate as acceptable for all piping, modified in proportion to their failure rate based on diameter and length. This defined more reasonably a design spill than the traditional worst case scenario but let intact the consequence to be avoided and the restriction on using only passive mitigations.

A consequence-based approach is very difficult for complex processing operations where releases cannot be mitigated by passive means. For example, a high pressure release at elevation where there is no liquid pool formation (and therefore no possibility for impoundment of the spill) does not lend itself well to a consequence-based approach.

In addition, the restriction to allow only passive mitigations, such as impoundment and vapor fences, imposes a potential negative impact on design and facility siting when trying to mitigate releases from LNG export facilities.

Conclusions

Natural gas liquefaction and export is a safe and proven technology and poses less hazard than many other chemical manufacturing processes. However, conversion of LNG Import Terminals to natural gas liquefaction and purification facilities requires more complex processing and involves significant inventories of compounds that pose somewhat greater hazards than LNG Import Terminals. Therefore, the careful application of industry best practices in the conversion of LNG Import Terminals to LNG Export is essential.

Better regulations, based on dialog with regulatory agencies and a shift from consequence-based regulations to the risk-based approach to process safety of natural gas liquefaction plants advocated by Woodward and Pitblado (1) would be helpful.

In addition, improved modeling tools and a better understanding of potential effects on the community are needed.

Acknowledgements

Constructively critical reviews by Kamal Shah and Frank Del Nogal are appreciated.

References

1 Woodward, John L. and Robin M. Pitblado, "LNG Risk Based Safety-Modeling and Consequence Analysis", John Wiley & Sons, Inc., Hoboken, New Jersey, 2010.

2 NFPA 59A, "Standard for the Production, Storage, and Handling of Natural Gas (LNG)", National Fire Protection Association, Quincy, MA, 2001 and 2006.

3 Volume 49 Code of Federal Regulations, Part 193, "Liquefied Natural Gas Facilities: Federal Safety Standards", United States, Washington, DC, 2013.

4 Volume 33, Code of Federal Regulations, Part 127, "Facilities Handling Liquefied Natural Gas and Liquefied Hazardous Gases", United States, Washington, DC, 2013.

5 Kohout, Andrew, "U. S. Regulatory Framework and Guidance for Siting of Liquefied Natural Gas Facilities – A Lifecycle Approach", paper presented at 15th International Symposium, Mary Kay O'Connor Process Safety Center, College Station, TX, October 23-25, 2012.

6 FERC and Cheniere Meeting Summary, Corpus Christi Project, Docket PF12-3.000, May 7, 2012

Biography

Jack Chosnek, Ph. D., P. E., Knowledge*One*, www.knowledge1.net Jack is a consultant in Process Safety Management and Knowledge Management and has over thirty five years of experience in the petrochemical industry. He has consulted for companies in the chemical, oil and gas, offshore, waste management and mining industries, developing policies and implementing process safety management systems, facilitating PHAs, writing operating procedures and conducting incident investigations and process safety audits. He has developed commercial, full-featured software for PHA facilitation and Management of Change (MOC). Jack consulted as the principal risk engineer for two multi-billion dollar projects carried out by large engineering firms.

Before becoming a consultant Jack worked for Celanese Corp. for 25 years in various positions in process safety management, technology management, operations, process engineering, piloting, tolling and R&D. Jack was a member of a project team that developed one of the first commercial scale-ups in the biotechnology industry.

Jack is the Chairman of the Process Safety Workshops at the South Texas Section (STS) of AIChE. He chaired the Process Plant Safety Symposium (PPSS) at the 4th Global Congress on Process Safety, and has chaired and co-chaired other process safety sessions in other years. He is the chair of the Technical Advisory Committee at Texas A&M's Mary Kay O'Connor Process Safety Center.

Jack is the holder of three patents related to chemical production. He has a BS and MS from the Technion—Israel Institute of Technology, and a PhD from the University of Missouri at Rolla, all in Chemical Engineering, as well as an MBA from Texas A&M—Corpus Christi.

Victor H. Edwards, Ph. D., P. E. is Director of Process Safety for IHI Engineers and Constructors International, 1080 Eldridge Parkway, Houston, TX 77077 (Telephone: 1-832-379-7742; e-mail: vic.edwards@ihi-ec.com). Responsible for health, safety, and environment in design, Vic has worked for IHI and predecessor companies for 29 years. His experience includes process engineering, process safety management, and process, biochemical, environmental, and mineral processing technologies. His project assignments have included biochemical and chemical processes, biomass energy resources, coal gasification, coal liquefaction, liquefied natural gas, chemical weapons demilitarization, hazardous waste disposal, offshore hydrocarbons facilities, and extraction of metals.

Dr. Edwards has received three DuPont awards for safety and environmental engineering excellence and two DuPont awards for health, safety, and environmental excellence during 28 years as an alliance engineering contractor. In 1998, Kvaerner Engineers and Constructors named him employee of the year. In 2007, Vic's Houston HSE team was recognized by Aker Kvaerner with the Process and Construction Excellence Award for HSE in Design.

Vic's earlier experience included Assistant Professor of Chemical Engineering at Cornell University, Program Manager at the National Science Foundation (on leave from Cornell), Research Fellow at Merck, alternate energy research at United Energy Resources, Visiting Professor of Environmental Engineering at Rice University, and process engineering at Fluor.

Vic has contributed more than sixty publications to the engineering literature, including several book chapters. He chaired the 1992 Process Plant Safety Symposium, the 1997 International Plant Operations and Design Conference, and the 2013 Global Congress on Process Safety, as well as numerous technical sessions at other conferences. Vic received the 2003 Service Award from the Process Safety Center at Texas A & M University, where he chaired the Technical Advisory Committee from 2004 to 2011. In 2009, Vic was elected to the Centennial Council of the Department of Chemical and Biomolecular Engineering at Rice University. He also serves on the Editorial Advisory Board of CHEMICAL PROCESSING magazine. Edwards earned his B. A. from Rice University and his Ph. D. from the University of California at Berkeley; both degrees in chemical engineering. A registered professional engineer in Texas, he is an AIChE Fellow, and a member of ACS, AAAS, NFPA, NSPE, and the New York Academy of Sciences.

The mitigation conundrum: when reducing one hazard may increase another

Filippo Gavelli, Ph.D., P.E.
Principal Engineer, Head of Dispersion Consulting, GexCon US

Scott Davis, Ph.D., P.E.
Principal Engineer, President, GexCon US

AIChE Spring Meeting
13th Topical Conference on Natural Gas Utilization
San Antonio, April 28-May 2, 2013

Abstract

Siting regulations for onshore LNG facilities in the United States require the consequences of credible accidental scenarios to be confined within the facility's boundaries – that is, no harm should occur to the public or public property from the worst credible events. Compliance with regulations based on worst credible events is often challenging and may require mitigation methods. One example is that of vapor barriers being used to control the dispersion of a flammable cloud from a release within the facility, in order to keep the cloud from extending beyond the property boundaries before dissipating below ½-LFL.

However, using barriers to keep the flammable cloud from dispersing too far downwind is likely to result in a larger volume of gas remaining near the release location and possibly within congested areas inside the facility. This may therefore increase the hazard footprint from the worst credible overpressure scenarios. Similarly, the flow obstruction posed by the barriers may decrease the air flow through the area increasing the overpressure hazard from smaller releases. Since small releases tend to occur more frequently, increasing their potential severity is likely to increase the overall risk to which facility personnel are exposed. Fortunately, not all barrier installations will have an adverse impact on overpressure hazard and/or personnel risk.

This paper will examine the impact of vapor barriers on overpressure hazards for different barrier configurations. The case study will apply CFD modeling to realistic scenarios, inside a detailed 3D model of a typical LNG liquefaction facility.

Introduction

The siting of LNG facilities in the United States is regulated by prescriptive requirements for the hazard analysis. The tacit rationale behind prescriptive (or deterministic) regulations is that there may be no hazard to the public or public property in the event of an accident. Since it would be effectively impossible to protect the public against cataclysmic events to nearly any kind of facility (the example of ground-splitting earthquakes has been used in the past to argue against the safety of LNG terminals), one of the critical parts of prescriptive regulations is the definition of the maximum credible event (MCE - even though that term is not explicitly used in the regulatory language). Facility developers need to demonstrate that the hazard footprint, to specified threshold levels, for any accidental scenario of magnitude up to the MCE.

Over the last several years, and particularly since the recent market changes led to numerous LNG export facilities being proposed in the United States, the criteria for the selection of the MCE have evolved significantly: in addition to the "design spill" defined in NFPA 59A, which was the only scenario required for LNG import facilities until a few years ago, "flashing and jetting" release scenarios for both LNG and flammable refrigerants (if used) are now also required to be evaluated. Accidental leaks selection criteria have not yet been codified in the regulations, therefore, analysts currently need to rely upon practical guidance from regulatory staff.

Practical Scenarios

Even though every facility is different, flashing and jetting MCEs in excess of 100 kg/s are rather common. For these scenarios, the hazard distances for unimpeded vapor cloud dispersion is likely to extend beyond the property boundaries. In these cases, prescriptive regulations allow the use of passive mitigation measures in order to control the vapor cloud dispersion and thus ensure that the hazard footprint from the MCEs will not extend beyond areas under the facility owner's control.

The most common vapor dispersion mitigation measure adopted is that of vapor barriers. Therefore, most LNG facilities being proposed in the United States currently require one or more sets of vapor barriers to ensure public safety according to the regulations.

In certain cases, facilities have been able to demonstrate that installing a single row of vapor barriers along, or just inside, the property boundaries is sufficient to meet the regulatory requirements for vapor dispersion hazards. This is generally the case when there is sufficient distance between the accidental leakage source (e.g., high-pressure piping within the liquefaction area) and the boundary, so that the flashing jet release loses much of its momentum before reaching the barrier; the low-momentum, dense vapor cloud is unable to “climb” the barrier and instead spreads laterally along the length of the fence, remaining within the facility’s boundaries. Figure 1 shows such an example: the vertical cut plane through the vapor cloud shows a small increase in thickness as the cloud reaches the barrier, pushed primarily by the wind (as for a cloud produced by vaporization of a liquid spill).

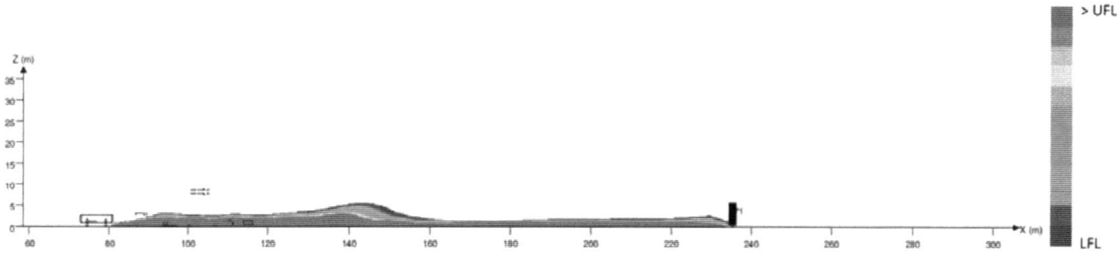

Figure 1. Low-momentum vapor cloud, contained by fenceline barrier

There are, however, situations in which the flashing jet release from an accidental leakage source reaches the vapor barrier with significant residual momentum. In these cases, the vapor cloud will manage to "jump" the barrier and spill over the opposite side, as shown in Figure 2.

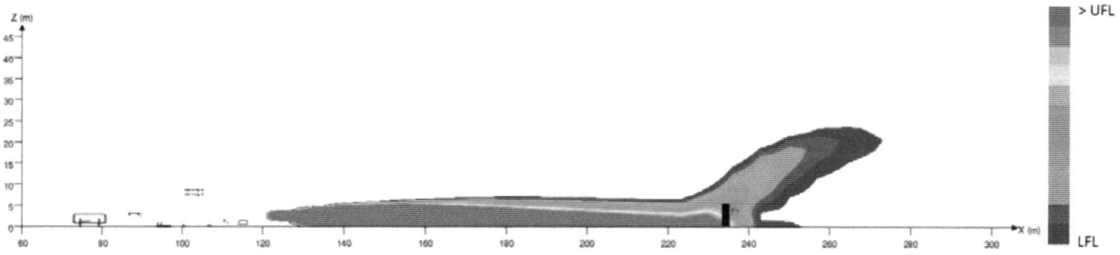

Figure 2. High-momentum vapor cloud, not contained by fenceline barrier

If the barrier is located along, or just inside, the property line, the barrier will be inadequate to control the vapor cloud dispersion in compliance with regulatory requirements. Potential solutions include: a) raising the vapor barrier height, or b) adding a second row of barriers between the leakage source and the property line. The barrier height necessary to control a momentum-driven dense vapor cloud can exceed 30 ft and become very expensive, due the need to sustain high wind loading, earthquake vibrations, etc. Therefore, adding a second row of barriers is often considered the most practical solution: the inner row intercepts the vapor cloud from a flashing jet release and reduces its momentum by forcing it to "jump" over the barrier; the inner row also reduces the mass flux of vapor that reaches the fenceline barrier. When the cloud reaches the fenceline barrier, it is smaller and with less momentum, making it possible for the barrier to contain it; in some cases, proper placement of the first barrier may render the fenceline barrier unnecessary, as shown in Figure 3.

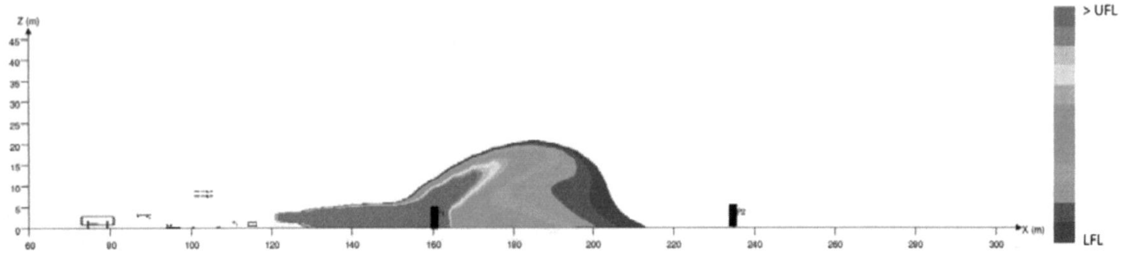

Figure 3. High-momentum vapor cloud, contained by two rows of barriers (the fenceline barrier is not strictly necessary in this case)

The conundrum

As shown in the examples above, by design vapor barriers tend to contain the vapor within a smaller area. This however may have an unintended adverse effect, particularly for the more reactive flammable refrigerants, as it may result in larger flammable volumes near the release location, which is often within or close to a congested area (e.g., the gas processing or liquefaction areas). Therefore, the use of vapor barriers to mitigate vapor dispersion hazards can – at least in principle – increase a facility's overpressure hazards. In practice, as will be discussed shortly, vapor barriers are not likely to result in increased overpressure hazards for the larger, less probable scenarios that must be evaluated for facility siting purposes; they are, however, likely to increase the hazard footprint of the smaller, more probable accidents.

Figure 4 shows an LNG liquefaction train from a small-scale LNG liquefaction facility. For the purpose of this example, the rest of the facility has been removed. Three different cases are considered, to evaluate the effect of vapor barrier location on potential overpressure hazards:

1. The reference case, without any vapor barriers (shown in Figure 4);
2. Vapor barriers on three sides, approximately 25 m away from the edges of the liquefaction train (shown in Figure 5);
3. Vapor barriers on three sides, approximately 10 m away from the edges of the liquefaction train (shown in Figure 6).

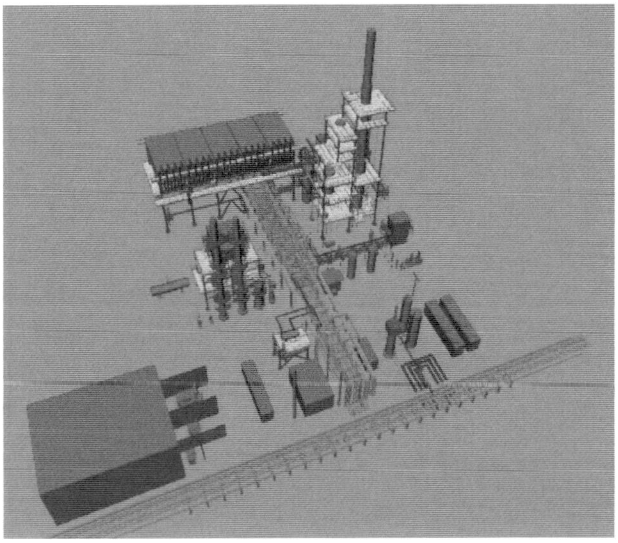

Figure 4. Reference case – no barriers

Figure 5. Wide barrier spacing (25 m outside process area)

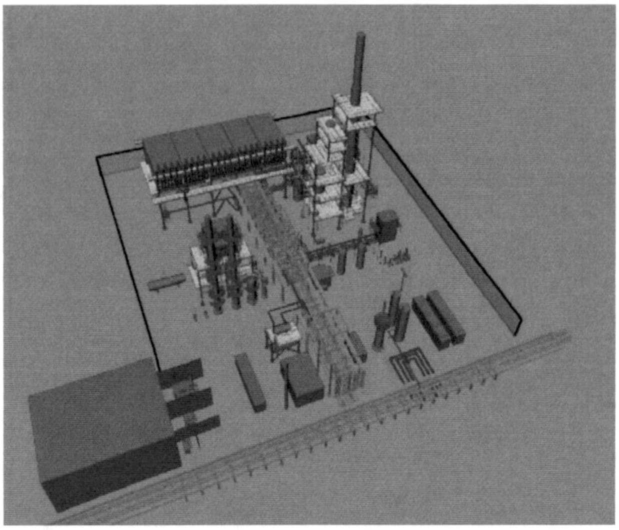

Figure 6. Tight barrier spacing (10 m outside process area)

A frequently used measure of the propensity for gas cloud accumulation within a space is the ventilation rate for a given wind speed and direction; this is often expressed in air changes per hour (ACH). Figure 7 shows the wind velocity vectors for the three barrier configurations, for an external wind at 2 m/s from the west. The vector plots show that the presence of the barriers affects the air flow through the liquefaction area, causing near-stagnant regions; these low-flow areas are more pronounced, as expected, for the tighter fence layout. The ACH rates for these scenarios are listed in Table 1; the values are normalized to the reference case (no barriers).

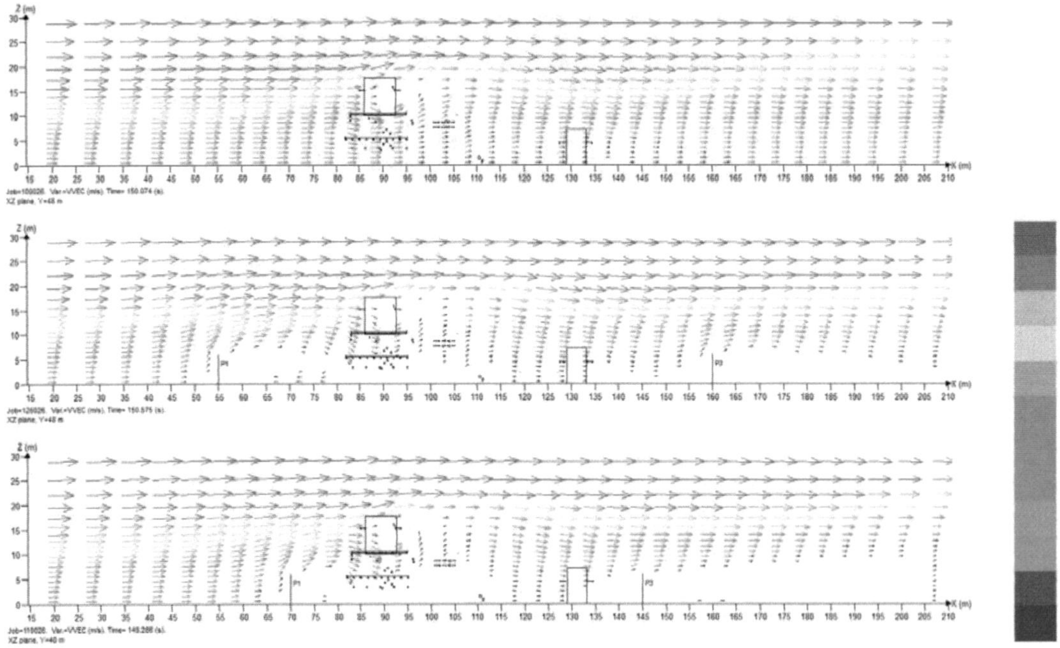

Figure 7. Wind velocity vectors for the three barrier configurations.

Table 1. Ventilation rates for different barrier spacings

Barrier gap	ACH (normalized)
No barriers	100%
25 m	91%
10 m	85%

The lower ACH due to the presence of vapor barriers will result, for any given release, in slower dilution of the vapor cloud. This, however, does not necessarily mean that overpressure hazards form the ignition of a vapor cloud will always be worse as ventilation is reduced. For example, Table 2 shows the max flammable volume obtained for a large release (over 100 kg/s) of a refrigerant within the liquefaction train. The results show that adding barriers increases the size of the flammable cloud in the congested area near the release, as the barriers limit the downwind dispersion of the cloud. However, the tighter barrier scenario (10 m out) results in a smaller flammable volume than the wider barrier gap (25 m); this occurs because, for a large release, the lower ventilation rate results in a richer cloud and therefore a smaller portion in the flammable range. In a prescriptive regulatory environment, the largest flammable volume determines the worst-case conditions for overpressure hazards and therefore the 10 m barrier distance would result in lesser consequences than the 25 m case.

Table 2. Max. flammable volumes for a large release

Barrier gap	Max. cloud volume (normalized)
No barriers	100%
25 m	119%
10 m	108%

Failure data for these facilities indicate that very large releases are very unlikely to occur; the more likely failure scenarios instead are small releases – e.g., from valve stem packing failure – on the order of 1 kg/s or less. While these releases do not pose a threat to the public, they may pose a threat to personnel safety. Table 3 shows the max flammable volume obtained for a 1 kg/s release of a refrigerant within the liquefaction train. The results show that as ventilation in reduced (tighter barrier spacing) the maximum flammable volume increases and therefore so does the overpressure hazard. This occurs because, for a small release, only a small portion of the cloud remains above the flammable range.

Table 3. Max. flammable volumes for a small release

Barrier gap	Max. cloud volume (normalized)
No barriers	100%
25 m	112%
10 m	127%

Conclusions

This paper examined the potential unintended effects of placing vapor barriers around congested areas of an LNG facility, to control large vapor cloud releases in order to comply with U.S. federal regulations. The analysis showed that, as expected, adding vapor barriers reduces the ventilation rates within the fenced-in area. The ventilation reduction generally results in larger flammable cloud sizes for small releases; the effect on large releases, however, is more difficult to predict and may actually result in smaller flammable volumes as much of the vapor cloud may remain above the flammable range.

Since U.S federal regulation require facility siting based on large releases and, for a given fuel and congested area, overpressures tend to increase with the size of the flammable cloud, the addition of vapor barriers to a facility does not necessarily result in a larger overpressure footprint for the LNG facility. However, the overpressure hazards for smaller releases are more likely to be increases, therefore, personnel hazards may be increased. Similarly, if a facility siting were subject to risk-based regulations, the increased consequences from the more frequent, smaller releases may indeed result in a larger risk footprint.

Given the large number of parameters that affect ventilation, gas dispersion and overpressures, the reader is cautioned to remember that the results presented in this paper are case-specific. As such, the numerical values and trend reported above should not be applied to any facility in lieu of a site-specific analysis.

Author Biographies

Dr. Filippo Gavelli is the Head of the Dispersion Consulting group at GexCon and is responsible for GexCon's LNG safety consulting business worldwide. He specializes in the analysis of heat transfer and fluid flow phenomena, including multiphase flows and cryogenic fluids. He has over 10 years of engineering consulting experience and 20 years of experience in computational fluid dynamics (CFD) modeling, using several research and commercial codes. He applies his expertise to modeling the atmospheric dispersion of hazardous gaseous releases, and has extensive experience modeling hazard scenarios for Liquefied Natural Gas (LNG) facilities, including vapor cloud dispersion, pool fires and vapor cloud deflagrations; his experience includes more than

20 LNG installations worldwide, including onshore, offshore and floating (FLNG) facilities for LNG import, export (liquefaction) and peak shaving.

Dr. Gavelli is a member of the technical committees for NFPA 59A (LNG facilities standard) and NFPA 2 (hydrogen technologies code) and was the lead in the model validation effort that led to FLACS' approval by the U.S. DOT under 49 CFR 193. Dr. Gavelli is the lead author of several gas dispersion and explosion safety-related papers and has been a regular contributor to LNG-related technical committees and expert panels for several years.

He is a member of the GexCon docents group, which develops and delivers safety seminars to facility owners and operators, safety engineers, and regulatory agencies, on the hazards associated with gas explosions, dust explosions and LNG. Dr. Gavelli is also responsible for software technical support and training for FLACS customers in North America and has provided training to over 100 users, including staff from regulatory and government agencies.

Dr. Scott Davis is the President at GexCon US and specializes in the engineering analysis and testing of combustion, thermal, and fluid processes. Dr. Davis is responsible for fire and explosion related activities, which include post-incident investigative work, worldwide training and experimentation, as well as performing risk assessments and safety studies for offshore and floating oil & gas installations, petrochemical facilities, and various other industries. These studies include explosion risk assessment, blast and venting analyses, assessment of combustible dust explosions, toxic/flammable gas releases and dispersion, hydrogen safety, ventilation, detector placement, and carbon monoxide dispersion with the assistance of the world-leading FLACS software developed by GexCon.

Dr. Davis serves on the committee responsible for NFPA 720 *Standard for the Installation of Carbon Monoxide (CO) Detection and Warning Equipment* and has served on that for NFPA 921 *Guide for Fire and Explosion Investigations.* He is also a member of GexCon's docents group, which develops and delivers worldwide industrial seminars to owners, operators, safety engineers, and regulatory agencies, on the hazards associated with gas explosions, dust explosions and LNG.

(137a) Catalytic Conversion Wood Syngas to 'Wide Cut' Diesel Over A Multifunctional Catalyst

Qiangu Yan, Department of Agricultural & Biological Engineering, Mississippi State University, Mississippi State University, MS and Fei Yu, Department of Agricultural & Biological Engineering, Mississippi State University, Mississippi State, MS

A continuous process involving gasification, syngas cleaning, and Fischer–Tropsch (FT) synthesis was developed to efficiently produce 'wide cut' diesel. Oak-tree wood chips were first gasified to syngas over a commercial pilot plant downdraft gasifier. The raw wood syngas contains about 47% N_2, 21% CO, 18% H_2, 12% CO_2, 2% CH_4 and trace amounts of impurities. A purification reaction system was designed to remove the impurities in the syngas such as moisture, oxygen, sulfur, ammonia, and tar. The purified syngas meets the requirements for catalytic conversion to liquid fuels. A multi-functional catalyst was developed and tested for the catalytic conversion of wood syngas to diesel. It was demonstrated that liquid fuels similar to commercial diesel was successfully synthesized from bio-syngas.

MODELO DE CÁLCULO DE LA PERMITIVIDAD DE CARBONES

Peña Benavides, Manuel G. y Vidal Teherán, Dairo A.
Universidad del Atlántico, Barranquilla, Colombia.

Resumen
El carbón posee muchas propiedades que, según su estado, lo hace aplicable para un uso específico. Una de esas propiedades es la permitividad eléctrica, que es la capacidad que posee un material de polarizarse. Esta propiedad es la que fija el comportamiento que van a tener los carbones cuando sean expuestos a una señal electromagnética, marca como se propaga la energía en el material, como se transforma la energía de microondas en calor al interior del mismo y, finalmente, nos indica si un material es susceptible de ser tratado por la vía de microondas. El método de medición de la potencia absorbida, es el que se propone en el presente estudio; este consiste en aplicar al material de análisis una onda electromagnética de una cierta potencia conocida como potencia suministrada para medir la cantidad de esa potencia que absorbe y posteriormente, a través de un modelo matemático, calcular el valor de la permitividad eléctrica.

Palabras claves: *Propiedad dieléctrica, permitividad, microondas, guía de ondas, carbón.*

1. Introducción

Permitividad

La capacidad que posee un material para calentarse en presencia de la acción de las microondas está en función de la permitividad. La unidad de medida en el Sistema Internacional SI, es el faradio por metro (F/m).

Tomada en función de la frecuencia, puede tomar valores tanto reales como complejos, ecuación (1). Por lo general no es constante, debido a que varía con la posición del medio, la frecuencia del campo que se aplique, la humedad del material y la temperatura.

$$\varepsilon = \varepsilon' - i\varepsilon'' \qquad (1)$$

De donde ε' es la constante dieléctrica o permitividad real y, ε'' es el factor de pérdidas dieléctricas o permitividad imaginaria.

Un factor de pérdidas alto indica que el material es susceptible de ser tratado por la vía de microondas.

En un medio que no sea lineal, la permitividad depende de la magnitud del campo eléctrico. En electromagnetismo al campo de desplazamiento eléctrico se le define como D, que representa como un campo eléctrico E influye en la organización de las cargas eléctricas del medio. La relación de estos campos con la permitividad se define con la siguiente expresión:

$$D = \varepsilon \cdot E \qquad (2)$$

En donde el campo de desplazamiento D se mide en culombios por metro cuadrado (C/m^2), mientras que el campo eléctrico E se mide en voltios por metro (V/m) y la permitividad eléctrica ε en faradio por metro (F/m).
En el caso en que ese campo eléctrico se dé en el vacío, la ecuación queda de la siguiente forma [7]:

$$D = \varepsilon_0 \cdot E \qquad (3)$$

De donde, ε_0 es la permitividad del vacío. Dicha permitividad está relacionada con la permeabilidad en el vacío μ_o y la velocidad de la luz C, por la siguiente expresión:

$$\varepsilon_o C^2 \mu_o = 1 \quad (4)$$

Despejando la permitividad en el vacío,

$$\varepsilon_o = \frac{1}{C^2 \mu_o} \quad (5)$$

De donde C es la velocidad de la luz ($3*10^8$ m/s) y μ_o es la permeabilidad magnética del vacío ($4\pi*10^{-7}$ Tm/A), definidas en unidades del SI. Se obtiene un valor de la permitividad del vacío de aproximadamente

$$\varepsilon_o = 8{,}8541878176 * 10^{-12}\ F/m$$

Efecto de las microondas sobre la materia

La interacción de las partículas cargadas que poseen ciertos materiales con el campo eléctrico que es generado a causa de la radiación electromagnética que provocan las microondas, hace que estos materiales se calienten.

La irradiación de microondas interactúa con la materia como una fuente interna de energía, haciendo que se caliente directamente desde la profundidad. La eficacia de esta técnica de calentamiento, depende, en gran medida, de las propiedades dieléctricas de la materia.

En el caso de moléculas polares, el campo eléctrico de las microondas hace que los dipolos permanentes presentes en dichas moléculas y los que son inducidos roten al intentar alinearse con el campo oscilante [9]. Para el caso en estudio en el presente trabajo, como se trabaja a una frecuencia de 2,45 GHz, la molécula rota 2450 millones de veces por segundo, y se encontró que a esta frecuencia es ideal para trabajar, debido a que la molécula de agua resuena aproximadamente en ese valor [28].

El carbón y las microondas

La mayoría de los materiales de carbono son excelentes absorbentes de microondas. Las microondas inducen una corriente en fase con el campo electromagnético del material. Los electrones encuentran resistencia al libre desplazamiento por el material, haciendo que cierta cantidad de la energía se disipe en forma de calor.

En los carbones puede ocurrir un interesante fenómeno adicional, la energía cinética de algunos electrones puede llegar a ser suficiente para que salten fuera del material provocando una ionización de la atmósfera que lo rodea [9]. Este fenómeno se puede apreciar con la aparición de chispas o arcos eléctricos.

Los carbones poseen pérdidas eléctricas elevadas y por ese motivo pueden ser sometidos a calentamiento por microondas. Además se pueden utilizar para recibir las microondas y calentar indirectamente otros materiales que no son buenos receptores de microondas, utilizándose en algunos procesos como la pirolisis de biomasa y residuos orgánicos o en reacciones catalíticas heterogéneas.

Guía de onda rectangular

Para la realización del calentamiento de los carbones se hizo uso de una guía de onda rectangular, diseñada y construida especialmente para este estudio.

Tienen varias aplicaciones como en sistemas de radiofrecuencia, microondas terrestres y satelitales. Son fáciles de fabricar y presentan varias ventajas puesto que poseen un gran ancho de banda y presentan pocas pérdidas.

Las corrientes inducidas en sus paredes generan un aumento en las pérdidas de la potencia suministrada, y para contrarrestar esto, la resistencia de las paredes de la guía se debe hacer lo más pequeño posible.

Las paredes deben ser especialmente laminadas, debido a que las corrientes tienden a concentrarse cerca del interior de la superficie de dichas paredes. Estas deben construirse con un material que sea un conductor perfecto.

2. Modelo Matemático para el Cálculo de la Permitividad en Carbones

En el modelo que se plantea con este estudio se realiza un calentamiento por vía de microondas al carbón y se compara con el agua calentada de la misma manera. Se registran los datos de temperatura y tiempo que se grafican y se aplican en el modelo, para el cálculo de la permitividad de los carbones.

Toma como punto de partida el balance de energía del sistema que, en general, está dado por la siguiente ecuación:

$$\{Flujo\ de\ calor\ por\ el\ ingreso\ de\ masa\} - \{Flujo\ de\ calor\ por\ salida\ de\ masa\} + \left\{\begin{matrix} Generación\ de\ energía\ térmica\ por \\ fricción\ molecular \end{matrix}\right\} - \{Calor\ por\ conducción\} = \{Acumulación\ de\ energía\}$$

Reemplazando los términos respectivo

$$F_0 C_P T_0 - F C_P T + 2\pi f V \varepsilon_0 E^2 \varepsilon_c^{''} + 2\pi f V_c \varepsilon_0 E^2 \varepsilon_v^{''} - UA(T - T_a) = M C_P \frac{dT}{dt} \quad (6)$$

Pero como no hay flujo de masa que ingrese ni que salga del sistema, en el caso en estudio, porque se da un calentamiento tipo batch:

$$F_0 = F = 0 \quad (7)$$

El calor generado por la fricción molecular del vidrio, provocado por la acción de las microondas, se puede despreciar debido a que es insignificante con respecto al generado por el carbón. La permitividad eléctrica relativa del vidrio de borosilicato 3.3 que es el tipo del vidrio Duran, es de 0,01702 [38].

El término de acumulación de energía térmica, en función del carbón y del vidrio queda expresado de la siguiente manera:

$$(M_v C_{Pv} + M_c C_{Pc}) \frac{dT}{dt} \quad (8)$$

Por todo lo anterior la ecuación del modelo matemático resulta de la siguiente forma:

$$(M_v C_{Pv} + M_c C_{Pc}) \frac{dT}{dt} = 2\pi f V_c \varepsilon_0 E^2 \varepsilon_c^{''} - UA(T_C - T_a) \quad (9)$$

De donde,
M_v= Masa del vidrio del tubo de ensayo que contiene la muestra del carbón y que se encuentra expuesto a la acción de las microondas, [Kg]
C_{Pv}= Capacidad calorífica del vidrio, [J/Kg °C]
M_c= Masa del carbón, [Kg]
C_{Pc}= Capacidad calorífica del carbón, [J/Kg °C]
$\frac{dT}{dt}$= La variación de la temperatura con respecto al tiempo, [°C/s]
f= Frecuencia de las microondas, [Hz]
V_c= Volumen de carbón, [m^3]
ε_0= Permitividad en el vacío, [F/m]
E^2= Campo eléctrico, [V/m]
$\varepsilon_c^{''}$= Permitividad relativa del carbón
UA= Coeficiente externo de transferencia de calor, [J/s °C]
T_C= Temperatura del carbón, [°C]
T_a= Temperatura ambiente, [°C]

Para el agua se realiza el mismo análisis que se hizo para el carbón obteniéndose la siguiente expresión:

$$(M_v C_{Pv} + M_{H_2O} C_{PH_2O}) \frac{dT}{dt} = 2\pi f V_{H_2O} \varepsilon_0 E^2 \varepsilon_{H_2O}^{''} - UA(T_{H_2O} - T_a) \quad (10)$$

De donde,
M_{H2O}= Masa del agua, [Kg]
C_{PH2O}= Capacidad calorífica del agua, [J/Kg °C]
V_{H2O}= Volumen de agua, [m^3]
$\varepsilon_{H2O}^{''}$= Permitividad relativa del agua
T_{H_2O}= Temperatura del agua, [°C]

La anterior ecuación posee dos condiciones en sus extremos:

1) Cuando t = 0, entonces, $T_{H_2O} = T_a$

Por consiguiente el segundo término de la derecha de la ecuación (10) se hace cero (0) y se obtiene la siguiente expresión para el t = 0:

$$(M_v C_{Pv} + M_{H_2O} C_{PH_2O}) \frac{dT}{dt}\Big|_{t=0} = 2\pi f V_{H_2O} \varepsilon_0 E^2 \varepsilon''_{H_2O} \qquad (11)$$

2) Cuando t = $t_{equilibrio}$, entonces, $\frac{dT}{dt} = 0$

Por consiguiente el término de la izquierda en la ecuación (10) se hace cero (0) y se obtiene la siguiente expresión para el t = $t_{equilibrio}$:

$$2\pi f V_{H_2O} \varepsilon_0 E^2 \varepsilon''_{H_2O} = UA(T_{H_2O} - T_a) \qquad (12)$$

De la ecuación (11) se despeja el campo eléctrico y se calcula, ya que los demás términos de la ecuación se pueden determinar o ya se conocen.

$$E^2 = \frac{(M_v C_{Pv} + M_{H_2O} C_{PH_2O})}{2\pi f V_{H_2O} \varepsilon_0 \varepsilon''_{H_2O}} \frac{dT}{dt} \qquad (13)$$

De la ecuación (12) se despeja la temperatura de equilibrio del agua y reemplazando el valor del campo eléctrico calculado con la ecuación (13) y haciendo las respectivas operaciones se obtiene el valor de dicha temperatura.

$$T_{H_2O} = \frac{2\pi f V_{H_2O} \varepsilon_0 E^2 \varepsilon''_{H_2O}}{UA} + T_a \qquad (14)$$

Esta temperatura del agua debe ser más alta que la temperatura de equilibrio del carbón, debido a que las propiedades dieléctricas del agua hacen que la acción de las microondas sea más fuerte en ella que en el carbón, por esta misma razón se espera que el valor de la permitividad del carbón sea menor que la del agua.

Cálculo de la permitividad del carbón

Hasta el momento se han determinado las ecuaciones que explican el sistema, pero aún falta determinar el valor de la permitividad del carbón.

Para el carbón se dan las mismas condiciones que para el agua, ecuaciones (11) y (12), que para el carbón quedan expresadas de la siguiente manera:

En el arranque $(T_c - T_a) = 0$

$$(M_v C_{Pv} + M_c C_{Pc}) \frac{dT}{dt}\Big|_{t=0} = 2\pi f V_c \varepsilon_0 E^2 \varepsilon''_c \qquad (15)$$

En el equilibrio $\left(\frac{dT}{dt} = 0\right)$

$$2\pi f V_c \varepsilon_0 E^2 \varepsilon''_c = UA(T_c - T_a) \qquad (16)$$

Despejando de la ecuación (15) la *permitividad en el arranque* del carbón, resulta la siguiente expresión:

$$\varepsilon''_c\Big|_{t=0} = \frac{(M_v C_{Pv} + M_c C_{Pc}) \frac{dT}{dt}\Big|_{t=0}}{2\pi f V_c \varepsilon_0 E^2} \qquad (17)$$

De donde,

$\varepsilon''_c\Big|_{t=0} = permitividad\ relativa\ en\ el\ arranque$

Despejando de la ecuación (16) la *permitividad en la temperatura de equilibrio* del carbón, resulta la siguiente expresión:

$$\varepsilon''_c\Big|_{e.e} = \frac{UA(T_c - T_a)}{2\pi f V_c \varepsilon_0 E^2} \qquad (18)$$

De donde,

$\varepsilon''_c\Big|_{e.e} = permitividad\ relativa\ en\ la\ temperatura\ de\ equilibrio$

Para poder resolver la ecuación anterior es necesario hacer una condición, que el campo eléctrico se mantiene constante para el carbón y para el agua, ya que la potencia suministrada es mínima, tan solo el 10% de la potencia total del equipo, así mismo, la potencia absorbida por el carbón es pequeña.

3. Gráficas Obtenidas

Se realizaron tres pruebas al agua y a cada uno de los carbones estudiados, bituminoso y antracítico.

Para el Agua

Gráfica 1. Agua Desionizada.

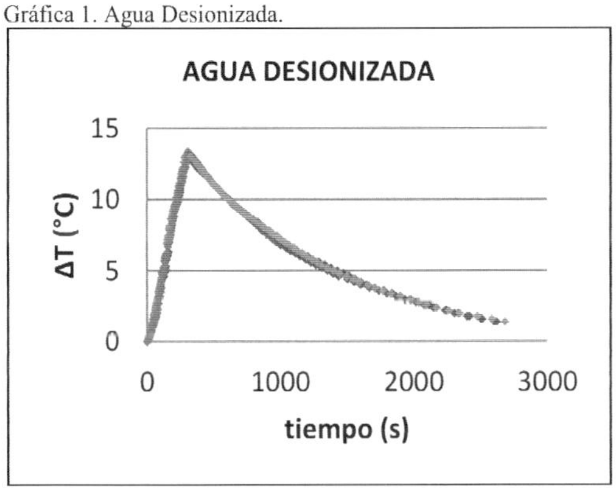

En cada una de las pruebas el calentamiento se realizó durante 5 minutos y se dejó enfriar hasta que el agua alcanzara su temperatura inicial. Se puede observar como el calentamiento se da en forma uniforme para las tres pruebas y que el enfriamiento se da de manera exponencial guardando el mismo perfil.

Para el Carbón Bituminoso

Gráfica 2. Carbón Bituminoso.

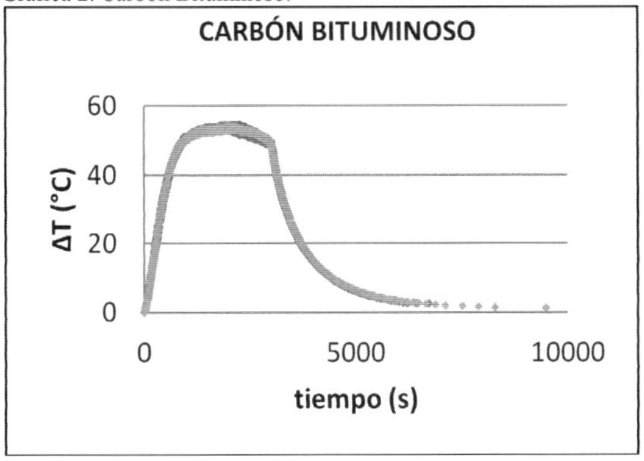

En la gráfica se puede observar como el calentamiento se mantiene constante y después de un tiempo se mantiene a la temperatura de equilibrio, a una diferencia de temperatura entre el carbón y el ambiente se encuentra alrededor de 53°C, y el carbón a 81°C, aproximadamente.

Para el Carbón Antracítico

Gráfica 3. Antracita.

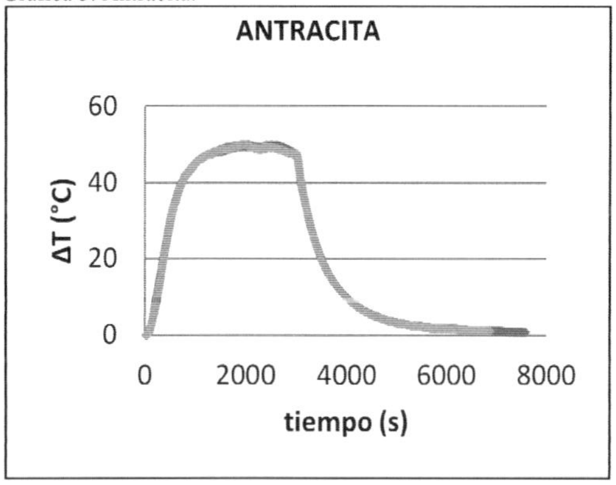

Se puede observar que el calentamiento es de pendiente constante hasta que se comienza a llegar a la temperatura de equilibrio, en donde se mantiene por unos minutos. Pasado un tiempo, el carbón empieza a sufrir daños, haciendo que la temperatura descienda aún con el calentamiento de las microondas.

El enfriamiento se da de manera casi exacta para las tres pruebas, conservando el mismo perfil con buena precisión.

4. Resultados

Luego de que se realizaron las pruebas, se elaboraron las gráficas, se analizaron, se extrajo información de ellas y se aplicó el modelo matemático se obtuvieron los siguientes resultados:

Carbón Bituminoso. Prueba 1.

Campo eléctrico, ecuación (13)

$$E = 370{,}5\,\frac{V}{m}$$

Temperatura de equilibrio del agua, ecuación (14)

$$T_{H_2O} = 90{,}21740144\,°C$$

Permitividad del carbón en el arranque, ecuación (17)

$$\varepsilon_c''|_{t=0} = 7{,}09196$$

Permitividad del carbón en el estado estacionario, ecuación (18)

$$\varepsilon_c''|_{e.e} = 2{,}19229$$

Carbón Bituminoso. Prueba 2.

Campo eléctrico, ecuación (13)

$$E = 379{,}8\,\frac{V}{m}$$

Temperatura de equilibrio del agua, ecuación (14)

$$T_{H_2O} = 94{,}74207057\,°C$$

Permitividad del carbón en el arranque, ecuación (17)

$$\varepsilon_c''|_{t=0} = 7{,}02674$$

Permitividad del carbón en el estado estacionario, ecuación (18)

$$\varepsilon_c''|_{e.e} = 2{,}10455$$

Carbón Bituminoso. Prueba 3.

Campo eléctrico, ecuación (13)

$$E = 376{,}6\,\frac{V}{m}$$

Temperatura de equilibrio del agua, ecuación (14)

$$T_{H_2O} = 92{,}2738915\,°C$$

Permitividad del carbón en el arranque, ecuación (17)

$$\varepsilon_c''|_{t=0} = 6{,}91878$$

Permitividad del carbón en el estado estacionario, ecuación (18)

$$\varepsilon_c''|_{e.e} = 2{,}02441$$

Antracita. Prueba 1.

Campo eléctrico, ecuación (13)

$$E = 370{,}5\,\frac{V}{m}$$

Temperatura de equilibrio del agua, ecuación (14)

$$T_{H_2O} = 90{,}21740144\,°C$$

Permitividad del carbón en el arranque, ecuación (17)

$$\varepsilon_c''|_{t=0} = 4{,}83344$$

Permitividad del carbón en el estado estacionario, ecuación (18)

$$\varepsilon_c''|_{e.e} = 2{,}54616$$

Antracita. Prueba 2.

Campo eléctrico, ecuación (13)

$$E = 379{,}8\,\frac{V}{m}$$

Temperatura de equilibrio del agua, ecuación (14)

$$T_{H_2O} = 94{,}74207057\,°C$$

Permitividad del carbón en el arranque, ecuación (17)

$$\varepsilon_c''|_{t=0} = 4{,}69888$$

Permitividad del carbón en el estado estacionario, ecuación (18)

$$\varepsilon_c''|_{e.e} = 2{,}41124$$

Antracita. Prueba 3.

Campo eléctrico, ecuación (13)

$$E = 376{,}6\,\frac{V}{m}$$

Temperatura de equilibrio del agua, ecuación (14)

$$T_{H_2O} = 92{,}2738915\ °C$$

Permitividad del carbón en el arranque, ecuación (17)

$$\varepsilon_c''|_{t=0} = 4{,}74887$$

Permitividad del carbón en el estado estacionario, ecuación (18)

$$\varepsilon_c''|_{e.e} = 2{,}51280$$

5. Conclusiones

Fue necesario plantear un modelo matemático, que permitiera plasmar la influencia de la energía transmitida por las microondas a la muestra en la temperatura de esta en el tiempo para poder observar el comportamiento de las mismas, y realizar una deducción de las permitividades de los carbones.

Al comparar los resultados obtenidos con el equipo y modelo matemático planteado, con los obtenidos con anteriormente para el agua, se puede afirmar la validez de los resultados obtenidos en el presente trabajo, ya que se puede apreciar la consistencia de la relación de las permitividades en el estado estacionario e inicial y se mantiene la tendencia a disminuir la permitividad, con un aumento de temperatura a la frecuencia trabajada.

Aunque los resultados obtenidos por medio del equipo construido y el modelamiento matemático planteado, pueden ser analizados y comparados con una tendencia parecida a la del agua, el margen de error que pueda existir está delimitado por la ausencia de un cálculo detallado –con modelos no ideales en el comportamiento del agua en el aumento de temperatura pero esto implica mayor complejidad y tiempo de cálculo.

6. Referencias

[1] ABDELGHANI-IDRISSI, M. A. Experimental investigations of occupied volume effect on the microwave heating and drying kinetics of cement powder in a mono-mode cavity. 2000.

[2] AGUILAR, J. A., GARCIA, R., VALDEZ, Z. Simulación del calentamiento de manganitas mediante microondas. 2007.

[3] AJIACO, F. Evaluación del comportamiento térmico de carbones del Cerrejón, carbones coquizantes y sus mezclas en la producción de coque metalúrgico. Universidad Nacional de Colombia. Bogotá, 2011.

[4] AKHTAR, M. J., FEHER, L., THUMM, M. A Multi-Layered Waveguide Technique for Determining Permittivity and Conductivity of Composite Materials. 2005.

[5] ALMANSA MORALES, Alejandro. Estudio experimental para la optimización energética y termográfica en aplicaciones de calentamiento por microondas. Universidad Politécnica de Cartagena. 2007, 19.

[6] ARIZTIMUÑOZ, A. GONZALEZ, F. RISUEÑO, A. El carbón como materia prima. Escuela Universitaria de Ingeniería Técnica Industrial.

[7] ASTIGARRAGA, Julio. Hornos de alta

frecuencia y microondas. Mc Graw Hill. Bogotá, 1999.
[8] BAKER-JARVIS, J., JANEZIC, M., GROSVENOR, J., GEYER, R. Transmission/Reflection and Short-Circuit Line Methods for Measuring Permittivity and Permeability. National Institute of Standards and Technology. Boulder, Colorado. 1993.
[9] BEVERLY MICROWAVE DIVISION. Magnetron theory of operation. Massachusetts, Estados Unidos.
[10] CARMONA MARTINEZ, F. Análisis y optimización de circuitos de microondas en guía de onda. Universidad Autónoma de Madrid. Madrid. 2009.
[11] CONSTANTINI, S. Guías de ondas. Universidad Metropolitana de Telecomunicaciones.
[12] DILDINE, Bob. Measuring complex permittivity and permeability at RF and microwave frequencies. Hewlett Packard. Yokogawa, Japón. 1989.
[13] FERNÁNDEZ, Y. *et al.* Procesos Térmicos Asistidos por Microondas cobre Materiales Carbonosos. Grupo de Materiales Carbonosos para Aplicaciones Tecnológicas. Oviedo, España. 2010.
[14] GRUPO ESPAÑOL DEL CARBÓN. Boletín del Grupo Español del Carbón. N°16, Mayo 2010.
[15] GUZMAN RABANILLO, D. Sistema de de deposición química en fase vapor para crecimiento homoepitaxial de diamante. Universidad de Cádiz. 2004.
[16]http://translate.google.com.co/translate?hl=es&langpair=en%7Ces&u=http://www.radartutorial.eu/08.transmitters/tx08.en.html
[17] http://www.catamutun.com/produc/carbon/
[18]http://www.lacomunidadpetrolera.com/showthread.php/2046-Carbonificaci%C3%B3n.
[19] http://www.metalloyd.co.uk/spanish/products_raw_materials.html
[20]http://www.oviedocorreo.es/personales/carbon/carbon_mineral/carbon%20mineral.htm
[21] http://www.pdv.com/lexico/museo/minerales/carbon.htm
[22] JOHNSON, M., HANSEN, W., RATL, W. Wave guide construction. United States Patent Ofice. Patent dec. 30, 1947.
[23] KEYSON, D. *et al.* Domestic Microwave Oven Adapted for Fast Heat Treatment of $Ba_{0.5}Sr_{0.5}(Ti_{0.8}Sn_{0.2})O_3$ Powders. Journal of Materials Processing Technology 189 (2007). Pág. 316 - 319.
[24] KLINBUN, W. RATTANADECHO, P. PAKDEE, W. Microwave heating of saturated packed bed using a rectangular waveguide (TE10 mode): Influence or particle size, sample dimension, frequency, and placement inside the guide. International Journal of Heat and Mass Transfer. Vol. 54. 2011. Pág. 1763 – 1774.
[25] LEON, Elard F. La importancia del carbón mineral en el desarrollo. Revista del Instituto de Investigaciones FIGMMG. Vol. 9, N° 18, 91 – 97. 2006. San Marcos, Perú.
[26] NAKAMURA, T., NIKAWA, Y. Precise measurement of complex permittivity of materials for telecommunications devices. Journal of Telecommunications and Information Technology. 2001.
[27] NEBOT, Eduardo. Caracterización de dieléctricos a frecuencia de microondas. Universidad de Zaragoza. 2004, 3.
[28] PICERNO, Alberto. Todo sobre hornos de microondas. Saber electrónica. Edición internacional N° 207. México, 2007. Páginas 5 – 22.
[29] PIRA, C. Densificazione di scariche a bagliore in dispositivi magnetron sputtering tramite accoppiamiento con sorgenti Hollow Cathode. Universitá degli Studi de Padova. 2007.
[30] RATTANADECHO, P. The simulation of microwave heating of wood using a rectangular wave guide: Influence of frequency and sample size. Chemical Engineering Science. Vol. 61. 2006. Pág. 4798 – 4811.
[31] RODRIGUEZ, J.M. et al. Medidas de propiedades dieléctricas de materiales de construcción utilizando una guía rectangular. Universidad Politécnica de Madrid. 2008.
[32] ROGOV, I.A. GORBATOV, A.V. Métodos físicos de procesamiento de productos

alimenticios. Industria Alimenticia. Moscú. 1974.
[33] STANLEY, R. Wave coupling arrangement. United States Patent Office. 1959
[34] STURM, G.S.J. , STEFANIDIS, G.D. , VERMEIJ, M.D. , VAN GERVEN, T.D.T. , STANKIEWICZ, A.I. Design principles of microwave applicators for small-scale process equipment. Chemical Engineering and Processing: Process Intensification. 2010. Pág 912 - 922.
[35] UNIDAD DE PLANEACION MINERO ENERGETICA. La cadena del carbón. Ministerio de Minas y Energías. República de Colombia. 2005. Bogotá, Colombia.
[36] USLU, T. ATALAY, U. AROL, A.I. Effect of Microwave Heating on Magnetic Separation of Pyrite. Colloids and Surfaces A: Physicochem. Eng. Aspects 225 (2003). Pág. 161 - 167.
[37] ZHENFENG, L. RAGHAVAN, G.S.V. ORSAT, V. Temperature and power control in microwave drying. Journal of Food Engineering 97. 2010. Pág. 478 - 483.
[38] http://www.vidrasa.com/esp/productos/duran/duran_pf.html
[39] J. B. Hasted, Liquid water: Dielectric properties, in *Water A comprehensive treatise*, Vol 1, Ed. F. Franks (Plenum Press, New York, 1972) Pág. 255-309.
[40] http://www.lsbu.ac.uk/water/microwave.html
[41] I.Takei, Dielectric relaxation of ice samples grown from vapor-phase or liquid-phase water, in *Physics and Chemistry of Ice*, ed. W. Kuhs (Royal Society of Chemistry, Cambridge, 2007) Pág. 577-584

(137c) Novel Warm Gas Multi-Contaminant Removal System

Gokhan Alptekin, Ambalavanan Jayaraman, Margarita Dubovik and Mike Cesario, TDA Research Inc., Wheat Ridge, CO

Gasification technologies convert coal and biomass feedstocks into synthesis gas feed streams that can either be used as a fuel for highly efficient power generation cycles or converted into value-added chemicals and transportation fuels. However, the coal/biomass-derived synthesis gas contains a myriad of trace contaminants (e.g. mercury, arsenic, selenium) and ammonia (NH_3) that must be removed to eliminate power plant emissions; NH_3 will convert into NO_xin the gas turbine combustor (an acid rain precursor), while Hg is designated as an hazardous air pollutant and being regulated by the U.S. EPA. These contaminants also poison the catalysts used in the manufacturing processes that converts synthesis gas into fuels or chemicals.

Although conventional clean-up technologies can effectively remove these contaminants, they typically require aqueous quenching and cooling of the syngas to around 100°F, followed by scrubbing with chemical or physical solvents, and finally absorption/adsorption of trace contaminants on solid sorbents. However, cooling the syngas and condensation of steam significantly penalizes the cycle efficiency. The power plant efficiency can be improved 3 to 4% if the synthesis gas is not cooled below its dew point because the energy in high pressure steam can be recovered by expanding it steam in the gas turbine and eliminating the energy penalty associated with the re-heating of the synthesis gas. Warm gas clean-up also eliminates the need for the expensive heat exchangers (to cool the synthesis gas to the operating temperature of the clean-up system and then to re-heat it back to before the combined cycle) and the gray water treatment problems associated with processing the large amounts of condensate.

In a DOE funded SBIR project, TDA Research, Inc. (TDA) is developing a chemical absorbent-based gas clean-up technology to remove the NH_3 and hydrogen cyanide (HCN) as well as all trace metal contaminants (e.g., Hg, As, Se and Cd) from the synthesis gas at high temperatures (500°F) in a single process step. The sorbent can be operated in a regenerable manner to remove NH_3 and Hg, while irreversibly absorbing all other contaminants. The regenerable operation for the Hg (which has very low affinity to any surface at high temperature) and NH_3 (which is present in very high concentration in the gas) ensures high utilization of the sorbent, while its high capacity to other contaminants and low cost ensures its economical use as an expendable sorbent. We will present the results from this sorbent development effort at the meeting.

(137d) "Clean" and "Dirty" Tar Reforming of Biomass Gasification Gas

Klas J. Andersson, Reforming Department, Haldor Topsøe A/S, Lyngby, Denmark and Poul Erik Højlund Nielsen, Chemicals Department, Haldor Topsøe A/S, Lyngby, Denmark

A prerequisite for the establishment of an efficient biomass-to-chemicals production or CHP applications utilizing engines is a good thermodynamic utilization of the hot synthesis gas exiting a biomass gasifier. A second and important criteria is the extraction or conversion of tars (polyaromatic hydrocarbons), which, if left untreated, leads to fouling of downstream equipment and issues around waste water treatment. Both above criteria are met by steam reforming the tar components at high temperatures in the sulfur-laden exit gas from the biomass gasifier. We discuss the benefits of different approaches depending on pressurized or near atmospheric gasification, leading to the concepts of "clean" and "dirty" tar reforming.

The basic difference between the two approaches is how dust is handled. In the dirty tar reforming processes the residual dust in the gas - after having passed cyclones - is removed downstream the tar reformer. In the clean tar reformer the dust removal takes place upstream the tar reformer and preferentially the dust removal takes place at temperatures in the vicinity of the gasifier exit temperature to minimize loss of exergy. The catalyst and processes are widely different. In the dusty tar reformer monolithic catalysts are a necessity whereas the clean reformer can use pelletized catalysts with a much higher content of active material.

We will discuss the virtues of the different approaches and look into the requirements for producing CHP, SNG or fuels like gasoline or diesel, supplemented by our achievements in clean and dirty tar reforming in industrial and pilot settings.

CFD-Based Probabilistic Explosion Hazard Analysis As an Early Tool to Improve FLNG Design

Scott Davis, Remi Martini and Filippo Gavelli

GexCon US

Abstract

Process areas on floating LNG facilities are typically designed to be very compact, due to the high costs of "floating real estate". Given the high congestion levels of FLNGs, explosion design accidental loads (DALs) can become very large and are very sensitive to the layout of piping and equipment, the size of safety gaps, specifics of the liquefaction process, etc. For these reasons, the analysis of explosion hazards on FLNGs is a critical step that should accompany the facility development from the very earliest design phase. This paper will discuss how a probabilistic explosion hazard analysis methodology based on computational fluid dynamics (CFD) modeling can be used to optimize the layout of an FLNG facility with respect to DALs. The methodology uses a detailed 3D model of the FLNG geometry (down to 1-2 inch pipe diameters) within which gas dispersion and vapor cloud explosion scenarios are simulated using a CFD tool; leak rate probability data and the TDIIM ignition model are then used to calculate probability of exceedance curves and to determine the DALs. The paper will present results from a study, which compared the effectiveness of a number of design alternatives, such as:

- Grated vs. solid decks;
- Weak barriers vs. blast walls;
- Safety gap size;
- Different liquefaction processes.

The paper will demonstrate how this methodology can provide FLNG designers with critical safety-related design implications previously not available.

Introduction

The advanced methodology implemented in the present work, which incorporates advanced techniques to supplement an "incomplete" model of the FLNG geometry in Concept or FEED phase to levels anticipated in the "as-built" condition, provides a powerful tool to assess FLNG layouts/technologies in the earliest design phase when evaluating *Inherently Safer Designs*. This paper will provide some case studies on how valuable insights were gained by applying a detailed 3D model in early Concept phase of the design.

Plated vs. Grated process deck

This section demonstrates the hazard of introducing grating on the process deck of on FLNG when using heavier-than-air gases/refrigerants in the liquefaction process. When some typical heavier-than-air refrigerants (e.g., mixed refrigerants) are released in the liquefaction area on a FLNG vessel with a completely plated process deck (solid deck above the hull deck), the released vapour tends to either remain on the process deck and its modules or falls overboard. However, when using grated sections between the liquefaction modules on the process deck, this heavy vapor tends to disperse and fall into the hull deck, which is a confined and is not well ventilated.

Even if the hull deck height up to the process deck is only between 5 and 8m, the flammable vapour can extend all along the length and width of the vessel, allowing very large flammable vapour clouds to form from such releases. Figure 1 and Figure 2 show that by changing the process deck from partly grated to fully plated, the size of the flammable vapour clouds are significantly reduced due to very limited quantities of flammable gas migrating to the hull deck area when the process deck is plated. When these results are combined with leak frequency data and statistical wind data, the exceedance frequency of large clouds is drastically reduced, as shown in Figure 3.

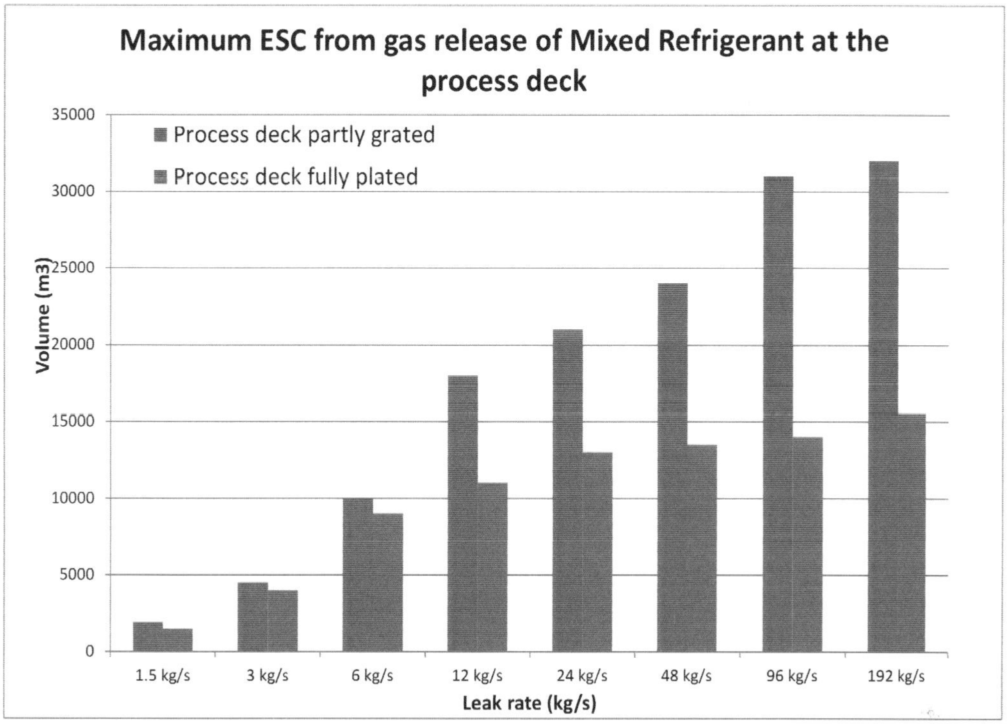

Figure 1: Maximum Equivalent Stoichiometric Cloud (ESC) from a release of mixed refrigerant at the process deck

Regarding explosion consequences, while it is intuitive to think that with increased grating on the process deck the explosion consequences in the hull deck will be reduced due to the increased venting. This however is not the case for the present study as the cloud size resulting from releases on the process deck result in clouds that occupy volumes both on the process deck and hull deck. The grating therefore does not the effect of venting for two reasons: when the flammable vapour cloud is below and above the grating, this allows the flame to accelerate through upper congested liquefaction modules, and into the confined hull deck, resulting in minimal benefits of venting through the grating (see); and in the case of having the flammable cloud only upon the process deck, the hull deck is typically too confined to adequately vent the explosion due to its height (5 to 8 meters).

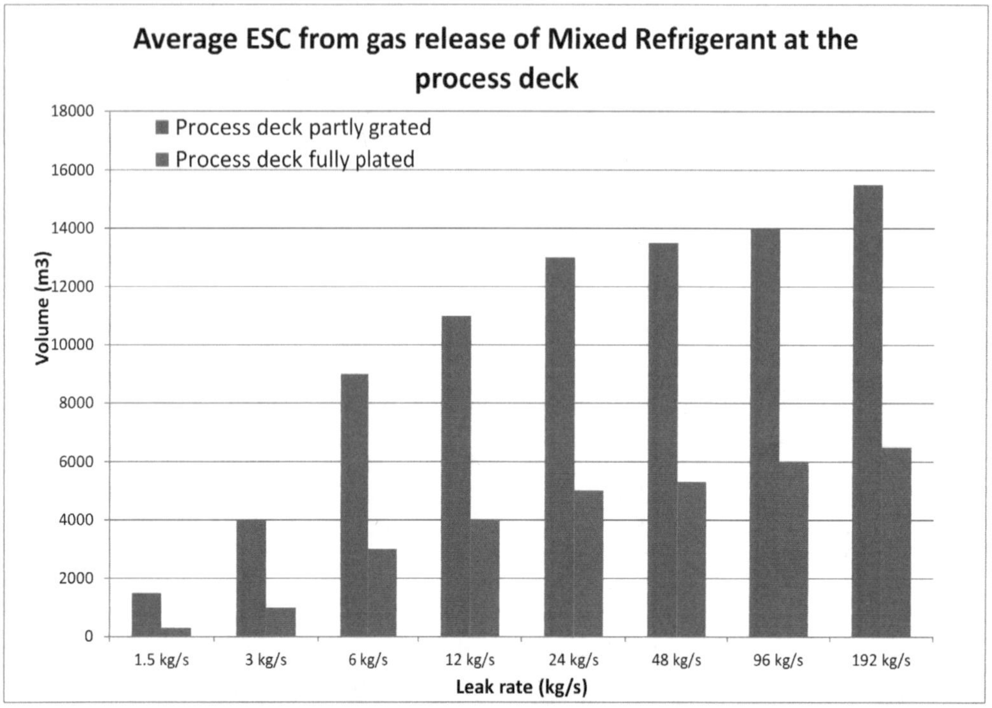

Figure 2: Average Equivalent Stoichiometric Cloud (ESC) from a release of mixed refrigerant at the process deck

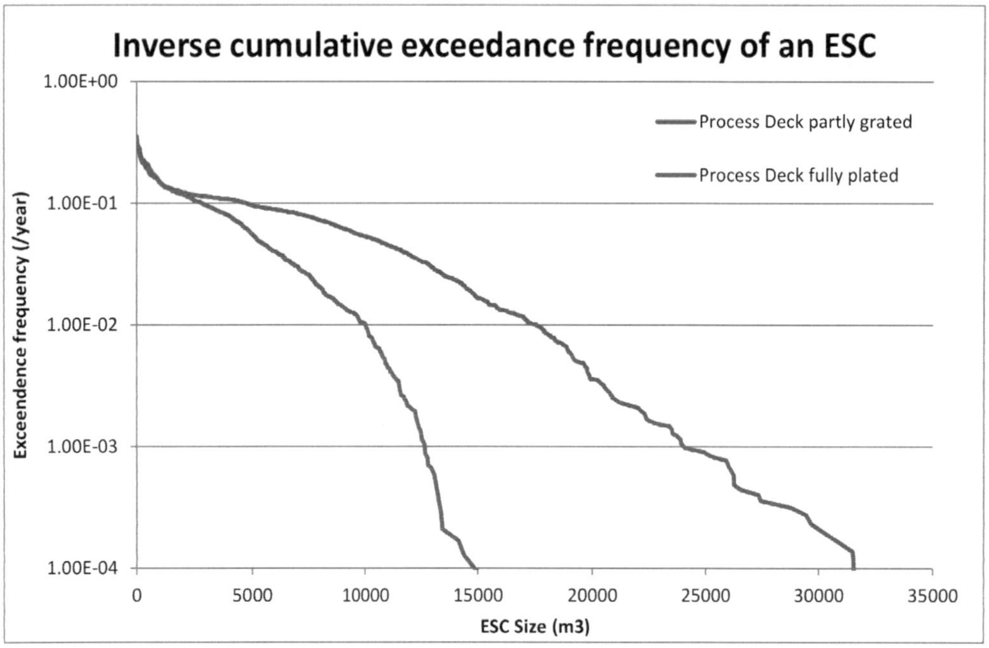

Figure 3: Inverse cumulative exceedance frequency of an explosion gas cloud

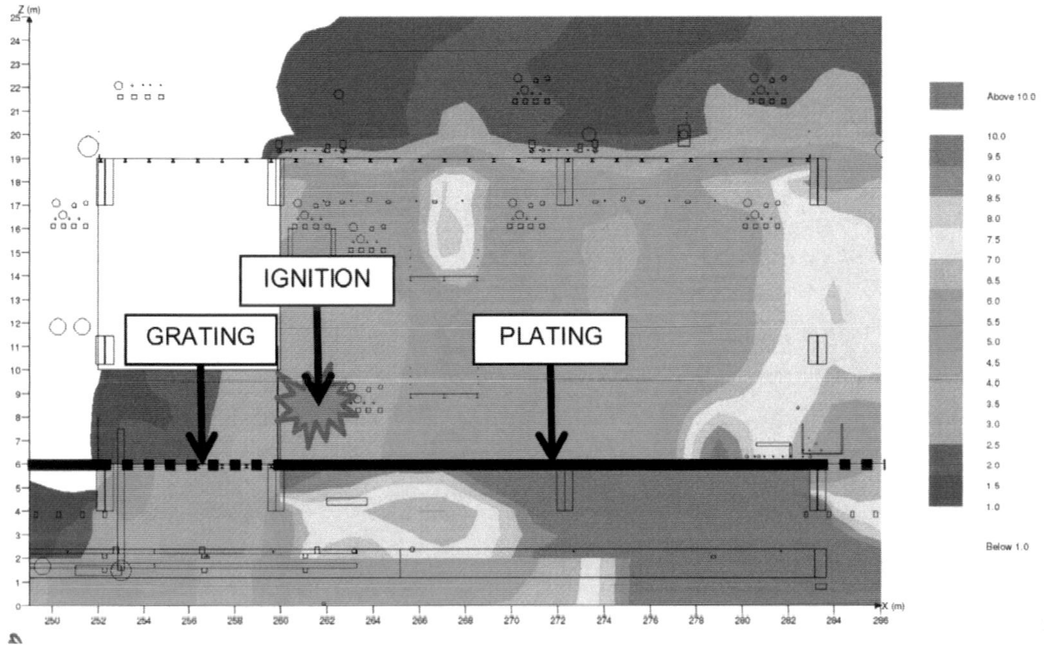

Figure 4: Example and an explosion simulation given more than 10 barg overpressure on the process deck

When assessing explosion risk, an ignition probability model is used to calculate the explosion frequency of a gas mixture. When using a time dependant ignition model, part of the ignition probability is calculated based on area (m2) in contact with the flammable mixture. Due to the presence of a flammable mixture in the hull deck, the whole deck area has to be considered as potential ignition source for the risk calculation. Even if the ignition sources within the hull deck are limited, they will slightly increase the ignition probability.

While a fully plated process deck does not have a significant impact on reducing the likelihood of an explosion (mainly due to the limited number of ignition sources at the hull deck), it significantly reduces the magnitude of the overpressures occurring with a frequency of $1x10^{-4}$/[year], due to the limited size and location of the gas cloud as shown in Figure 5.

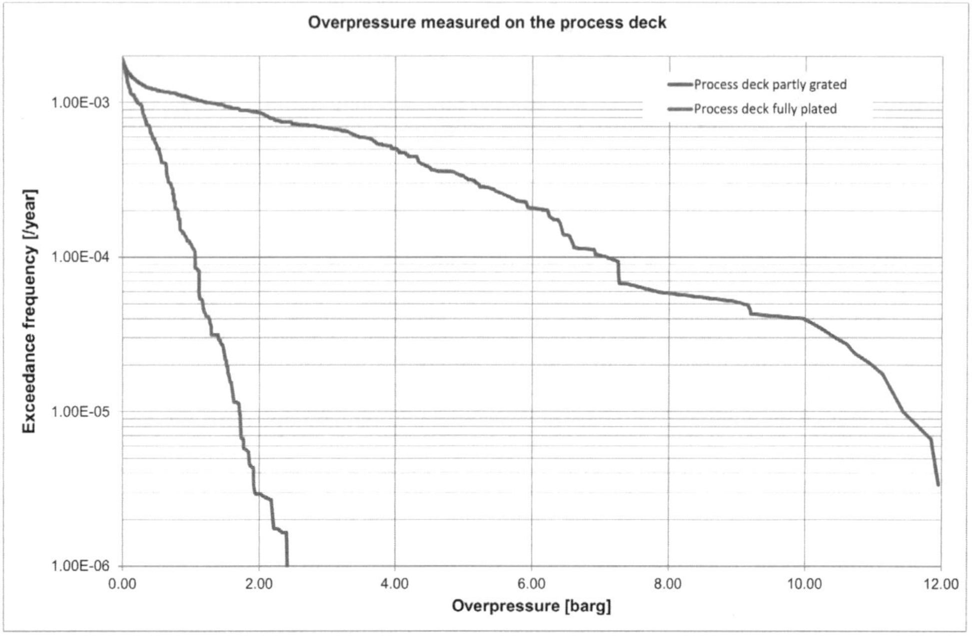

Figure 5: Inverse cumulative exceedance frequency of an explosion gas cloud

Gas Detection

Gas detection is a critical safety device that can initiate certain mitigation measures. The time to detect a leak will largely affect two parameters: the size of the gas cloud due to initiation of ESD and blowdown; and the time of exposure to ignition sources (ignition source isolation after confirmed gas detection). Studies have shown that using an efficient gas detector layout, an accidental release can be detected within few seconds. This present case study only compares generic detection times for all leak rates and assumes that all accidental releases are detected. The difference in the two detection philosophies are that the confirmed gas alarm is reach in 30 seconds for the base case and 15 seconds in the optimized case.

Two detection philosophies are compared:

	TIME (s)				
	Leak starts	Low alarm	High Alarm*	ESDV closed	Blowdown
Base Case	0	15	30	60	150
Optimized	0	10	15	45	135

*Start of Emergency ShutDown (ESD)

Despite its generic nature, this example enlightens the importance of detection time in terms of explosion risk reduction. For the present case study, the explosion loads occurring once every 10,000 years (10^{-4}) were reduced from 2.2 barg to 1.7 barg.

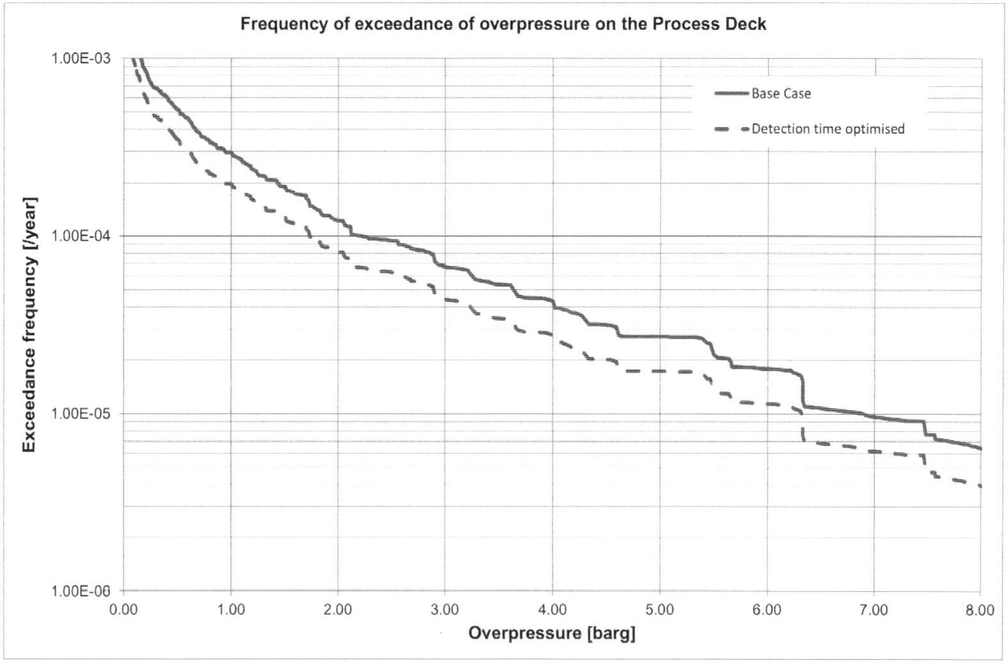

Figure 6: Inverse cumulative exceedance frequency of an explosion gas cloud

Safety Gaps

This case study evaluated the effectiveness of two different safety gaps (12.5 meters and 20 meters) between the liquefaction modules (see).

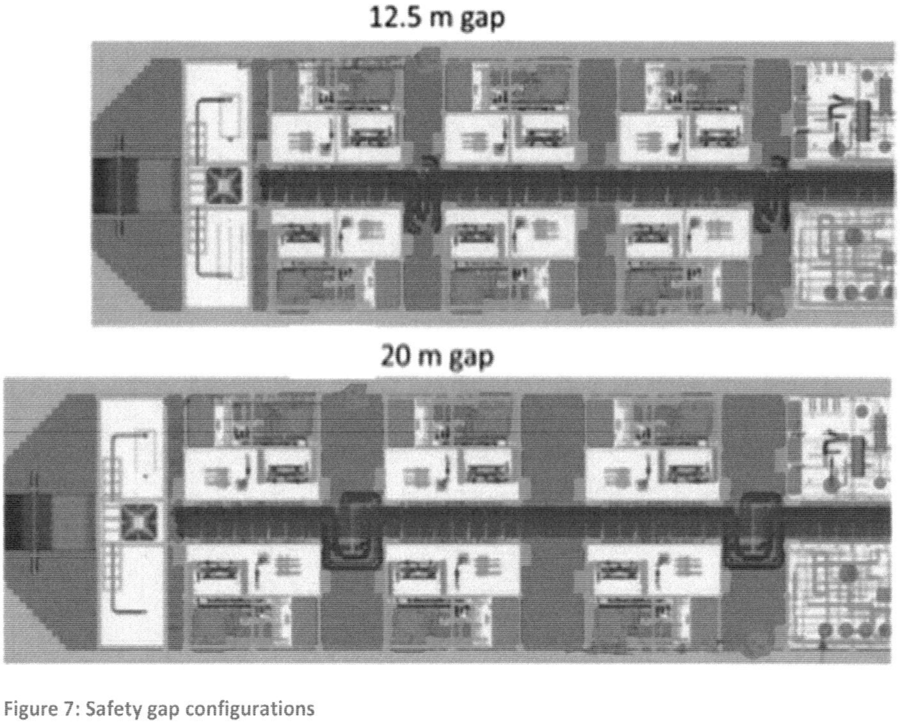

Figure 7: Safety gap configurations

The results showed that the ventilation rate within the module (measured by ACH) remained essentially unchanged as the safety gap was increased from 12.5 to 20 meters. Therefore, increasing the gap size from 12.5 to 20 meters did not significantly reduce the size of a gas cloud that remained inside the module where the leak originated. In contrast, significant reductions in cloud sizes were observed for clouds that migrated out of the origin module and into neighboring modules (see Figure 8). This was significant with respect to overall gas cloud size and ignition frequency as discuss in a previous section.

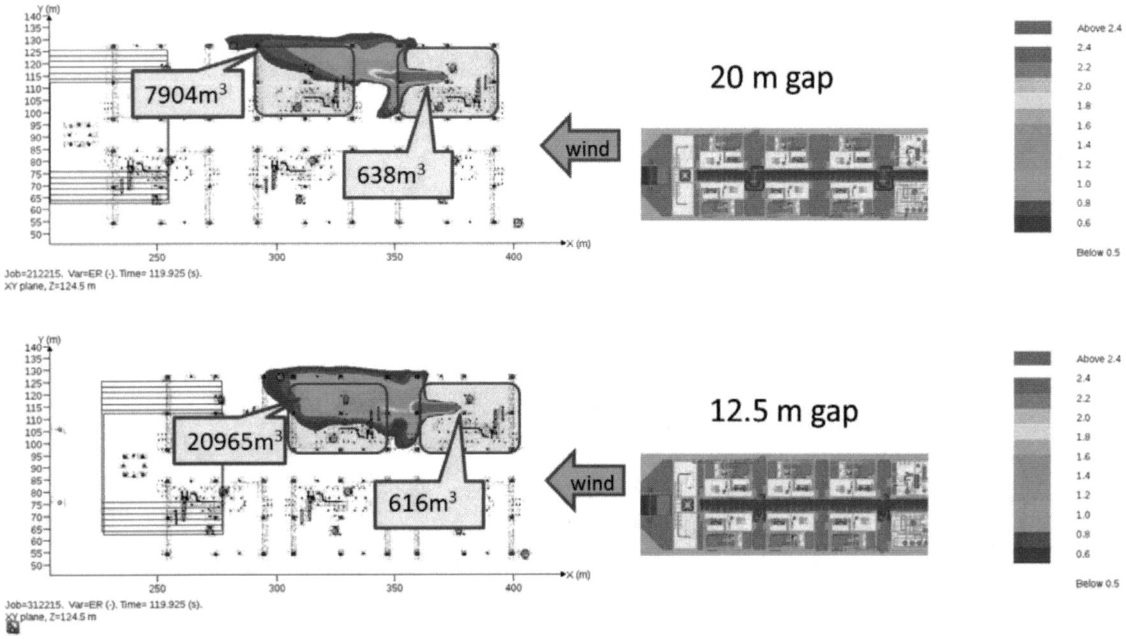

Figure 8: Dispersioin example for the 20 meter gap and 12.5 meter gap

Conclusions

This paper examined the benefits of using advance CFD simulations for explosion risk in early design phase of FLNG vessels. Many insights were obtained early enough as to positively influence the design towards an inherently safer design.

Forced Dispersion Analysis of LNG Vapor Mitigation Using Fire Dynamic Simulator

*Wilson Y. Molina-Torres, Byung Kyu Kim, Ray A. Mentzer and M. Sam Mannan**
Mary Kay O'Connor Process Safety Center, Artie McFerrin Department of Chemical Engineering,
Texas A&M University System, College Station, TX
*Corresponding author. Tel.: +1 979 862 3985. E-mail address: mannan@tamu.edu (M.S. Mannan)

Abstract

With the expected demand of liquefied natural gas (LNG) to double by year 2020 world-wide, the mitigation techniques for an LNG spill emergency require further investigation to provide accurate information in evaluating the consequences of an LNG spill and the potential impact on the public safety and security. This work applies the Fire dynamic simulator (FDS) to set up and simulate LNG forced dispersion using a water spray application. A comparison study with a previous methodology has been conducted and potential application of FDS in the LNG safety research is discussed.

1. Introduction

The demand of the natural gas in the US had exceeded the production level for last couple of the decades [1]. However, with the increases of natural gas production in US from the development of the shale gas industry, supplies of natural gas is expected to surpass the domestic demands. Once a liquefied natural gas (LNG) importer, the US natural gas industry is in the phase of transiting to LNG exporter, and expected to play a significant role in redefining the global natural gas market [2].

With the expectation of the LNG production in US, it becomes critical to understand all the hazards related to the liquefaction and transportation of the LNG to ensure public safety. It was concluded that the safety mitigation measures require further study to minimize the uncertainty in determining the influence on reducing the hazards imposed to the society [3]. The Fire dynamic simulator (FDS) is an open source computational fluid dynamics (CFD) model of fire-driven fluid flow developed by the National Institute of Standards and Technology (NIST). This work applies the FDS code to evaluate the LNG forced mitigation phenomena. A preliminary comparison analysis had been conducted using the Fluent model. The potentials of applying FDS and further work to minimize the uncertainties of the vapor prediction are discussed.

2. Numerical modeling with FDS code

The FDS code is a publically available CFD model developed by the NIST. FDS was developed as software to model fire-driven fluid flow with emphasis on smoke and heat transport [4]. The FDS code is commonly used in the fire protection engineering industry and the hydrodynamic model embedded in the code can be used to model the low-speed fluid flow, such as vapor dispersion. FDS provides solutions for the Navier Stokes equations of motion, energy and species concentration in space and time. The vertical profiles in plume properties (velocity, concentration, species concentration, etc.) are not simplified or averaged, and the turbulence model components are provided by adjusting the damping effects of thermal or density stratification.

The FDS dispersion code had been evaluated for LNG vapor dispersion [5]. FDS provided fairly reasonable dispersion prediction. A safety factor of 2 was recommended to predict the unobstructed flow fields, and factor of 3 for the obstructed flow field from the results.

2.1 LES turbulence modeling

The FDS codes apply the Large Eddy Simulation (LES) to evaluate the turbulent flow induced within the vapor fields [4]. The LES model provides a qualitative description of the flow physics and applies the quantitative model to produce large-scale motions directly [6]. LES requires a flow field where only the large-scale components are present, and this process can be achieve through filtering process. With the available sub-grid scale models, large-scale fields are solved, while the smaller scale motions are approximated through localized filter functions of either a Gaussian or top-hat filter.

While the Reynolds Averaged Navier-Stokes (RANS) approach approximates and averages out the fluid fluctuations for all turbulent length scales, LES provides only the turbulent fluctuations that are resolvable by its mesh. This enables better fidelity of fluid flow behavior from turbulent fluctuations. Theoretically, this approach may assist in reducing the potential hazards that might rise from approximation or averaging the concentration. Although, LES is more computationally expansive, the model can provide an accurate turbulent vapor behavior [5].

2.2 Lagrangian spray model

The Lagrangian particles are utilized in FDS to simulate the fine droplets for the water sprinkler or spray [4]. A prescribed number of droplets are introduced from the nozzle every second, and the droplet characteristics, such as the density, specific heat, diameter, temperature, and mass flow rate can be defined. The force balance of an individual droplet is evaluated by the governing equation (1) and overall momentum imparted from the droplets to the gas flow in a grid cell is calculated using equation (2) [7].

$$\frac{du_d}{dt} = g - \frac{1}{2}\frac{\rho C_D \pi r_d^2}{m_d}(u_d - u)|u_d - u| \qquad (1)$$

$$f_b = \frac{1}{V}\sum_{d \in V}\left[\left[\frac{1}{2}\rho C_D \pi r_d^2 (u_d - u)\right]|u_d - u| - \frac{dm_d}{dt}(u_d - u)\right] \qquad (2)$$

u_d is the velocity of the droplet, $C_d$the drag coefficient, $r_d$is the droplet radius, u is the velocity of the gas, ρ is the density of the gas, V is the volume of the grid cell, m_d is the mass of the droplet, and g is the gravity constant. The initial droplet trajectory is evaluated with the predefined droplet characteristics and the overall momentum effects of overall droplets in a computational grid cell are calculated to determine the level of interaction of droplet/vapor system.

In additional to the momentum effects, the FDS solver also considers the heating and evaporation of the droplet particles.

$$m_L C_L \frac{dT_L}{dt} = Ah(T_g - T_L) + Ah_s(T_s - T_L) + \dot{q_r} + \frac{dm_L}{dt}h_V \qquad (3)$$

A is the surface area of the liquid droplet, h_m is the mass transfer coefficient, ρ is the density of the gas, $C_L$is the liquid specific heat, h is the heat transfer coefficient between the liquid and wall, $m_L$is the mass of the liquid droplet, h_S is the heat transfer coefficient between the liquid and solid surface, $h_V$is the latent heat of vaporization of the liquid, and $\dot{q_r}$is the radiative heating of the droplet. As the droplets travel in the gas flow, the droplet temperatures are constantly evaluated by considering the convective thermal transfer with the gas phase. The heat transfer with the ground is not considered in this work as the droplet interactions get terminated once the droplets travel back to ground. Also, the radiative heat exchange is neglected, as the heat transfer in this work does not involve high thermal source, such as fire or flame. The Lagrangian spray model also considers the evaporation effects of the droplets to the surroundings.

$$\frac{dm_i}{dt} = -Ahm\rho(Y_L - Y_g) \quad (4)$$

where the liquid equilibrium vapor mass fraction Y_L, the local gas phase vapor mass fraction Y_g, the assumed uniform liquid temperature T, and the local temperature T_g. However, considering the short duration of the droplets traveling in the atmosphere, the effects of evaporation will be relatively less significant compared to the heat transfer effects described as equation 3.

3. Simulation specifications

3.1 Computational domain and water spray modeling

The water spray settings and the computational domain are described in Fig. 1.

Parameter	Setting
Nozzle size (m)	1.27×10^{-3}
Cone angle (deg)	60
Droplet material	Water
Droplet diameter (mm)	0.935
Temperature (k)	300
Mass flow rate (kg/s)	4.6

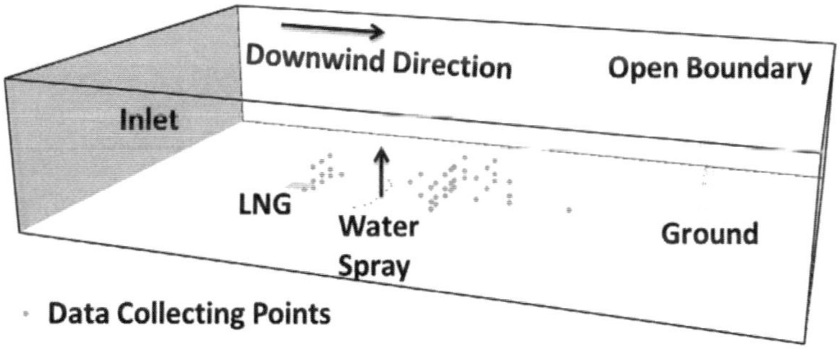

Fig. 1. Water spray applied for FDS modeling and computational domain setup

The computational domain was created in FDS platform with the downwind direction in the x direction. The size of the domain was 50 m x 30 m x 10 m. The For the simulation grid size the cell size was refined to acquire grid-independent setting to a 0.2-meter in a cube-like cell. A steady state result was achieved for the concentration profiles at different distances and approximately 1,875,000 grids were created. The simulation starts running without the LNG release for 300 s in order to obtain a constant (steady) wind profile on the domain. The natural dispersion was set for 600 s, and forced dispersion of LNG vapors with water sprays was applied for 500 s. These scenarios are based on the experimental work from March 2009 test [9].

For the LNG inlet boundary condition, the LNG source term was set using a circular inlet condition at the ground level. The methane was injected at its boiling temperature of 111 K. The size of the inlet was set at 0.857 m with an average mass flux rate set at 0.2 kg/(m^2-s). The atmospheric conditions used for the simulation were adopted from the experimental data [8].

The data collecting points and spray location, LNG pool size was adopted from the March 2009 MKOPSC LNG spill experiments [9]. The spray settings were also adopted from the experimental setup. A total of 100 droplets are injected each time step, which was set as 1 s, per nozzle. The droplet type is set as water droplets, which takes into account of the evaporation, and radiative absorption. However, these functions are negligible, as the droplet travel time is relatively short and temperature applied in the simulation is not high enough for the radiative heat transfer to be dominant. The coalescence and collision of the droplets are not considered in this modeling and the droplets are discharged in the shape of upward-oriented full cone type nozzle.

3.2 Wind Profile

The FDS is capable of simulating the atmospheric condition using the steady or unsteady wind profiles [4]. The magnitude of velocity profile and direction can be defined by specifying the velocity along the x, y, and z directions within the domain boundaries [5]. The FDS has two different functions available to describe the atmospheric condition. The ramp function can be used to make changes in the velocity component in the normal direction. As the ramp function can only simulate the wind field in the normal direction, it is difficult to simulate the transient wind. Another option is to apply the vent condition in the inlet and create corresponding ramp function. The wind profile can be described using a parabolic wind profile or a typical atmospheric wind profile in normal direction. For this study, a parabolic wind profile combined with the vent function was applied to provide a closer approximation of the wind behavior presented in the field experimentation. The parabolic wind profile used in this work is as follow:

$$u = u_O\left(\frac{z}{z_O}\right)^p \quad (5)$$

where u is the wind velocity at height z , u_O is the wind velocity known at an specific height z_O and p is an empirical constant that varies depending on the stability of the atmosphere.

It is recommended to use low wind speed with stable atmospheric condition, as it will result in longer distances to downwind concentration. The requirement outlined in 49 CFR Part 193.2059 is the lowest wind speed that occurs 90% of time for the area or 2 m/s, to replicate the worst-case scenario. It is expected that 2 m/s wind speed will demonstrate the worst-case approach, with the farthest downwind concentration.

4. Model comparison

To evaluate the LNG force dispersion modeling of FDS, the simulation predictions were compared with the Fluent results. The physical parameters for Fluent were adopted from the simulation work done by Kim et al. (2012) and the experimental data from MKOPSC LNG spill experiments were also compared [8, 10].

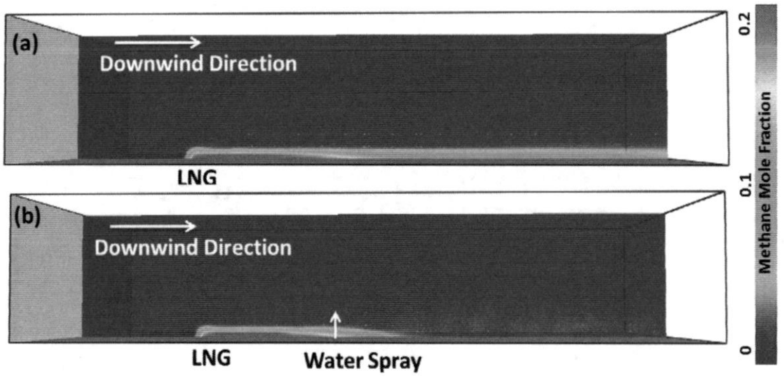

Fig. 2. Vapor contour simulated for (a) LNG natural dispersion and (b) forced dispersion with water spray

After acquiring the steady state solution of the wind profile, LNG vapors were injected to the domain as shown in Fig. 2 (a). The LNG vapors dispersed in the prevailing downwind direction, and diluted as results of heat transfer and air entrainments into the vapor clouds. The water spray application on the LNG vapors is illustrated in Fig. 2 (b). As the water spray starts to interact with LNG vapors, the droplets start lifting the LNG vapors to a higher elevation, increasing the buoyancy of the cloud. The momentum and heat transfer effects from the water spray resulted in decrease of the vapor concentration in the post-spray region.

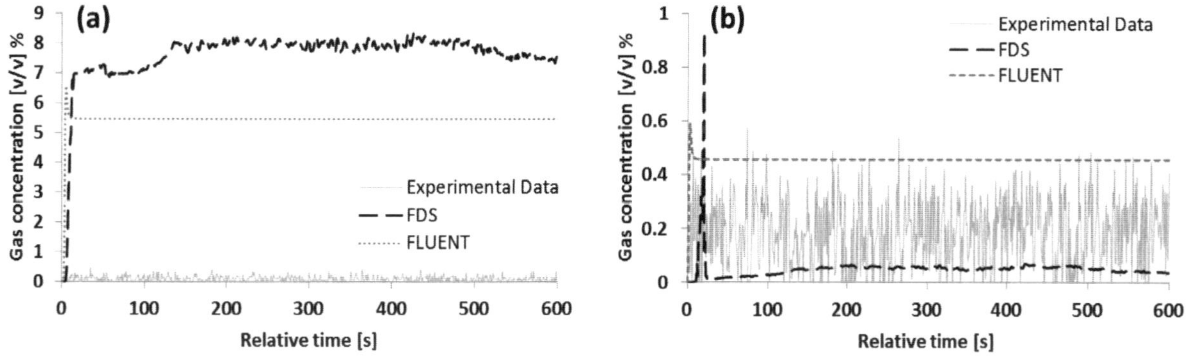

Fig. 3. Comparison of LNG natural dispersion prediction (FDS/Fluent) with experimental data at downwind distance 0 m at (a) 0.5 m elevation and (b) 1.2 m elevation

Fig. 3 shows the comparison of the LNG natural dispersion with the experimental data to the simulation prediction (FDS/Fluent) at 0 m distance. At the lower elevation, the results obtained by both FDS and Fluent over-predicted the LNG vapor concentration as shown in Fig. 3 (a). For the 1.2 m elevation case, Fluent showed agreeable prediction to the experimental result, while FDS provided an under-predicted vapor concentration as shown in Fig. 3 (b). This is mainly due to the limited vapor movement in the FDS. The FDS solver was capable to disperse the LNG vapors only up to approximately 1 m elevation. Further investigation is required to implement better heat transfer model and grid setting to improve the vapor movement prediction using FDS. The high peak observed during the initial dispersion of the LNG vapors shown in Fig. 3 (b) was caused by the LNG vapor behavior. When LNG was injected into the domain, the prevailing wind dominated the LNG vapor movement, as there was only a small amount of LNG vapors initially. Although heavier than air, the LNG introduced into the domain during the initial stage might had been carried by the wind as bulk of vapor, which might had caused the concentration to increase for short amount of time. As more vapors are introduced to the domain, the LNG vapor concentration reached steady state after approximately 200 s.

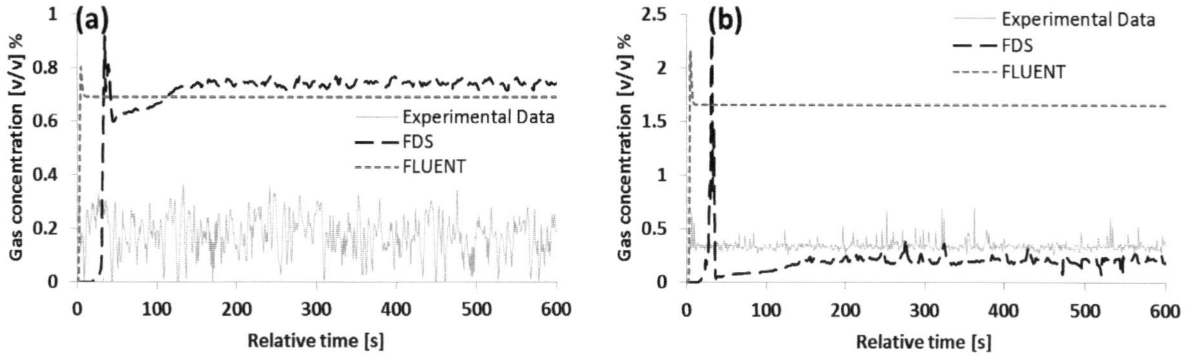

Fig. 4. Comparison of LNG natural dispersion prediction (FDS/Fluent) with experimental data at downwind distance 9.7 m at (a) 0.5 m elevation and (b) 1.2 m elevation

Fig. 4 shows the vapor concentration plotted for 9.7 m away from the LNG source at two elevations. These data will show the LNG vapor behavior after certain amount of time, as the vapors had traveled 9.7 m from the LNG source. Similar to Fig. 3, both FDS and Fluent over-predicted the experimental data at 0.5 m elevation. However, the FDS under-predicted the experimental data at 1.2 m elevation, as shown in Fig. 4 (b). Further investigation will address the issue of limited heat transfer and other essential physical parameters to improve the LNG natural dispersion using FDS.

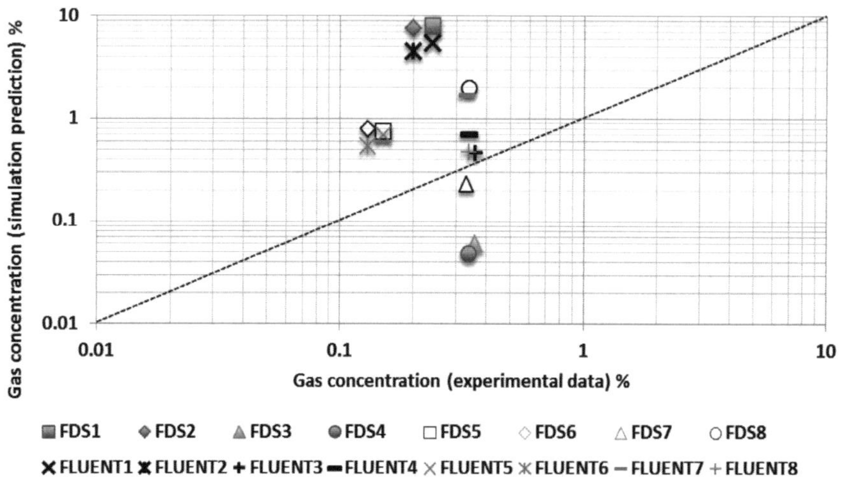

Fig. 5. Comparison of concentration prediction (FDS/Fluent)

Fig. 5 compares the FDS and Fluent prediction. It can be seen from these predictions that FDS provided under-prediction, which may require further safety precaution in real applications. Kohout (2010) recommended to implement the safety factor of 2 for an unobstructed flow fields [5]. Further study will address the degree of under-prediction quantitatively to provide detail guidelines on developing and implementing the safety factors for the application of FDS.

Fig. 6 shows the overall concentration prediction of LNG natural dispersion and forced mitigation using the water spray application. Both FDS and Fluent achieved a steady state solution after 600 s, and when water spray was applied, the water droplets influenced the vapor behavior. The FDS prediction showed less significant changes of vapor concentration at 0.5 m elevation; however, the vapor concentration at 1.2 m elevation showed approximately 5 times more vapor increased at this level. This indicates that the FDS predicted the LNG vapors to travel towards higher elevation (1.2 m elevation) after the water spray application, pushing the vapor cloud beyond the 0.5 m elevation. On the other hand, Fluent predicted the vapor behavior in the opposite trend, where the vapor concentration in both elevations decreased.

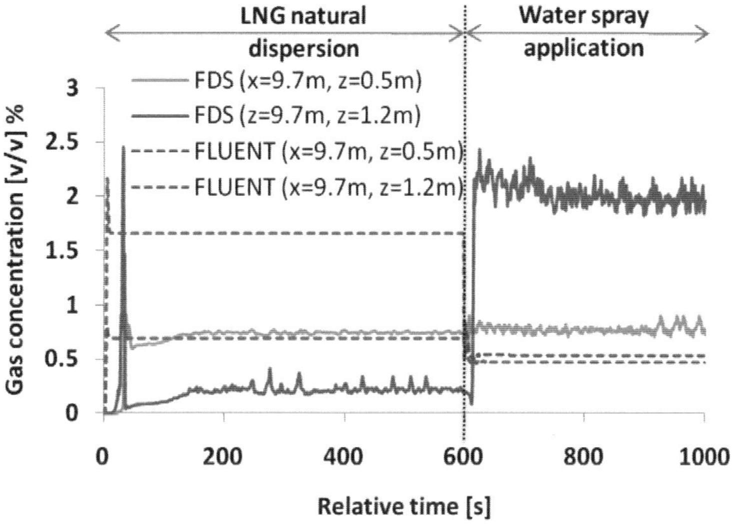

Fig. 6. Vapor concentration of LNG natural dispersion and forced mitigation using water spray application

The opposite trend can be explained by the two different turbulence models applied for FDS and Fluent. The LES predicts the vapor behavior different from the k- ε turbulence model, which was used for Fluent application. It can be seen from the Fluent results that the solver predicted the vapors to travel beyond the 1.2 m elevation, while the FDS solver predicted that the vapors would travel up to 1.2 m elevation. Currently, further investigation is planned to compare with the experimental results to determine the accuracy of both turbulence model.

5. Conclusion

This work presented preliminary results of applying FDS in setting up and simulating the LNG forced dispersion phenomena using the water spray application. The results of LNG vapor dispersion using FDS showed under-prediction of the vapor concentration at higher elevation. Further analysis will determine the exact level of the under-prediction and implementation of the safety factors. The forced dispersion simulated using the FDS produced opposite trend compared to the Fluent prediction. This is mainly because of the discrepancy in describing the turbulent flow induced from the interaction of water droplets. Detail study is required to determine the accuracy of the turbulence model, which could provide better description of droplet/vapor interaction system.

FDS can provide a rigorous prediction of the complex behavior of LNG vapor dispersion. Currently, FDS requires further analysis on the following to improve the vapor dispersion simulation: grid sensitivity, implementation of more specific models for heat transfer model, and the description of a specific fully developed atmospheric profile (wind speed and temperature).

References

1. EIA, *Annual Energy Outlook 2012 with Projections to 2035*, 2012, US Department of Energy.
2. IEA, *World Energy Outlook 2011: Are we entering a golden age of gas?* , 2011, IEA: Paris, France.
3. USGAO, *Public Safety Consequences of a Terrorist Attack on a Tanker Carrying Liquefied Natural Gas Need Clarification (Maritime Security GAO-07-316), United States Government Accountability Office*, 2007: Washington, D.C.
4. McGrattan, K.F., et al., *Fire Dynamics Simulator (Version 5) Use's Guide (Special Publication 1019-5)*, 2007, National Institute of Standards and Technology: Gaithersburg, MD.
5. Kohout, A.J., *Evaluation of Fire Dynamics Simulator for liquefied natural gas vapor dispersion hazards*, in *Fire Protection Engineering*2011, University of Maryland: College Park, MD.
6. Tu, J., G. Yeoh, and C. Liu, *Computational fluid dynamics: a practical approach* 2008, Boston, MA: Butterworth-Heinemann.
7. NIST, *Fire dynamics simulator technical reference guide (Vol 1: Mathematical model-NIST Special Publication 1018)*, 2012, National Institute of Standards and Technology: Gaithersburg, MD.
8. MKOPSC, *Data Report of MKOPSC LNG Spill Tests: 2005-2009*, M.S. Mannan, Editor 2010, Mary Kay O'Connor Process Safety Center, Texas A&M Univeristy: College Station, TX.
9. Rana, M., *Forced dispersion of liquefied natural gas vapor clouds with water spray curtain application, Chemical Engineering, Texas A&M University,* ***PhD Dissertation***, 2009. p. 232.
10. Kim, B.K., et al., *Modeling of water spray application in the forced dispersion of LNG vapor cloud using a combined Eulerian-Lagrangian approach.* Industrial & Engineering Chemistry Research, 2012. **51**: p. 13803-13814.

HSE Evolution in Conversion of LNG Import Terminals to Export

Jack Chosnek, Ph. D., P. E.
President and Principal
Knowledge*One*

Filippo Gavelli, Ph. D., P.E.
Head of Dispersion Consulting
GexCon

Victor H. Edwards, Ph. D., P. E.(TX)
Director of Process Safety
IHI E&C International Corporation

AIChE Spring Meeting
13th Topical Conference on Natural Gas Utilization
San Antonio, April 28-May 2, 2013

Introduction

With the abundance of natural gas (NG) in the U.S. in recent years, companies that had built marine terminals for the import of Liquefied Natural Gas (LNG) and subsequent re-gasification for pipeline transportation are now considering converting those facilities for liquefaction of pipeline gas in order to export LNG. This has brought challenges in the Health, Safety & Environment (HSE) area as well as in the regulatory arena regarding plot plan development and the handling of high pressure liquefied hydrocarbons. Besides the need to address the increased hazards of NG liquefaction versus LNG re-gasification, government agencies are faced with application of regulations and codes to the liquefaction of NG that were intended mainly to address LNG re-gasification terminals earlier. As the regulations try to apply the consequence-based codes to the variety of materials used in the liquefaction process, such as refrigerants that may include propane, ethane, ethylene, butanes, and other heavier hydrocarbons that may also be present in the NG, the need for tools and techniques to evaluate safe alternatives becomes evident. In addition, the environmental perspective also changes from regasification to the liquefaction. This paper will describe the methodology used in an NG liquefaction project to address these needs, which among other aspects include a combination of traditional vapor dispersion, vapor cloud explosion, and thermal radiation modelling tools, with computational fluid dynamics (CFD) modelling. The environmental factors, as well as the use of barriers to screen equipment noise and to limit the extent of vapor clouds will also be discussed.

Many US LNG Import Facilities are Being Converted to LNG Export

Domestic supplies of natural gas are growing rapidly due to availability of extensive shale gas resources. As a result, US natural gas prices have dropped to a fraction of their peak several years ago and US LNG imports have decreased dramatically.

At the same time, LNG prices in Japan, in other parts of the Asia Pacific region, and in Europe have remained high. Contributing factors are the Japanese decision to shut down nuclear power generation after the Fukushima tragedy, and the growing global demand for natural gas. Yet another factor is the environmental advantages of natural gas as a fuel when compared with coal.

Consequently, many US LNG import facilities are being converted to either LNG export facilities or to bidirectional LNG facilities.

Conversion from LNG Import to LNG Export poses Increased Hazards

Chosnek and Edwards (1) discussed the hazards of conversion of LNG Import Facilities to LNG Export. These hazards include the following:

- Natural gas liquefaction facilities required for LNG export are more complex and congested due to higher number of processing equipment involving different processing steps with multiple products, and require more space than corresponding LNG regasification facilities.
- Plot space at existing LNG import sites for new LNG export facilities may be limited.
- The risks of fire and explosion may be increased by the presence of more reactive chemicals (e.g., propane, ethylene) and higher congestion typical of liquefaction and other gas processing facilities.
- These risks are also increased by the use of more reactive heavy hydrocarbon streams as by-products from natural gas purification.
- Liquefaction often requires large inventories of refrigerants such as ethane, ethylene, propane, propylene, and butanes or their mixtures, at comparatively high pressures.
- Hydrogen sulfide streams may result from purification of natural gas prior to liquefaction, although the primary impact is on potential emissions of sulfur oxides resulting from oxidation of the hydrogen sulfide by-product.

Current US LNG Regulations are Consequence-based

Chosnek and Edwards (1) highlighted that current US LNG regulations are consequence-based instead of a preferred risk-based approach (2). Summarized below are the current criteria (3-6):

- Exclusion Zones (at property line) from a release caused by a "design spill"

  - ½ Lower Flammable Limit for flammable vapor dispersion

  - 1,600 BTU/hr-ft^2 Thermal Radiation Limit for pool fires

  - 1 psi Overpressure Limit for Vapor Cloud Explosions (although not in the regulations and never an issue for LNG import facilities, this is now an important criterion for liquefaction facilities in the US)

These existing US regulations were adequate for LNG import, although the constraint that only passive mitigations (impoundments, vapor fences) could be used to reduce the hazard footprint was a disadvantage.

Consequence modeling of existing US import terminals was done in most cases using the following two software products:

- LNGFIRE3 for Thermal Radiation
- DEGADIS for Flammable Vapor Dispersion

These tools model the design spill as an LNG pool on an otherwise empty horizontal plane, which is a reasonable approximation when the major hazard is an LNG spill that can be controlled by impoundment.

Advanced Consequence Modeling is needed to Safely Design LNG Export Facilities

As mentioned above, natural gas liquefaction facilities introduce new hazards, such as overpressures from vapor cloud explosions, because of C2+ hydrocarbons and increased congestion.

Comparatively recently (October 2011), FERC approved two additional software packages for LNG vapor dispersion hazard analyses: PHAST™ from DNV, an analytical tool described by Witlox, *et al.* (7), and FLACS from GexCon, a CFD tool described by Hansen, *et al.* (8).

The two models now available for LNG hazard analysis offer a number of advantages over the previously allowed DEGADIS:

- Model verification and extensive validation with field data

- Wide variety of source terms

- Mixtures of gases can be considered

- Both vapor dispersion and vapor cloud explosion can be modeled

-Tabular and graphic outputs can be generated (2D in the case of PHAST, 2D or 3D in the case of FLACS)

The main differences between the two models are as follows:

- PHAST is based on analytical data and correlations (Unified Dispersion Model), which allows for rapid calculations; FLACS is a 3D CFD model that solves the Navier-Stokes and turbulence closure equations in each of several hundred thousand computational grid cells, therefore requires longer times to complete a calculation;

- The correlations implemented in PHAST do not account for the presence of obstacles or terrain features, therefore its results are less accurate and it does not allow all mitigation measures to be evaluated; FLACS can include any obstacle or obstruction (as long as they are built into the geometry model used for the simulations) and therefore provides more realistic results.

While both tools have pros and cons, an efficient and effective approach can be achieved by combining the use of PHAST and FLACS in order to take advantage of each model's strengths. This approach can make thorough and detailed consequence modeling possible within the typical time frame of LNG facility design development, leading to a safer and regulatory-compliant plot plan:

- Use PHAST to rapidly screen a large number of scenarios, involving releases of hazardous fluids from different process streams and locations (approximately 100 scenarios can be examined for a single LNG project).
- Rearrange equipment on the plot plan to minimize adverse consequences of releases of hazardous fluids.
- Validate the safer plot plan using FLACS.
- Establish and evaluate passive protections, such as vapor fences, using FLACS.
- Quantify location specific overpressures from vapor cloud explosions using FLACS.

Atmospheric Emissions will Increase and Amended Permits may Face Stricter Regulations

Natural gas liquefaction requires energy, so greenhouse gas (CO_2 and CH_4) emissions may increase if power is generated onsite.

- Emissions of ozone precursors (NOx and VOCs) may increase.
- Greater fugitive emissions may result due to increased process size and complexity.
- To a smaller extent in the US pipeline gas, emissions of SOx may increase.
- Amendment of existing permits may require meeting new and more demanding regulations.
- Best Available Control Technology (BACT) emission controls may be required.

Modest Increases in Aqueous Discharges

- LNG import terminals have few aqueous streams to treat
    - Storm water with potential lube oil contamination
    - Demineralization by-product water
    - Fire water system test water
    - Hydro-test water
    - Sanitary waste
- LNG export terminals have a few more:
    - All of the above aqueous streams
    - Higher potential for lube oil contamination of storm water (larger plot plan and significantly more rotating equipment)
    - Storm water with potential amine contamination
    - Amended permit may need to achieve newer and stricter regulations

It should be possible to minimize storm water contamination through segregation, containment and rigorous maintenance and housekeeping practices.

Noise Sources and Mitigation

New export facilities add new piping and equipment to the existing site that are new noise sources. Examples include air-cooled heat exchangers, compressors, and turbines. FERC has very strict limitations on noise from LNG facilities.

In addition to buildings, enclosures, silencers, and sound insulation, a noise barrier wall may be required. Computer modeling of noise from equipment and piping will be required during both FEED and Detailed Design. Noise modeling during FEED can give guidance on plot plan modification to minimize noise exposure to any nearby noise sensitive areas.

Design should also incorporate features to permit easy provision of additional noise controls if as-built facilities are noisier than predicted during design phase. Examples include building foundations for noise barrier walls strong enough for an increased wall height and providing spool pieces in compressor piping to allow easy provision of silencers if needed.

Conclusions

LNG import and export facilities are among the safest and most environmentally friendly fossil fuel processes. Conversion of a number of US LNG import facilities to either export or bi-directional facilities is being proposed. Fitting major new natural gas liquefaction facilities into an existing site can be challenging.

Challenges include the following:

- Limited plot space
- Increased fire and explosion hazards
- Increased emissions and stricter regulations
- Increased noise levels

This paper discusses how these challenges can be overcome.

Acknowledgments

We gratefully acknowledge constructively critical reviews by Kamal Shah and Frank Del Nogal.

References

1. Chosnek, Jack and Victor H. Edwards, "LNG Imports to Exports: Process Safety and Regulatory Challenges", paper presented at the AIChE Spring Meeting, 13th Topical Conference on Natural Gas Utilization San Antonio, April 28-May 2, 2013

2. Woodward, John L. and Robin M. Pitblado, "LNG Risk Based Safety-Modeling and Consequence Analysis", John Wiley & Sons, Inc., Hoboken, New Jersey, 2010.
3. NFPA 59A, "Standard for the Production, Storage, and Handling of Natural Gas (LNG)", National Fire Protection Association, Quincy, MA, 2001 and 2006.
4. Volume 49 Code of Federal Regulations, Part 193, "Liquefied Natural Gas Facilities: Federal Safety Standards", United States, Washington, DC, 2013.
5. Volume 33, Code of Federal Regulations, Part 127, "Facilities Handling Liquefied Natural Gas and Liquefied Hazardous Gases", United States, Washington, DC, 2013.
6. Kohout, Andrew, "U. S. Regulatory Framework and Guidance for Siting of Liquefied Natural Gas Facilities – A Lifecycle Approach", paper presented at 15th International Symposium, Mary Kay O'Connor Process Safety Center, College Station, TX, October 23-25, 2012.
7. H.W.M. Witlox, Mike Harper, and Robin Pitblado, "Validation of PHAST Dispersion Model As Required for USA LNG Siting Applications", LNG Plant Safety and Applications, AIChE Spring Meeting, April 2012.
8. Hansen, O. R., Gavelli, F., Ichard, M., Davis, S. G., Validation of FLACS against experimental data sets from the model evaluation database for LNG vapor dispersion, J. Loss Prev. Proc. Ind., 23:867-877, 2010.

Biography

Jack Chosnek, Ph. D., P. E., Knowledge*One*, www.knowledge1.net Jack is a consultant in Process Safety Management and Knowledge Management and has over thirty five years of experience in the petrochemical industry. He has consulted for companies in the chemical, oil and gas, offshore, waste management and mining industries, developing policies and implementing process safety management systems, facilitating PHAs, writing operating procedures and conducting incident investigations and process safety audits. He has developed commercial, full-featured software for PHA facilitation and Management of Change (MOC). Jack consulted as the principal risk engineer for two multi-billion dollar projects carried out by large engineering firms.

Before becoming a consultant Jack worked for Celanese Corp. for 25 years in various positions in process safety management, technology management, operations, process engineering, piloting, tolling and R&D. Jack was a member of a project team that developed one of the first commercial scale-ups in the biotechnology industry.

Jack is the Chairman of the Process Safety Workshops at the South Texas Section (STS) of AIChE. He chaired the Process Plant Safety Symposium (PPSS) at the 4th Global Congress on Process Safety, and has chaired and co-chaired other process safety sessions in other years. He is the chair of the Technical Advisory Committee at Texas A&M's Mary Kay O'Connor Process Safety Center.

Jack is the holder of three patents related to chemical production. He has a BS and MS from the Technion—Israel Institute of Technology, and a PhD from the University of Missouri at Rolla, all in Chemical Engineering, as well as an MBA from Texas A&M—Corpus Christi

Filippo Gavelli, Ph. D., P. E.

Dr. Filippo Gavelli is the Head of the Dispersion Consulting group at GexCon and is responsible for GexCon's LNG safety consulting business worldwide. He specializes in the analysis of heat transfer and fluid flow phenomena, including multiphase flows and cryogenic fluids. He has over 10 years of engineering consulting experience and 20 years of experience in computational fluid dynamics (CFD) modeling, using several research and commercial codes. He applies his expertise to modeling the atmospheric dispersion of hazardous gaseous releases, and has extensive experience modeling hazard scenarios for Liquefied Natural Gas (LNG) facilities, including vapor cloud dispersion, pool fires and vapor cloud deflagrations; his experience includes more than 20 LNG installations worldwide, including onshore, offshore and floating (FLNG) facilities for LNG import, export (liquefaction) and peak shaving. Dr. Gavelli is a member of the technical committees for NFPA 59A (LNG facilities standard) and NFPA 2 (hydrogen technologies code) and was the lead in the model validation effort that led to FLACS' approval by the U.S. DOT under 49 CFR 193. Dr. Gavelli is the lead author of several gas dispersion and explosion safety-related papers and has been a regular contributor to LNG-related technical committees and expert panels for several years. He is a member of the GexCon docents group, which develops and delivers safety seminars to facility owners and operators, safety engineers, and regulatory agencies, on the hazards associated with gas explosions, dust explosions and LNG. Dr. Gavelli is also responsible for software technical support and training for FLACS customers in North America and has provided training to over 100 users, including staff from regulatory and government agencies.

Victor H. Edwards, Ph. D., P. E. is Director of Process Safety for IHI E&C International Corporation, Houston, TX 77077 (832-379-7742; e-mail: vic.edwards@ihi-ec.com). Responsible for health, safety, and environment in design, Vic has worked for IHI and predecessor companies for 29 years. His experience includes process engineering, process safety management, and process, biochemical, environmental, and mineral processing technologies. His project assignments have included biochemical and chemical processes, biomass energy resources, coal gasification, coal liquefaction, liquefied natural gas, chemical weapons demilitarization, hazardous waste disposal, offshore hydrocarbons facilities, and extraction of metals.

Dr. Edwards has received three DuPont awards for safety and environmental engineering excellence and two DuPont awards for health, safety, and environmental excellence during 28 years as an alliance engineering contractor. In 1998, Kvaerner Engineers and Constructors named him employee of the year. In 2007, Vic's Houston HSE team was recognized by Aker Kvaerner with the Process and Construction Excellence Award for HSE in Design.

Vic's earlier experience included Assistant Professor of Chemical Engineering at Cornell University, Program Manager at the National Science Foundation (on leave from Cornell), Research Fellow at Merck, alternate energy research at United Energy Resources, Visiting Professor of Environmental Engineering at Rice University, and process engineering at Fluor.

Vic has contributed more than sixty publications to the engineering literature, including several book chapters. He chaired the 1992 Process Plant

Safety Symposium, the 1997 International Plant Operations and Design Conference, and the 2013 Global Congress on Process Safety, as well as numerous technical sessions at other conferences. Vic received the 2003 Service Award from the Process Safety Center at Texas A & M University, where he chaired the Technical Advisory Committee from 2004 to 2011. In 2009, Vic was elected to the Centennial Council of the Department of Chemical and Biomolecular Engineering at Rice University. He also serves on the Editorial Advisory Board of CHEMICAL PROCESSING magazine. Edwards earned his B. A. from Rice University and his Ph. D. from the University of California at Berkeley; both degrees in chemical engineering. A registered professional engineer in Texas, he is an AIChE Fellow, and a member of ACS, AAAS, NFPA, NSPE, and the New York Academy of Sciences.

9TH GLOBAL CONGRESS ON PROCESS SAFETY

Guidelines for Relative Hazard Ranking of Refrigerants and Siting Considerations for LNG Liquefaction Units

Ryan J. Hart, Ph.D., P.E.
Exponent, Inc.
4580 Weaver Parkway, Suite 100, Warrenville, Illinois 60555
rhart@exponent.com

Delmar "Trey" Morrison III, Ph.D., P.E.
Exponent, Inc., Warrenville, Illinois 60555

Alfonso F. Ibarreta, Ph.D., P.E. and Harri K. Kytömaa, Ph.D., P.E.
Exponent, Inc., Natick, Massachusetts

Prepared for Presentation at
American Institute of Chemical Engineers
2013 Spring Meeting
9th Global Congress on Process Safety
San Antonio, Texas
April 28 – May 1, 2013

UNPUBLISHED

Guidelines for Relative Hazard Ranking of Refrigerants and Siting Considerations for LNG Liquefaction Units

Ryan J. Hart, PhD
Exponent, Inc.
4580 Weaver Parkway, Suite 100
Warrenville, IL 60555
rhart@exponent.com

Delmar "Trey" Morrison, PhD
Alfonso F. Ibarreta, PhD
Harri K. Kytomaa, PhD
Exponent, Inc.

Keywords: LNG, refrigerant, exclusion zone, vapor cloud

Abstract

Recent growth in the availability of domestic natural gas has led the industry to push for the development of liquefaction facilities, at both existing import/regasification facilities and new locations. As a consequence of this evolving need for North American liquefaction capacity, the Federal Regulators (FERC) have added interpretations and expectations for new accidental release case studies that are specific to liquefaction facilities. The requirements for LNG facility siting and design are codified in the 49 CFR 193 and NFPA 59A, but many of the exclusion zone analysis requirements are stated broadly and involve some level of expert interpretation.

Accidental release case studies required for the Resource Report 11 filings must now include LNG and flammable refrigerants. The hazards posed by refrigerants had not been previously relevant for receiving terminals. One goal of the required analysis is to calculate the extent of flammable vapor clouds, to demonstrate and confirm that these do not cross the proposed facility's property boundaries. Given economic and geographic limitations, there is often a trade-off between liquefaction unit capacity, siting, and even technology. Flammable refrigerants also introduce the new risk of vapor cloud explosions and associated overpressures. As a result, refrigerant vapor dispersion analysis is also now required, as is the quantification of the associated overpressure from potential vapor cloud explosions.

However, the analysis for liquefaction units must include both the LNG and the flammable liquid refrigerants. Two primary liquefaction refrigeration technologies are typically considered: propane pre-cooled mixed refrigerant and mixed refrigerant (PRICO). We have found that different considerations may drive the flammable vapor exclusion zones due to liquid spills and jetting and flashing of pressurized releases. The combination of physical properties, mixture composition, inventory, and operating parameters can lead to widely different results. The tools that are available and approved for performing these analyses impose limitations upon the mixtures and the number of constituents that can be modeled.

This paper will provide case studies and discussion illustrating the dominant parameters for liquefaction unit accidental release modeling. Guidance for identifying and selecting refrigerant release scenarios will be provided. Of most interest may be the interaction of the relative hazard distances for LNG and refrigerant vapor dispersion with refrigerant overpressure analysis.

1 Introduction

The liquefaction of natural gas for storage and transportation is, historically, a unique process for facilities located in the United States. Outside of peakshaving facilities, which are primarily used to liquefy natural gas during low periods of demand and re-gasify the stored liquid during high demand periods, large scale liquefaction facilities are a rarity in the US natural gas infrastructure. This has been driven by the fact in recent US history the amount of natural gas produced domestically has trailed domestic demand considerably. The graph provided in Figure 1 illustrates this point by providing the import of natural gas (MMcf, million cubic feet) into the US dating back into the 1970's [1].

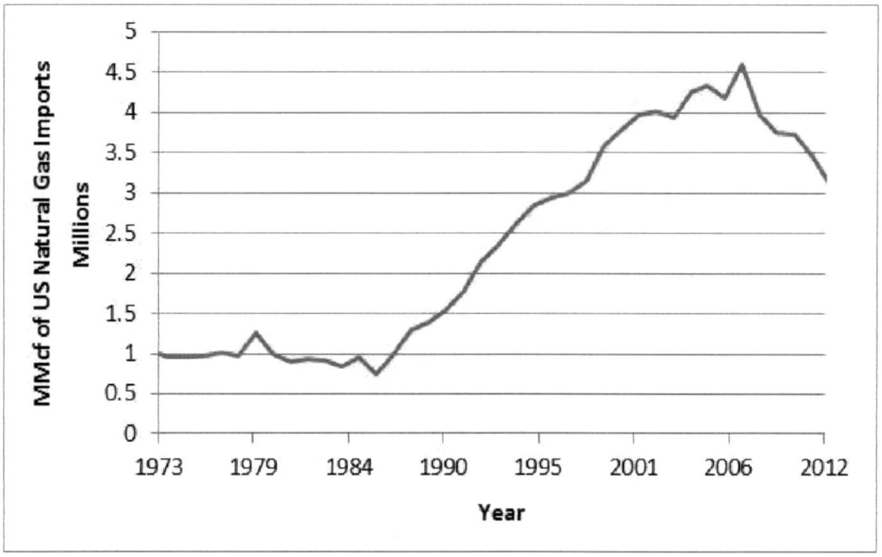

Figure 1. Historical US imports of natural gas (total, in millions of MMcf).

US imports of natural gas peaked in 2007 at just over 4,600,000 MMcf of natural gas. Since that time, the rate of natural gas imports has decreased by an average of 7.3% each year. This trend correlates with the rate of domestic natural gas production, which has been increasing steadily since 2006 (Figure 2, blue line, left axis) [2]. The increase in domestic production of natural gas is largely attributed to the increase in production from unconventional resources – mainly shale gas – as evidenced in Figure 2 (red line, right axis) [3]. In fact, the rate of shale gas production has been increasing at a rate of 44% per year since production began in 2007.

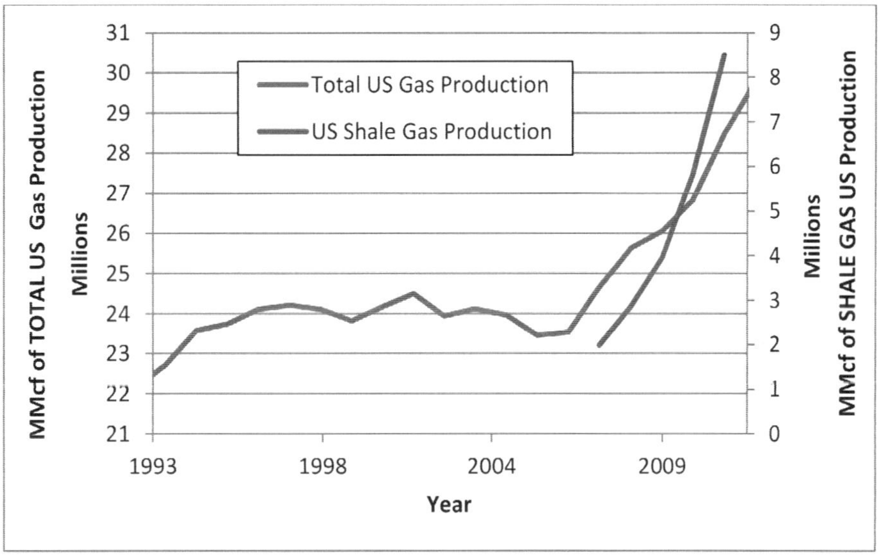

Figure 2. Historical US production of total natural gas (blue line, in millions of MMcf) and shale gas (red line, in millions of MMcf).

Because the domestic underground storage capacity of natural gas has remained relatively constant over the past two decades (between 8 and 9 million MMcf) [4], increases in domestic production can only be absorbed by the US infrastructure through decreasing imports, increasing exports, or converting our gas storage to liquid storage. If the rate of domestic production continues at its current pace, it is expected that production capacity will outpace domestic consumption at some point in the near future. This leaves only two options to handle increasing production rates: domestic storage of liquid natural gas (LNG) or the export and sale of gas. Exporting natural gas to countries not adjacent to the US (i.e. Canada and Mexico) is typically accomplished via trans-oceanic shipments of LNG. In fact, there have already been a number of proposed or potential LNG export facilities in the US, as shown in Figure 3 [5].

The codes and standards for which the facilities must be designed, constructed, and operated are found in 49 CFR Part 193 [6], and the authority having jurisdiction (AHJ) for these facilities is the Federal Energy Regulatory Commission (FERC). As part of the application process, the applicants (facility owners) must perform an environmental analysis in a series of thirteen (13) Resource Reports. Included in the assessment is a reliability and safety assessment (Resource Report 11) [7]. FERC then reviews the applications and must give approval prior to the commencement of any construction activities.

Resource Report 11 concerns reliability and safety of the facility, which includes a facility siting study. Although seemingly innocuous, the facility siting typically requires thorough analyses of potential loss of containment (LOC) events and the consequences of said events. 49 CFR Part 193, Subpart B addresses the siting requirements (including thermal radiation analysis and flammable vapor cloud analysis); however, the federal code references NFPA 59A (2001) throughout much of the subpart for specific guidance.

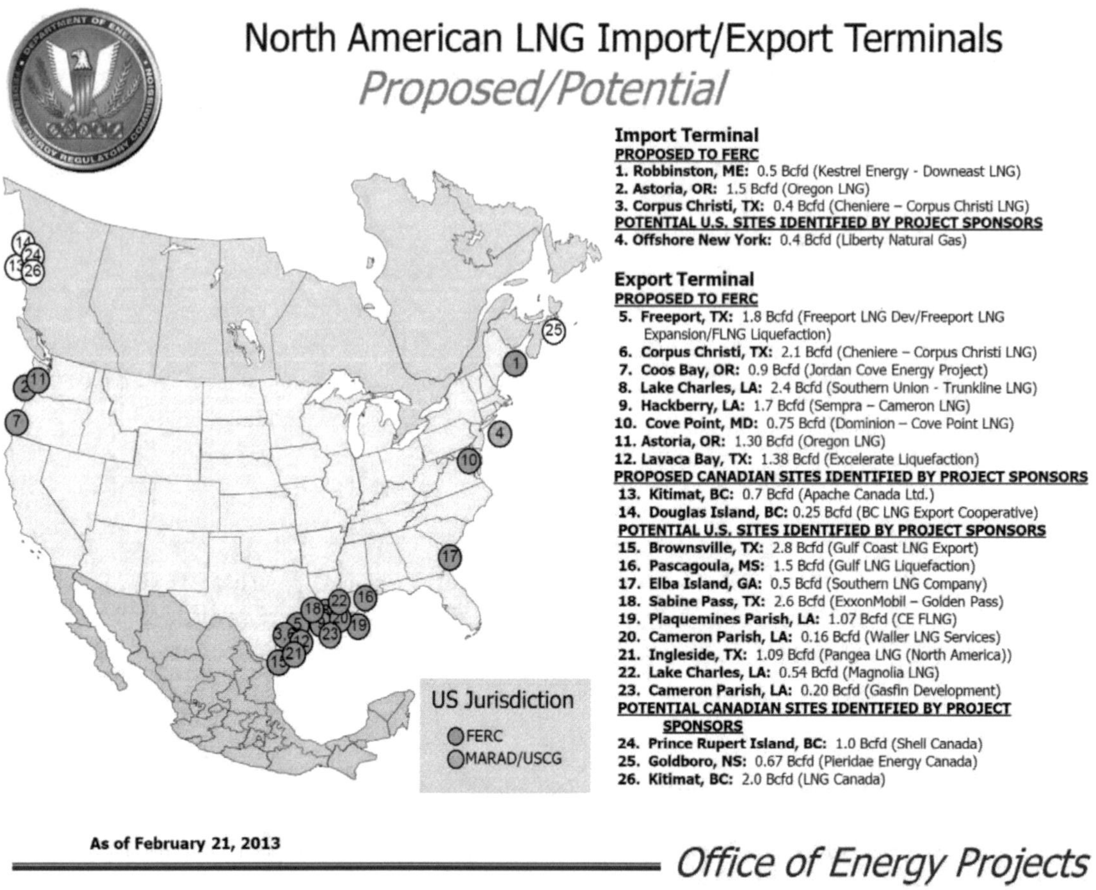

Figure 3. Proposed and potential liquefied natural gas import/export terminals, as identified by the US Department of Energy Federal Energy Regulatory Commission (FERC).

NFPA 59A is the "Standard for the Production, Storage, and Handling of Liquefied Natural Gas (LNG)" [8]. The standard provides general guidance for facility siting considering thermal and flammable vapor dispersion exclusion zones. The flammable vapor dispersion exclusion zones are found via calculation of the extent of flammable cloud vapor dispersion resulting from a LOC event. Although not required for import terminals, the exclusion zones at export terminals must consider the thermal radiation and flammable vapor dispersion exclusion zones for flammable refrigerants. Additionally, the AHJ (in this case FERC) has required that explosion overpressures be considered for the ignition of flammable refrigerant vapor clouds, based on 40 CFR 68 "Chemical Accident Prevention Provisions" requirements.

The exclusion zones related to refrigerant release scenarios may sometimes be larger than those associated with the traditional LNG-release exclusion zones. In the present analysis, we will use a FERC-approved commercial modeling tool, PHAST v6.7 [9], to evaluate the thermal radiation and flammable vapor dispersion exclusion zones associated with typical LNG liquefaction facility refrigerants. Additionally, the overpressures resulting from ignition of the flammable clouds will also be discussed. The results for the refrigerant vapor dispersion and thermal radiation analyses will be compared against LNG (methane).

2 LNG Liquefaction Technology

The transport of LNG is favorable due to the high density of liquids relative to gases and is a well understood process as a large portion of historical US imports has been received as LNG. However, unlike propane where the fluid can be liquefied by simply increasing the pressure, natural gas (primarily methane) must be refrigerated as part of its liquefaction process. The critical temperature of methane is -82.7°C (-117°F) and the critical temperature of propane is 96.6°C (206°F) [10]. Thus, as part of the liquefaction process for natural gas, the fluid must be refrigerated. The typical storage temperature of LNG is around -162°C (-260°F) [11], which is right at the normal boiling point for methane (-161.6°C, [10]).

The liquefaction of natural gas is accomplished using modified versions of the refrigeration cycles most commonly applied in residential appliances. All cycles use a working fluid that is expanded via Joules-Thomson (J-T) or isentropic means. The working fluid can vary greatly depending on the type of technology employed. The two primary technologies for the liquefaction of natural gas utilize a single-stage mixed refrigerant (such as the PRICO® process) or the two-stage propane/mixed refrigerant process [12]. More information on commercial LNG liquefaction technology is available elsewhere [13-15].

Either the two-stage or single-stage liquefaction process utilizes a mixed refrigerant as the working fluid at some point in the process. The mixed refrigerant composition can vary depending on the developer of the technology but is typically a mixture of the following components: methane, ethylene, propane, butane or higher molecular weight hydrocarbons, and nitrogen. As nitrogen is an inherently inert material, the contribution to thermal radiation and flammable vapor dispersion exclusion zones is non-existent. Selected properties of typical refrigerant materials can be found in Table 1 [16]. For comparison, the density of air is 0.075 lb/ft^3 (1.20 kg/m^3) at 68°F (20°C).

Table 1. Selected properties for the refrigerant components examined in this study. All properties were taken from the PHAST v6.7 materials database.

Material	Methane	Ethylene	Propane	n-Butane
Molecular Weight	16.0	28.1	44.1	58.1
Normal Boiling Point (n.b.p.)	-260°F (-162°C)	-155°F (-104°C)	-43.7°F (-42.0°C)	31.1°F (-0.50°C)
Liquid Density at n.b.p.	26.4 lb/ft^3 (423 kg/m^3)	35.5 lb/ft^3 (569 kg/m^3)	36.4 lb/ft^3 (583 kg/m^3)	37.6 lb/ft^3 (602 kg/m^3)
Vapor Density at 20°C	0.042 lb/ft^3 (0.668 kg/m^3)	0.073 lb/ft3 (1.17 kg/m3)	0.116 lb/ft^3 (1.86 kg/m^3)	0.155 lb/ft^3 (2.49 kg/m^3)
Sat. Pressure at 20°C	3630 psi (25.0 MPa)	910 psi (6.27 MPa)	121 psi (0.834 MPa)	30.2 psi (0.208 MPa)

3 Hazards of Refrigerants

The composition, temperature, and pressure of the mixed refrigerant may vary depending on the type of facility, liquefaction technology employed, and point along the refrigeration cycle. Therefore, a broad range of fluid temperatures and pressures must be considered in the exclusion zone analysis. While specific mixed refrigerant operating conditions and compositions are not made publicly available, all of these materials may be stored on-site as pure species (for refrigerant make-up) and may be present as a near pure species in the liquefaction cycle. Thus, for the exclusion zone analysis provided here we will examine pure component releases covering a range of temperatures and pressures. The specific species examined include: methane, ethylene, propane, and butane.

3.1 Thermal Radiation

The thermal radiation exclusion zone considers the ignition of a quantity of a spilled liquid material in an impoundment area. The size of the thermal radiation exclusion zone will therefore depend on the type of material and size of impoundment. Although the size of the impoundment area is unique to each and every facility, the resulting thermal radiation exclusion zone is expected to scale with the size of the impoundment.

The "Pool Fire" model in PHAST v6.7 was used to calculate the thermal radiation distances to the 1,600 BTU/ft^2.hr (5.0 kW/m^2) heat flux boundary for 50 ft (15 m) diameter circular pools of methane, ethylene, propane, and butane. The results are provided in Table 2. For all scenarios the wind speed was 4.5 mph (2 m/s), atmospheric stability class F, and the ambient temperature and surface temperature were 68°F (20°C).

Table 2. Pool fire thermal radiation distances to the 1,600 BTU/ft^2.hr (5.0 kW/m^2) heat flux boundary.

Material	Distance to 1,600 BTU/ft^2.hr
Methane	254 ft (77.3 m)
Ethylene	206 ft (62.8 m)
Propane	230 ft (70.0 m)
Butane	236 ft (71.8 m)

The results indicate that for equivalent pool diameters the methane liquid spills result in the farthest distance to the 1,600 BTU/ft^2.hr heat flux boundary. Propane and butane pool fires result in similar exclusion zones and ethylene resulted in the smallest exclusion zone.

3.2 Flammable Vapor Dispersion

The flammable refrigerants involved in the liquefaction cycle for LNG production pose a significant hazard in the event of a LOC. High pressure and possibly cold refrigerants may be discharged from piping or appurtenances in the event of a release. If the release involves the fluid in its liquid state, these liquids may flash or vaporize completely upon release, they may spill into collection areas (impoundments), or both may occur. Even if the fluid released is in the vapor state, a high momentum cloud will be formed (although at lower discharge rate for equivalent leak size) which can disperse consider distances if aligned with the prevailing wind direction.

For the flammable vapor dispersion exclusion zone analysis, two separate sets of scenarios are examined: (1) refrigerant releases under typical storage conditions and (2) refrigerant releases under "product LNG conditions" (i.e. conditions at which LNG is produced). For the typical storage conditions, it was assumed that the refrigerants were stored at their vapor pressure at 68°F (20°C); however, the pressure for methane was capped at 1000 psi. For the typical process conditions, it was assumed that the refrigerants were processed at -260°F (-162°C) and 100 psi (0.689 MPa); however, the process temperature for butane was selected just above its melting point at -216°F (-138°C). A summary of the release conditions examined are provided in Table 3. The LFL values reported are the PHAST v6.7 default values for the pure species.

Table 3. Release scenarios and fluid conditions considered for the flammable vapor dispersion exclusion zone analysis.

Material	Temperature	Pressure	LFL
	Refrigerant Storage Releases		
Methane	68°F (20°C)	1,000 psi (6.89 MPa)	4.4%
Ethylene	68°F (20°C)	910 psi (6.27 MPa)	2.3%
Propane	68°F (20°C)	121 psi (0.834 MPa)	2.0%
Butane	68°F (20°C)	30.2 psi (0.208 MPa)	1.5%
	Refrigerant Process Releases		
Methane	-260°F (-162°C)	100 psi (0.689 MPa)	4.4%
Ethylene	-260°F (-162°C)	100 psi (0.689 MPa)	2.3%
Propane	-260°F (-162°C)	100 psi (0.689 MPa)	2.0%
Butane	-216°F (-138°C)	100 psi (0.689 MPa)	1.5%

The model input parameters examined in the PHAST v6.7 vapor dispersion calculations are provided in Table 4. The surface roughness factor selected corresponds to dispersion along open, flat terrain. The leak elevation that was chosen corresponds to the height of a typical pipe rack. The model uses the orifice equation to calculate discharge rate, and a discharge coefficient of 0.6 corresponds to a typical square-edged orifice (what one might expect in the event of clean break of piping connection). The leak size was chosen to be 1 in, which is an expected piping connection for an instrument tap. All releases were allowed to continue until a full steady-state cloud was formed. The releases were in the horizontal orientation for all cases.

Table 4. PHAST v6.7 model input parameters for vapor dispersion exclusion zone analysis of the releases described in Table 3.

Parameter	Value
Ambient Temperature	68°F (20°C)
Surface Temperature	68°F (20°C)
Relative Humidity	70%
Wind Speed	4.5 mph (2 m/s)
Pasquill-Gifford Atmospheric Stability	Category F
Surface Roughness Factor (m)	12 in (0.03 m)
Leak Elevation	12 ft (3.7 m)
Discharge Coefficient	0.6
Leak Size	1 in (2.5 cm)

The results of the flammable vapor dispersion for the releases described above, reported as distances from the source to the edge of the ½-LFL vapor cloud, are provided in Table 5. A portion of the propane process and butane process releases rained out (liquid droplets reached the ground), yielding rainout values of 7.4% and 20%, respectively. For all other releases, there was no liquid that rained out.

Certain trends are observed in the calculated vapor dispersion results. Firstly, the flammable cloud dispersion distances correlate with the discharge rate (as one might expect) for all components. Secondly, it appears that the clouds with higher density vapors (which disperse closer to the ground where the wind speeds are lower) travel greater distances. Lastly, it appears as though within a given set of releases (i.e. storage vs. process) the materials with the higher molecular weights (and coincidentally higher densities and lower LFL values) travel farther distances. The only exception to this is the ethylene process release, which was found to travel the greatest distance of any of the materials.

Table 5. PHAST v6.7 calculated flammable vapor cloud release rates and dispersion distances to the ½-LFL for the high pressure storage and process releases described in Table 3.

Material	Discharge Rate	Distance to ½-LFL	Height at ½-LFL
	Refrigerant Storage Releases		
Methane	8.73 lb/s (3.96 kg/s)	108 ft (33.0 m)	12.2 ft (3.73 m)
Ethylene	16.6 lb/s (7.53 kg/s)	234 ft (71.4 m)	6.93 ft (2.11 m)
Propane	20.2 lb/s (9.16 kg/s)	244 ft (74.3 m)	0.545 ft (0.166 m)
Butane	10.8 lb/s (4.90 kg/s)	356 ft (109 m)	0 ft (0 m)
	Refrigerant Process Releases		
Methane	16.2 lb/s (7.36 kg/s)	699 ft (214 m)	0 ft (0 m)
Ethylene	20.5 lb/s (9.28 kg/s)	850 ft (259 m)	0 ft (0 m)
Propane	21.8 lb/s (9.89 kg/s)	755 ft (230 m)	0 ft (0 m)
Butane	22.3 lb/s (10.1 kg/s)	770 ft (235 m)	0 ft (0 m)

Although only two of the pressurized refrigerant releases resulted in appreciable rainout of liquid fractions, the collection of liquids in impoundments from low pressure releases should also be considered in flammable vapor dispersion exclusion zone analyses. Next we will consider the resulting vaporization, cloud formation, and flammable vapor dispersion from liquid spills of the four refrigerant materials into a circular impoundment area that is 50 ft (15 m) in diameter. The impoundment material chosen for this analysis was regular concrete, and the PHAST default thermal conductivity (κ = 1.21 W/m.K) and thermal diffusivity ($\alpha = 5.72 \times 10^{-7}$ m^2/s) were used. We have found that the choice of concrete thermal properties can significantly affect the rate of vaporization for the spilled cryogenic liquids, but we only provide the default values here. Each of the four materials was assumed to spill at a rate of 11.0 lb/s (5.00 kg/s) at their normal point boiling temperature (provided in Table 1). The weather conditions described in Table 4 were used and the spills occurred for 10 minutes.

Table 6. PHAST v6.7 flammable vapor dispersion distances to the ½-LFL boundary for liquid spills into a 50 ft (15 m) diameter regular concrete impoundment.

Material	Discharge Rate	Distance to ½-LFL	Height at ½-LFL
Methane	11.0 lb/s (5.0 kg/s)	1175 ft (358 m)	0 ft (0 m)
Ethylene	11.0 lb/s (5.0 kg/s)	674 ft (205 m)	0 ft (0 m)
Propane	11.0 lb/s (5.0 kg/s)	344 ft (105 m)	0 ft (0 m)
Butane	11.0 lb/s (5.0 kg/s)	328 ft (100 m)	0 ft (0 m)

Unlike the high pressure process and storage refrigerant release scenarios, the liquid spills into impoundment indicate that the materials with the lowest normal boiling points produce the longest dispersing clouds. In fact, an 11 lb/s (5 kg/s) liquid methane spill resulted in a cloud that travelled nearly the length of four football fields. With methane being the lone exception, the liquid refrigerant spills resulted in clouds that travelled less distance than the high pressure refrigerant process releases.

3.3 Overpressure Analysis

Several explosion overpressure calculation models exist and are available for determination of the flammable clouds described above. In PHAST v6.7, the available models include TNT, TNO Multi-energy (TNO), and Baker-Strehlow-Tang (BST). For the purposes of evaluating the resulting overpressure profiles of the releases described previously, we will be focusing on only the BST model results. Although all of the models have their advantages and disadvantages, the BST model is the only one of the three that allows for consideration of material reactivity. Since we are examining overpressures as the result of ethylene vapor cloud explosions, and ethylene is widely considered to be a highly reactive fuel (the others are generally accepted to have moderate to low reactivity), the BST model was chosen.

The BST model takes confinement, congestion, and material reactivity into consideration. However, unlike the TNO model the blast strength is not defined by the user. The following inputs need to be prescribed [17]:

> ***Material reactivity***: The material reactivity can be defined as Low, Medium or High. Low reactivity corresponds to components such as methane, carbon monoxide or ammonia, while High reactivity corresponds to very reactive materials such as hydrogen, ethylene or acetylene. The reactivity values for the refrigerant species are provided in Table 7, below.

Table 7. The material reactivities used for the Baker-Strehlow-Tang explosion model.

Refrigerant	Material Reactivity
Methane	Low
Ethylene	High
Propane	Medium
Butane	Medium

***Flame expansion*:** The flame expansion describes the number of dimensions in which the blast can propagate. For example, if a blast occurs within a pipe, the overpressures can only propagate in a single direction. For a blast that is confined between two floors, the overpressures can only propagate in 2 dimensions. In the case of most leaks at a liquefaction facility, the overpressures are able to propagate in all three dimensions. A flame expansion parameter of 3 was therefore selected.

***Obstacle density*:** The obstacle density is selected based on the blockage ratio in congested areas.

- A "low" obstacle density corresponds to a blockage ratio of less than 10%,
- A "medium" obstacle density corresponds to a blockage ratio between 10% and 40%,
- A "high" obstacle density corresponds to a blockage ratio above 40%.

At the typical liquefaction facility, most confined areas correspond to pipe racks which are not highly congested; pipes are typically aligned parallel to one another along pipe supports, rather than staggered. A "medium" obstacle density was selected for all scenarios.

***Ground reflection factor*:** The ground reflection factor is a number between 1 (no reflection) and 2 (perfect ground reflection). In order to provide for the most conservative overpressure calculations, a ground reflection of 2 was selected. This assumption corresponds to perfect ground reflection and highest possible overpressures.

***Confined volume*:** The confined volume corresponds to the total flammable cloud volume that exists in congested regions. For the analyses represented here, it was conservatively assumed that the entire flammable cloud was congested for each scenario. The flammable cloud volume is defined as the portion of the vapor cloud found between the LFL and UFL boundaries.

The resulting flammable vapor clouds described in Section 3.2 were examined using the methods discussed above. The ignition location was placed at the farthest LFL boundary, and the individual distances from the release source to the 1 psi overpressure boundary are provided in Table 8.

Table 8. PHAST v6.7 calculated distances to the LFL and BST model calculated distances to 1 psi overpressure boundary for the refrigerant storage, process, and liquid spill release scenarios described in Section 3.2.

Material	Discharge Rate	Distance to LFL	Distance to 1 psi Overpressure
	Refrigerant Storage Releases		
Methane	8.73 lb/s (3.96 kg/s)	56.3 ft (17.2 m)	60.9 ft (18.6 m)
Ethylene	16.6 lb/s (7.53 kg/s)	108 ft (33.0 m)	346 ft (106 m)
Propane	20.2 lb/s (9.16 kg/s)	113 ft (34.4 m)	289 ft (88.0 m)
Butane	10.8 lb/s (4.90 kg/s)	154 ft (46.8 m)	371 ft (113 m)
	Refrigerant Process Releases		
Methane	16.2 lb/s (7.36 kg/s)	312 ft (95.0 m)	443 ft (135 m)
Ethylene	20.5 lb/s (9.28 kg/s)	497 ft (152 m)	1381 ft (421 m)
Propane	21.8 lb/s (9.89 kg/s)	327 ft (99.8 m)	776 ft (237 m)
Butane	22.3 lb/s (10.1 kg/s)	336 ft (102 m)	832 ft (254 m)
	Refrigerant Liquid Spill Releases		
Methane	11.0 lb/s (5.0 kg/s)	613 ft (187 m)	852 ft (260 m)
Ethylene	11.0 lb/s (5.0 kg/s)	334 ft (102 m)	1307 ft (398 m)
Propane	11.0 lb/s (5.0 kg/s)	236 ft (71.8 m)	786 ft (240 m)
Butane	11.0 lb/s (5.0 kg/s)	232 ft (70.6 m)	806 ft (246 m)

Ignition of the flammable cloud for the low reactivity fuel, methane, resulted in BST calculated overpressure distances to the 1 psi boundary that didn't exceed the vapor cloud dispersion distance to the ½-LFL value. For the medium reactivity fuels propane and butane, the BST calculated distances to the 1 psi overpressure boundary were similar to their ½-LFL vapor dispersion distances (butane was slightly higher and propane was slightly lower for most cases). However, for the high reactivity fuel ethylene the BST calculated distance to the 1 psi boundary greatly exceed the vapor dispersion to the ½-LFL boundary.

4 Conclusions

The hazard consequence of LOC events related to refrigerant storage and processing and LNG liquefaction facilities can be successfully examined using modern software tools. The model tool PHAST v6.7 was used here to examine the thermal radiation, flammable vapor dispersion, and overpressure exclusion zones. Such information is not only required by the FERC for regulatory permit application and approval, but can provide useful information for facility siting.

The results indicate that for typical refrigerant materials the ½-LFL vapor cloud dispersion exclusion zone represents the largest expected exclusion zone. The only exception is for highly reactive fuels, such as ethylene, which may produce overpressures to the 1 psi overpressure boundary that greatly exceed the predicted ½-LFL vapor cloud extents. Thus, fuel characteristics and release conditions both can greatly influence the exclusion zone analyses.

Proposed liquefaction facilities would be benefited by screening the potential exclusion zones from anticipated LOC events at early stages in the design of the facility. Consider that some of the calculated exclusion zones for typical LOC events at a liquefaction facility may result in exclusion zones that extend well beyond 1300 ft. Such information will play a pivotal role in facility siting, and should be conducted prior to the full arrangement of the facility piping and equipment systems (at which point it may be impractical to move large structures or capital equipment).

5 References

[1] U.S. Natural Gas Imports (MMcf), released on February 28, 2013. Accessed via <www.eia.gov>.

[2] U.S. Natural Gas Gross Withdrawals (MMcf), released on February 28, 2013. Accessed via <www.eia.gov>.

[3] U.S. Natural Gas Gross Withdrawals from Shale Gas (Million Cubic Feet), released on February 28, 2013. Accessed via <www.eia.gov>.

[4] U.S. Total Natural Gas Underground Storage Capacity (MMcf), released on February 28, 2013. Accessed via <www.eia.gov>.

[5] North American LNG Import/Export Terminals – Proposed/Potential. US Department of Energy Federal Energy Regulatory Commission, February 21, 2013. Accessed via <ferc.gov/industries/gas/indus-act/lng.asp>.

[6] "Liquefied Natural Gas Facilities: Federal Safety Standards." Code of Federal Regulations Title 49, Pt. 193, 2011 edition.

[7] Federal Energy Regulatory Commission, "Guidance Manual for Environmental Report Preparation." FERC – Office of Energy Projects (August 2002).

[8] NFPA 59A, Standard for the Production, Storage, and Handling of Liquefied Natural Gas (LNG), 2001 Edition.

[9] PHMSA Docket No. 2011-0075, Final Decision on petition for approval for PHAST UDM Version 6.6 and 6.7, October 7, 2011.

[10] Air Liquide Gas Encyclopedia, Air Liquid (2009). Accessed via <encyclopedia.airliquide.com>.

[11] U.S. Department of Energy, "Liquefied Natural Gas: Understanding the Basic Facts." U.S. Department of Energy – Office of Fossil Energy (August 2005).

[12] Kidnay, AJ, WR Parrish, DG McCartney, "Fundamentals of Natural Gas Processing, 2nd ed." CRC Press (2011).

[13] Ransbarger, W, "A fresh look at LNG process efficiency." LNG Industry (Spring 2007).

[14] Shukri, T, "LNG technology selection," Hydrocarbon Engineering (February 2004).

[15] Shukri, T and M Barclay, "Single mixed refrigerant process has appeal for growing offshore market," LNG Journal (June 2007).

[16] PHAST v6.7 Material Properties, DNV Software.

[17] Guidelines for Vapor Cloud Explosions, Pressure Vessel Burst, BLEVE and Flash Fire Hazards. Second Edition. Wiley (2010).

AUTHOR INDEX

AUTHOR INDEX